Dynamics of One-Dimensional Maps

Mathematics and Its Applications

Volume 407

Dynamics of One-Dimensional Maps

by

A.N. Sharkovsky
S.F. Kolyada
A.G. Sivak
and
V.V. Fedorenko

Institute of Mathematics,
Ukrainian Academy of Sciences,
Kiev, Ukraine

KLUWER ACADEMIC PUBLISHERS
DORDRECHT / BOSTON / LONDON

A C.I.P. Catalogue record for this book is available from the Library of Congress.

ISBN 978-90-481-4846-2

Published by Kluwer Academic Publishers,
P.O. Box, 3300 AA Dordrecht, The Netherlands.

Kluwer Academic Publishers incorporates
the publishing programmes of
D. Reidel, Martinus Nijhoff, Dr. W. Junk and MTP Press.

Sold and distributed in the U.S.A. and Canada
by Kluwer Academic Publishers,
101 Philip Drive, Norwell, MA 02061, U.S.A.

In all other countries, sold and distributed
by Kluwer Academic Publishers,
P.O. Box 322, 3300 AH Dordrecht, The Netherlands.

This is a completely revised and updated translation of
the original Russian work of the same title,
published by Naukova Dumka, Kiev, 1989.
Translated by A.G. Sivak, P. Malyshev and D. Malyshev

Printed on acid-free paper

CONTENTS

INTRODUCTION

Last decades are marked by the appearance of a permanently increasing number of scientific and engineering problems connected with the investigation of nonlinear processes and phenomena. It is now clear that nonlinear processes are not exceptional; on the contrary, they can be regarded as a typical mode of existence of matter. At the same time, independently of their nature, these processes are often characterized by similar intrinsic mechanisms and admit universal approaches to their description.

As a result, we observe fundamental changes in the methods and tools used for mathematical simulation. Today, parallel with well-known methods studied in textbooks and special monographs for many years, mathematical simulation often employs the results of nonlinear dynamics—a new rapidly developing field of natural sciences whose mathematical apparatus is based on the theory of dynamical systems.

The extensive development of nonlinear dynamics observed nowadays is explained not only by increasing practical needs but also by new possibilities in the analysis of a great variety of nonlinear models discovered for last 20 years. In this connection, a decisive role was played by simple nonlinear systems, discovered by physicists and mathematicians, which, on the one hand, are characterized by quite complicated dynamics but, on the other hand, admit fairly complete qualitative analysis. The analysis of these systems (both qualitative and numerical) revealed many common regularities and essential features of nonlinearity that should be kept in mind both in constructing new nonlinear mathematical models and in analyzing these models. Among these features, one should, first of all, mention stochastization and the emergence of structures (the relevant branches of science are called the theory of strange attractors and synergetics, respectively).

The theory of one-dimensional dynamical systems is one of the most efficient tools of nonlinear dynamics because, on the one hand, one-dimensional systems can be described fairly completely and, on the other hand, they exhibit all basic complicated nonlinear effects. The investigations in the theory of one-dimensional dynamical systems gave absolutely new results in the theory of difference equations, difference–differential equations, and some classes of differential equations. Thus, significant successes were attained in constructing new types of solutions, which can be efficiently used in simulating the processes of emergence of ordered coherent structures, the phenomenon of intermittence, and self-stochastic modes. Significant achievements in this field led to the appearance of a new direction in the mathematical theory of turbulence based on the use of nonlinear difference equations and other equations (close to nonlinear difference equations) as mathematical tools.

It is clear that iterations of continuous maps of an interval into itself are very simple dynamical systems. It may seem that the use of one-dimensional dynamical systems substantially restricts our possibilities and the natural ordering of points in the real line may result in the absence of some types of dynamical behavior in one-dimensional systems. However, it is well known that even quadratic maps from the family $x \mapsto x^2 + \lambda$ may have infinitely many periodic points for some values of the parameter λ. Furthermore, for $\lambda = -2$, the map possesses an invariant measure absolutely continuous with respect to the Lebesgue measure, i.e., for this map, "stochastic" behavior is a typical behavior of bounded sequences of iterations. Actually, the trajectories of one-dimensional maps exhibit an extremely rich picture of dynamical behavior characterized, on the one hand, by stable fixed points and periodic orbits and, on the other hand, by modes which are practically indistinguishable from random processes being, at the same time, absolutely deterministic.

This book has two principal goals: First, we try to make the reader acquainted with the fundamentals of the theory of one-dimensional dynamical systems. We study, as a rule, very simple nonlinear maps with a single point of extremum. Maps of this sort are usually called *unimodal*. It turns out that unimodality imposes practically no restrictions on the dynamical behavior.

The second goal is to equip the reader with a more or less comprehensive outlook on the problems appearing in the theory of dynamical systems and describe the methods used for their solution in the case of one-dimensional maps.

To understand distinctive features of topological dynamics on an interval on a more profound level, the reader must not only study the formulations of the results but also carefully analyze their proofs. Unfortunately, the size of the book is limited and, therefore, some theorems are presented without proofs.

This book does not contain special historical notes; only basic facts given in the form of theorems contain references to their authors. Almost all results are achievements of the last 20–30 years. The interest to the qualitative investigation of iterations of continuous and discontinuous functions of a real variable was growing since 1930s when applied problems requiring the study of such iterations appeared. However, these investigations were not carried out systematically till 1970s. The results of many authors worked at that time are now well known. We would like to mention here less known works of Barna [1], Leonov [1–3], and Pulkin [1, 2], which also contain many important results.

In Chapter 1, following Sharkovsky, Maistrenko, and Romanenko [2], we give an elementary introduction to the theory of one-dimensional maps. This chapter contains an exposition of basic concepts of the theory of dynamical systems and numerous examples illustrating various situations encountered in the investigation of one-dimensional maps.

Chapter 2 deals with the methods of symbolic dynamics. In particular, it contains a presentation of the basic concepts and results of the theory of kneading invariants for unimodal maps.

In Chapters 3 and 4, we prove theorems on coexistence of periodic trajectories. The maps whose topological entropy is equal to zero (i.e., maps that have only cycles of periods $1, 2, 2^2, \dots$) are studied in detail and classified.

Various topological aspects of the dynamics of unimodal maps are studied in Chapter 5. We analyze the distinctive features of the limiting behavior of trajectories of smooth maps. In particular, for some classes of smooth maps, we establish theorems on the number of sinks and study the problem of existence of wandering intervals.

In Chapter 6, for a broad class of maps, we prove that almost all points (with respect to the Lebesgue measure) are attracted by the same sink. Our attention is mainly focused on the problem of existence of an invariant measure absolutely continuous with respect to the Lebesgue measure. We also study the problem of Lyapunov stability of dynamical systems and determine the measures of repelling and attracting invariant sets.

The problem of stability of separate trajectories under perturbations of maps and the problem of structural stability of dynamical systems as a whole are discussed in Chapter 7.

In Chapter 8, we study one-parameter families of maps. We analyze bifurcations of periodic trajectories and properties of the set of bifurcation values of the parameter, including universal properties such as Feigenbaum universality.

Unfortunately, in the present book, we do not consider the maps of a circle onto itself and the maps of the complex plane onto itself. Some results established for maps of an interval onto itself are related to the dynamics of rational endomorphisms of the Riemann sphere: The beauty of the dynamics of the considered maps of the real line onto itself from the family $x \mapsto x^2 + \lambda$, $\lambda \in \mathbb{R}$, becomes visible (in the direct meaning of this word) if we pass to the family $z \mapsto z^2 + \lambda$, where z is a complex variable and λ is a complex parameter (see Peitgen and Richter [1]).

We hope that our book will be useful for everybody who is interested in nonlinear dynamics.

1. FUNDAMENTAL CONCEPTS OF THE THEORY OF DYNAMICAL SYSTEMS. TYPICAL EXAMPLES AND SOME RESULTS

Dynamical systems are usually understood as one-parameter groups (or semigroups) f^t of maps of a space X into itself (this space is either topological or metric). If t belongs to $\mathbb{R}$ or $\mathbb{R}^+$, then a dynamical system is sometimes called a flow and if t belongs to $\mathbb{Z}$ or $\mathbb{Z}^+$, then this dynamical system is called a cascade. These names are connected with the fact that, under the action of f^t, the points of X "begin to move" $(x \mapsto f^t(x))$, and the space "splits" into the trajectories of this motion.

A pair (X, f), where f is a mapping of the space X into itself, defines a dynamical system with discrete time, i.e., a semigroup of maps $\{f^n, n \in \mathbb{Z}^+\}$, where $f^n = f \circ f^{n-1}$, $n = 1, 2, \ldots$, and f^0 is the identity map. If the space X is the real line $\mathbb{R}$ or an interval $I \subset \mathbb{R}$, then this dynamical system with one-dimensional phase space and discrete time is, in a certain sense, the simplest one; nevertheless, in many cases, it is characterized by very complicated dynamics. In some aspects, e.g., from the viewpoint of the descriptive theory of sets, one-dimensional dynamical systems can be as complicated as dynamical system on arbitrary compact sets.

1. Trajectories of One-Dimensional Dynamical Systems

The main object of the theory of dynamical systems is a trajectory or an orbit (in what follows, we use both these terms). The set

$$\mathrm{orb}\,(x) = \{x, f(x), f^2(x), \ldots\} = \bigcup_{n=0}^{\infty} f^n(x)$$

is called the *trajectory* of a dynamical system (X, f) passing through a point $x \in X$ (it is sometimes convenient to regard a trajectory as a sequence of points $x, f(x), f^2(x), \ldots$ but not as a set because this point of view is closer to the concept of motion along the trajectory governed by the map $n \mapsto f^n(x)$). The trajectory passing through a point x

is denoted either by the symbol $\mathrm{orb}\,(x)$ or by $\mathrm{orb}_f(x)$. In most cases, it is necessary to clarify the behavior of a trajectory (or a family of trajectories) on a bounded or unbounded time interval. In what terms and in what form one can answer this or similar questions ?

In the theory of dynamical systems, the asymptotic behavior of trajectories is usually characterized by ω-limit sets. A point $x' \in X$ is called an *ω-limit point* of a trajectory $\{x, f(x), \ldots , f^n(x), \ldots\}$ if, for any $n' > 0$ and any neighborhood U of x', there exists $n'' > n'$ such that $f^{n''}(x) \in U$ (i.e., there exists a sequence $n_1 < n_2 < \ldots \to \infty$ such that $f^{n_i}(x) \to x'$). The set of all ω-limit point of the trajectory passing through the point x is denoted by $\omega_f(x)$ or simply by $\omega(x)$. This set is closed. Moreover, if X is compact, it is invariant and nonempty (if X is not a compact set, then it is possible that $\omega(x) = \varnothing$, i.e., the trajectory eventually leaves X). Thus, if X is a compact set, then $\omega(x)$ is the smallest closed set such that any its neighborhood contains all points of the trajectory $\{f^n(x)\}$ beginning with some n (depending on the choice of a neighborhood).

The most simple behavior is exhibited by periodic trajectories or cycles. A point $x_0 \in X$ is called a *periodic point* with period m if $f^m(x_0) = x_0$ and $f^n(x_0) \neq x_0$ for $0 < n < m$. Each point $x_n = f^n(x_0)$, $n = 1, 2, \ldots , m-1$, is also a periodic point with period m, and the points $x_0, x_1, \ldots , x_{m-1}$ form a *periodic trajectory* or a *cycle* with period m. Periodic trajectories play an important role in the theory of dynamical systems. For one-dimensional dynamical systems, they are of particular importance.

The ω-limit sets of periodic trajectories coincide with these trajectories. Generally speaking, if the ω-limit set of a trajectory is a cycle, then this trajectory is either periodic or *asymptotically periodic,* i.e., it is *attracted by a periodic trajectory.*

There exists a simple graphic procedure for constructing trajectories of dynamical systems defined on an interval. This procedure can be employed, e.g., in studying the behavior of trajectories in the vicinity of a fixed point or a cycle.

Consider a mapping $x \mapsto f(x)$ defined on an interval I and a point $x_0 \in I$. The procedure of graphic representation of the trajectory of the point x_0 is called the *Königs–Lamerey diagram* and can be described as follows: In the plane (x, y), we draw the graphs of the functions $y = f(x)$ and $y = x$. The trajectory of the point x_0 is represented by a broken line $M_1N_1M_2N_2M_3N_3 \ldots$ whose chains are parallel to the coordinate axes (see Fig. 1). The abscissae of the points M_1, N_1 and M_2, N_2 and M_3, etc., are the successive iterations of the point x_0 equal to x_0, $x_1 = f(x_0)$, $x_2 = f(x_1), \ldots$ respectively. The ordinates of the points M_1 and N_1, M_2 and N_2, M_3 and N_3, etc., are equal to $x_1 = f(x_0)$, $x_2 = f(x_1)$, $x_3 = f(x_2), \ldots$, respectively. Thus, to construct the broken line $M_1N_1M_2N_2M_3N_3 \ldots$, one must start from the point x_0 and successively move along its trajectory.

The fixed points of the map f are associated with the points of intersection of the graphs of the functions $y = f(x)$ and $y = x$. In Fig. 1, these are the points β_0 and β_0'. Moreover, the point β_0 is repelling and the point β_0' is attracting, since the trajectories of the points close to β_0 recede from β_0, and the trajectories of points close to β_0' approach this point.

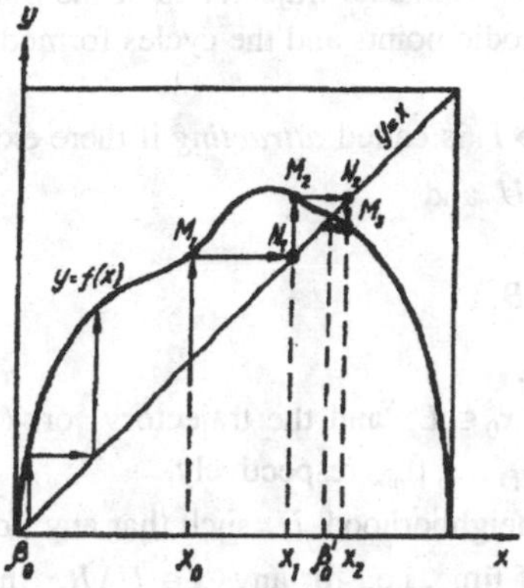

Fig. 1

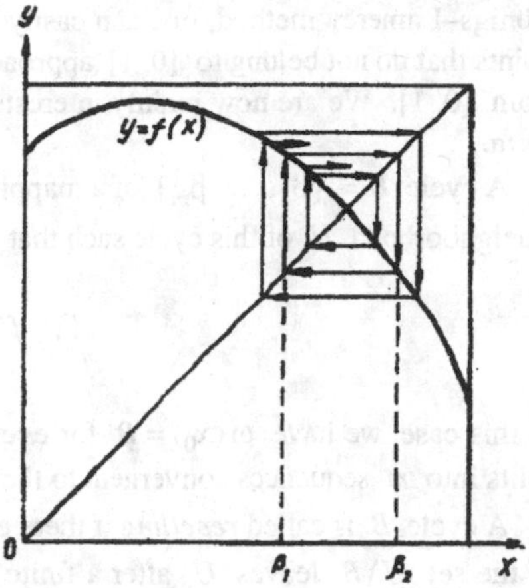

Fig. 2

The closed broken line $M_1N_1M_2N_2\ldots$, where $M_{n+1} = M_1$, corresponds to a cycle of period n. In Fig. 2, we present an example of a closed broken line with $n = 2$. It corresponds to a cycle of period 2 that consists of the points β_1 and β_2 such that $f(\beta_1) = \beta_2$ and $f(\beta_2) = \beta_1$. This cycle is attracting because broken lines close to the closed broken line corresponding to this cycle approach this line.

For the maps whose graphs are displayed in Fig. 1 and Fig. 2, the ω-limit set of every trajectory can be defined quite simply: Any trajectory is attracted either by a fixed point or by a cycle with period 2. If a map possesses a cycle with period greater than 2, then the behavior of trajectories near this cycle can be studied by using a computer. However, in many cases, both the Königs–Lamerey method and numerical simulation fail to detect any regularities in the behavior of the trajectories: Thus, one observes no convergence to fixed points or cycles; furthermore, the behavior of trajectories is completely different even if these trajectories correspond to initial points lying at very short distances from each other, etc. The reader can readily check this fact by analyzing (e.g., with a calculator) the trajectories of the maps

$$x \mapsto f_\lambda(x) = \lambda x(1-x) \tag{1}$$

for different values of the parameter $\lambda > 0$. It seems useful to choose $x_0 \in (0, 1)$ and successively consider the values $\lambda \in \{1.5;\, 2.9;\, 3.4;\, 3.57;\, 3.83;\, 4\}$.

The maps in family (1) are defined for $x \in \mathbb{R}$. Moreover, $f_\lambda(0) = f_\lambda(1) = 0$ and

$$\max_{x\in\mathbb{R}} f_\lambda(x) = f_\lambda\left(\frac{1}{2}\right) = \frac{\lambda}{4}$$

for $\lambda > 0$. Therefore, for $\lambda \in (0, 4]$, the interval $[0, 1]$ maps into itself. By using the Königs–Lamerey method, one can easily show that, in this case, the trajectories of the points that do not belong to $[0, 1]$ approach infinity. Consider trajectories of the points from $[0, 1]$. We are now mainly interested in periodic points and the cycles formed by them.

A cycle $B = \{\beta_1, \ldots, \beta_m\}$ of a mapping $f: I \to I$ is called *attracting* if there exists a neighborhood U of this cycle such that $f(U) \subset U$ and

$$\bigcap_{n \geq 0} f^n(U) = B.$$

In this case, we have $\omega(x_0) = B$ for every point $x_0 \in U$ and the trajectory $\mathrm{orb}(x_0)$ splits into m sequences convergent to the points $\beta_1, \ldots, \beta_m$, respectively.

A cycle B is called *repelling* if there exists its neighborhood U such that any point of the set $U \backslash B$ leaves U after a finite period of time, i.e., for any $x \in U \backslash B$, there exists $n = n(x)$ such that $f^n(x) \notin U$.

These definitions can also be used in the case of an arbitrary topological space.

If f is differentiable, then one can use the following simple sufficient conditions that enable one to distinguish between attracting and repelling cycles: It is necessary to compute the quantity

$$\mu(B) = \frac{d}{dx} f^m(x) \bigg|_{x = x_0 \in B} = f'(\beta_1) \cdot \ldots \cdot f'(\beta_m),$$

which is called the *multiplier* of a cycle B. If $|\mu(B)| < 1$, then B is an attracting cycle and if $|\mu(B)| > 1$, then B is a repelling cycle. For $|\mu(B)| = 1$, the cycle B is called nonhyperbolic. In this case, it may be either attracting or repelling. One can also observe a more complicated behavior of trajectories in its neighborhood.

The examples presented below illustrate the changes in the behavior of trajectories of a map f_λ from family (1) for various values of the parameter λ. In these examples, we write f instead of f_λ wherever this does not lead to any ambiguities.

1. $0 < \lambda \leq 1$. In this case, the interval $I = [0, 1]$ contains a single fixed point $x = 0$ and this point is attracting. Since $f(x) < x$ for $x \in I \backslash \{0\}$, we can write

$$\bigcap_{n=0}^{\infty} f^n(I \backslash \{0\}) = \{0\},$$

i.e., for any point $x \in I \backslash \{0\}$, we have $f^n \to 0$ as $n \to \infty$. Hence, every trajectory $\mathrm{orb}(x_0)$ is attracted by the fixed point $x = 0$ (Fig. 3).

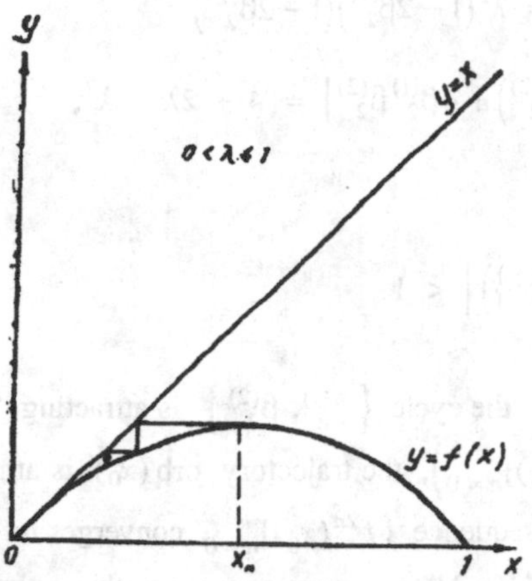

Fig. 3

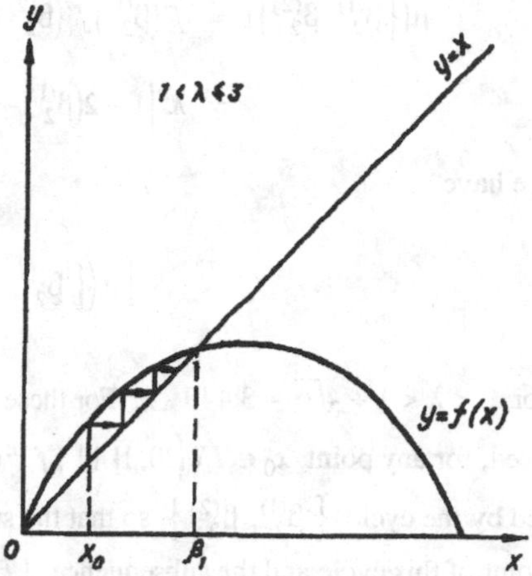

Fig. 4

2. $1 < \lambda \leq 3$. For $\lambda > 1$, the fixed point $x = 0$ becomes repelling $(f'(0) > 1)$ and a new fixed point $\beta_1 = 1 - 1/\lambda$ appears in the interval I (Fig. 4). Since $f'(x) = \lambda(1 - 2x)$, the multiplier $\mu(\beta_1)$ is equal to $2 - \lambda$ and, therefore, the fixed point $x = \beta_1$ is attracting for $1 < \lambda < 3$. For any point $x_0 \in (0, 1)$, we have $f^n(x_0) \to \beta_1$ as $n \to \infty$. Note that $\mu(\beta_1) > 0$ for $1 < \lambda < 2$ and the trajectory $\operatorname{orb}(x_0)$ monotonically approaches β_1. For $2 < \lambda < 3$, we have $\mu(\beta_1) < 0$ and the trajectory $\operatorname{orb}(x_0)$ approaches β_1 oscillating about this point and taking, in turn, values greater and lower than β_1.

For $\lambda = 3$, the fixed point $x = \beta_1$ is still attracting although, in this case, $|\mu(\beta_1)| = 1$.

3. $3 < \lambda \leq 1 + \sqrt{6}$. As the parameter λ becomes greater than $\lambda_1 = 3$, we observe the appearance of a new bifurcation, namely, the fixed point $x = \beta_1$ becomes repelling $(|\mu(\beta_1)| > 1$ for $\lambda > 3)$ and generates a new attracting cycle with period 2. The changes in the behavior of the map f in the vicinity of the point $x = \beta_1$ are displayed in Fig. 5, where we present the graphs of the function $y = f(f(x))$ for the parameter λ crossing the value $\lambda_1 = 3$.

A cycle of period 2 (Fig. 6) is formed by the points

$$\beta_2^{(1),\,(2)} = \frac{\lambda + 1 \pm \sqrt{\lambda^2 - 2\lambda - 3}}{2\lambda}.$$

The values $\beta_2^{(1)}$ and $\beta_2^{(2)}$ are defined as the roots of the equation $f^2(x) = x$ that differ from the roots of the equation $f(x) = x$ that defines the fixed points of f. Thus, for $\beta_2^{(1)}$ and $\beta_2^{(2)}$, we arrive at the equation $\lambda^2 x^2 - \lambda(\lambda + 1)x + (\lambda + 1) = 0$.

Since

$$\mu(\{\beta_2^{(1)}, \beta_2^{(2)}\}) = f'(\beta_2^{(1)})f'(\beta_2^{(2)}) = \lambda^2(1-2\beta_2^{(1)})(1-2\beta_2^{(2)})$$

$$= \lambda^2\left[1-2(\beta_2^{(1)}-\beta_2^{(2)})+4\beta_2^{(1)}\beta_2^{(2)}\right] = 4 + 2\lambda - \lambda^2,$$

we have

$$\left|\mu(\{\beta_2^{(1)}, \beta_2^{(2)}\})\right| < 1$$

for $3 < \lambda < 1 + \sqrt{6} \approx 3.449\ldots$. For these λ, the cycle $\{\beta_2^{(1)}, \beta_2^{(2)}\}$ is attracting. Indeed, for any point $x_0 \in I \setminus \left(\{0, 1\} \cup \{f^{-n}(\beta_1)\}_{n=0}^{\infty}\right)$, the trajectory $\operatorname{orb}(x_0)$ is attracted by the cycle $\{\beta_2^{(1)}, \beta_2^{(2)}\}$ so that the subsequence $\{f^{2n}(x_0)\}_{n=0}^{\infty}$ converges to one point of this cycle and the subsequence $\{f^{2n+1}(x_0)\}_{n=0}^{\infty}$ converges to another point of this cycle.

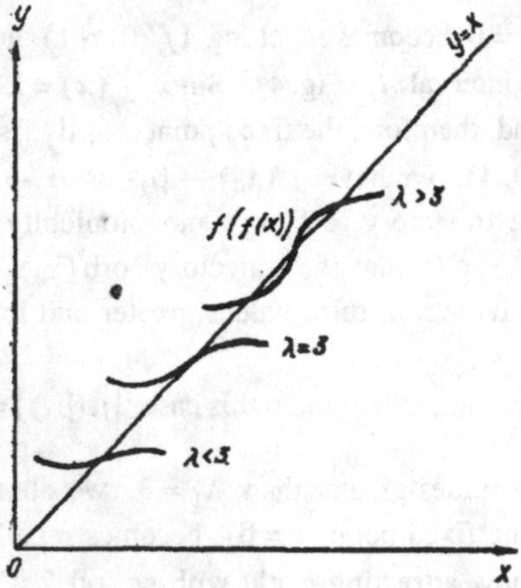

Fig. 5

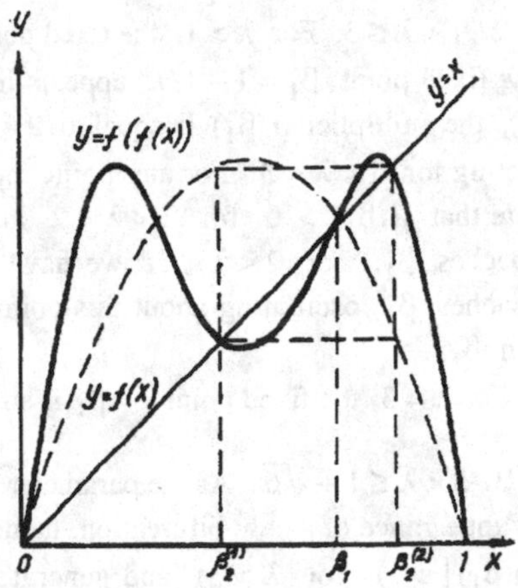

Fig. 6

We can specify the character of convergence of a trajectory to the cycle by using the multiplier $\mu(\{\beta_2^{(1)}, \beta_2^{(2)}\})$. As the parameter λ increases from 3 to $1+\sqrt{6}$, the multiplier increases from -1 to 1. Hence, for $3 < \lambda < 1 + \sqrt{5}$ and $\mu > 0$, the subsequences $\{f^{2n}(x_0)\}$ and $\{f^{2n+1}(x_0)\}$ are monotone beginning with certain n. Furthermore, one of them is increasing, while the other one is decreasing (since $f'(x) < 0$ for $x = \beta_2^{(1)}$ and $x = \beta_2^{(2)}$). For $1 + \sqrt{5} < \lambda < 1 + \sqrt{6}$, we have $\mu < 0$ and the subsequences $\{f^{2n}(x_0)\}$ and $\{f^{2n+1}(x_0)\}$ approach $\beta_2^{(1)}$ and $\beta_2^{(2)}$ oscillating about $\beta_2^{(1)}$ and $\beta_2^{(2)}$, respectively, so that the subsequences $\{f^{4n}(x_0)\}$, $\{f^{4n+2}(x_0)\}$, $\{f^{4n+1}(x_0)\}$, and $\{f^{4n+3}(x_0)\}$ are monotone.

4. $1+\sqrt{6} < \lambda < 3.569\ldots$. As the parameter λ crosses the value $\lambda_2 < 1+\sqrt{6} \approx 3.449\ldots$, we observe the appearance of the next bifurcation: The cycle $\{\beta_2^{(1)}, \beta_2^{(2)}\}$ becomes repelling (for $\lambda > 1+\sqrt{6}$, we have $|\mu(\{\beta_2^{(1)}, \beta_2^{(2)}\})| > 1$) and generates a new attracting cycle of period 4. This new cycle attracts all points of I except a countable set of points

$$\{0, 1\} \cup \left\{f^{-n}\left(\left\{\beta_1, \beta_2^{(1)}, \beta_2^{(2)}\right\}\right)\right\}_{n=0}^{\infty}.$$

If the parameter λ increases further, then, at $\lambda_3 \approx 3.54$, the cycle of period 4 also becomes repelling and generates an attracting cycle of period 8 (which attracts all points of the interval except countably many points). The process of consecutive doubling of the periods of attracting cycles occurs as the parameter λ increases to $\lambda = \lambda^* \approx 3.569\ldots$.

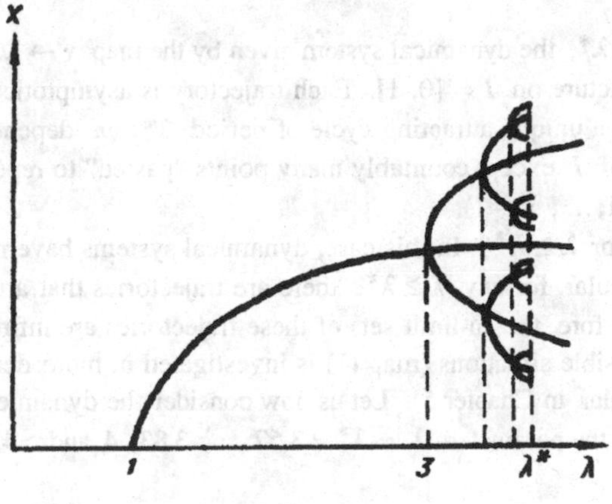

Fig. 7

5. There exists a convenient graphic representation of the qualitative reconstructions of cycles occurring as the parameter λ increases. It is called the bifurcation diagram (Fig. 7). The bifurcation curves of this diagram corresponding to $\beta_2^{(1)}$ and $\beta_2^{(2)}$ diverge as the branches of a parabola according to the formula for $\beta_2^{(1)}$ and $\beta_2^{(2)}$, namely,

$$|\beta_2^{(1)} - \beta_2^{(2)}| = O\left(\sqrt{|\lambda - \lambda_1|}\right) \quad \text{as} \quad \lambda \to \lambda_1 = 3.$$

At the same time, the fixed point β_1 drifts slower: $|\beta_1(\lambda) - \beta_1(\lambda_1)| = O(|\lambda - \lambda_1|)$.

A similar picture is also observed in the neighborhood of the subsequent bifurcation values $\lambda_2, \lambda_3, \ldots$.

As noted by Feigenbaum [3], if we compute the values λ_n with sufficiently high accuracy and construct the ratios

$$\delta_n = \frac{\lambda_n - \lambda_{n-1}}{\lambda_{n+1} - \lambda_n}, \quad n = 1, 2, \ldots,$$

then $\delta_n \to \delta = 4.66920\ldots$ as $n \to \infty$, i.e., the rate of appearance of cycles with doubled periods (as n increases) is characterized by the constant δ. There exists another constant $\alpha \approx 2.502\ldots$ that characterizes the sizes of emerging cycles. Let β'_{2^n} be the first point to the right of $x = 1/2$ belonging to a cycle with period 2^n (this point appears for $\lambda > \lambda_n$) and let $\beta''_{2^n} = f^{2^{n-1}}(\beta'_{2^n})$. Then

$$\frac{\beta'_{2^n} - \beta''_{2^n}}{\beta'_{2^{n+1}} - \beta''_{2^{n+1}}} \to \alpha = 2.502\ldots \quad \text{as} \quad n \to \infty.$$

6. For any $\lambda < \lambda^*$, the dynamical system given by the map $x \to \lambda x(1-x)$ has a relatively simple structure on $I = [0, 1]$. Each trajectory is asymptotically periodic. For any λ, there exists a unique attracting cycle of period 2^m (m depends on λ), which attracts all points of I except countably many points "pasted" to repelling cycles with periods 2^i, $i = 0, 1, \ldots, m-1$).

What happens for $\lambda \geq \lambda^*$? In this case, dynamical systems have more complicated structure. In particular, for any $\lambda \geq \lambda^*$, there are trajectories that are not attracted to any cycle and, therefore, the ω-limit sets of these trajectories are infinite. Here, we do not analyze all possible situations (map (1) is investigated in more details in what follows and, in particular, in Chapter 5). Let us now consider the dynamical system for the following values of the parameter: $\lambda = \lambda^* \approx 3.57\ldots$, 3.83, 4 and >4.

7. For $\lambda = \lambda^*$, map (1) already possesses cycles with periods 2^i, $i = 0, 1, 2, \ldots$ (all these cycles are repelling), but have no cycles with other periods. The set $K = (\mathrm{Per}(f))'$ of limiting points for the set of periodic points $\mathrm{Per}(f)$ is a nonempty nowhere dense perfect set, i.e., it is homeomorphic to the Cantor set. This set K does not contain periodic points, i.e., $K \cap \mathrm{Per}(f) = \varnothing$.

The dynamical system is minimal on K. Indeed, for any point $x \in K$, the trajectory $\mathrm{orb}(x)$ is dense in K, i.e., $\omega(x) = K$. The set K contains the point $x = 1/2$ (and, hence, $K = \omega(1/2)$). All points of the interval I, except the countable set

$$P = \bigcup_{i=0}^{\infty} f^{-i}(\mathrm{Per}(f)),$$

are attracted by the set K. Indeed, if $x \in I \backslash P$, then $\omega(x) = K$. We discuss the proofs of these statements in Chapter 5.

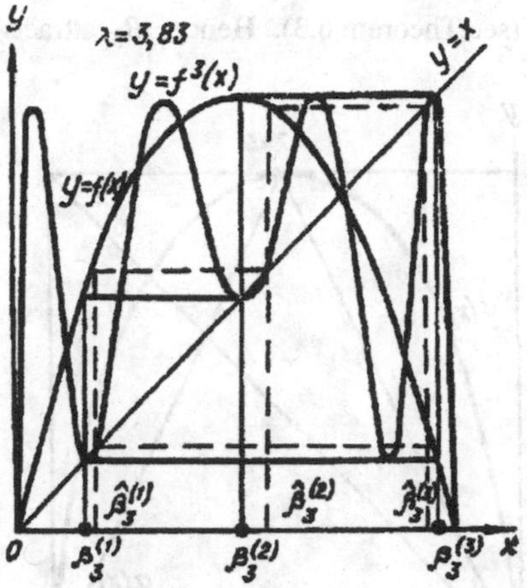

Fig. 8

8. $\lambda = 3.83$. As the parameter λ increases further, we observe the appearance of new cycles and, in particular, cycles whose periods are not equal to 2^i, $i = 0, 1, 2, \ldots$. For $\lambda = 3.83$, the map already has cycles of all periods $m \in \mathbb{N}$. The cycle B_3 of period 3 formed by the points $\beta_3^{(1)}, \beta_3^{(2)}$, and $\beta_3^{(3)}$ (Fig. 8) is attracting. In addition to the attracting cycle, there is a repelling cycle of period 3: $\left\{\hat{\beta}_3^{(1)}, \hat{\beta}_3^{(2)}, \hat{\beta}_3^{(3)}\right\}$ (points of these cycles can be computed as the roots of the following sixth-degree polynomial: $(f^3(x) - x)/(f(x) - x)$).

What points are attracted by the attracting cycle B_3? Let I_0 denote an open interval whose ends are the preimages of the point $\hat{\beta}_3^{(3)}$, i.e., the points $\hat{\beta}_3^{(2)}$ and $1 - \hat{\beta}_3^{(2)}$, $I_0 = (1 - \hat{\beta}_3^{(2)}, \hat{\beta}_3^{(2)})$. By using a computer, one can check that

(a) $f^3(I_0) \subset I_0$ (it suffices to show that $f^3(1/2) \in I_0$);

(b) the interval I_0 contains a single fixed point $\beta_3^{(2)}$ of the map f^3, and this point is attracting; the map $f^3|_{I_0}$ has no cycles of period 2.

Therefore, for any $x_0 \in I_0$, we have $f^{3n}(x_0) \to \beta_3^{(2)}$ as $n \to \infty$, i.e., the point x_0 is attracted by the cycle B_3 and the interval I_0 belongs to the basin of attraction of this cycle. It is clear that any trajectory attracted by the cycle B_3 also passes through the interval I_0. Hence, the set

$$P = \bigcup_{i=0}^{\infty} f^{-i}(I_0)$$

consists of the points of I attracted by the cycle B_3. The set P is open and dense in I and $\operatorname{mes} P = \operatorname{mes} I = 1$ (see Theorem 6.3). Hence, B_3 attracts almost all points of I.

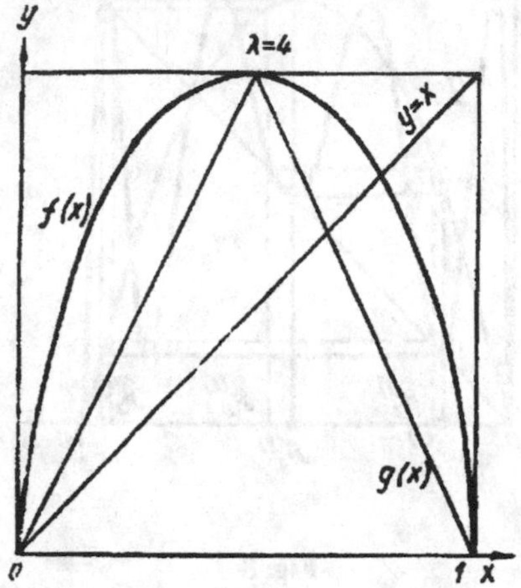

Fig. 9

The set $I \backslash P$ consists of the points that are not attracted by the cycle B_3. This is a perfect nowhere dense set, i.e., it is homeomorphic to the Cantor set. The fact that the set $I \backslash P$ is perfect follows from the fact that any distinct (maximal) open intervals which form $f^{-i}(I_0)$ have common ends neither for different nor for equal $i \geq 0$ (the same is true for the ends of the interval I (i.e., for the points 0 and 1)). We also note that the points x such that

$$\omega(x) = \hat{P} = (I \backslash P) \cap [f^2(1/2), f(1/2)]$$

and

$$\overline{\operatorname{Per}(f)} = \hat{P} \cup \{0\} \cup \{B_3\}$$

are everywhere dense in the set $I \backslash P$. This dynamical system is studied in more details in Chapter 5. The problem of the appearance of sets homeomorphic to the Cantor set is discussed below (see case 10 with $\lambda > 4$).

9. $\lambda = 4$. In this case, $\max_{x \in I} f(x) = f(1/2) = 1$ and, therefore, $f(I) = I$ (Fig. 9). In order to understand the properties of the dynamic system defined by the mapping

$$x \to f(x) = 4x(1-x), \qquad (2)$$

we use the fact that this mapping is topologically equivalent to the linear mapping

$$x \to g(x) = \begin{cases} 2x, & 0 \le x \le 1/2, \\ 2(1-x), & 1/2 < x \le 1. \end{cases} \tag{3}$$

Two maps $g_1 : X_1 \to X_1$ and $g_2 : X_2 \to X_2$ are called *topologically conjugate* or equivalent if there exists a homeomorphism $h : X_1 \to X_2$ such that the diagram

$$\begin{array}{ccc} X_1 & \xrightarrow{g_1} & X_1 \\ h \downarrow & & \downarrow h \\ X_2 & \xrightarrow{g_2} & X_2 \end{array}$$

is commutative, i.e., $h \circ g_1 = g_2 \circ h$.

For maps (2) and (3), we have $X_1 = X_2 = I$ and the conjugating homeomorphism $h : I \to I$ is given by the function $h(x) = \frac{2}{\pi} \arcsin \sqrt{x}$.

If two maps are conjugate, then the dynamical systems generated by these maps are also conjugate (or equivalent) (if $h \circ f = g \circ h$, then $h \circ f^n = g^n \circ h$ for any $n > 0$). Every trajectory of a dynamical system is associated with a trajectory of another dynamical system (this correspondence is established by the function h; the trajectory of the map f passing through the point x_0 is associated with the trajectory of g that passes through the point $h(x_0)$). The corresponding trajectories have the same asymptotic properties (the ω-limit sets of the trajectories $\{f^n(x_0)\}$ and $\{g^n(h(x_0))\}$ are homeomorphic; if one of these trajectories is attracted by a cycle, then the other is also attracted by a cycle, and so on).

Therefore, we can study the dynamics of map (3) instead of map (2) because this is much simpler.

Map (3) is expanding, i.e., it increases the distance between close points because the modulus of its derivative is everywhere greater than 1. This means that, for any open (in I) interval $\mathcal{J} \subset I$, there exists a number $m > 0$ such that $g^m(\mathcal{J}) = I$.

The proof of this fact is almost obvious: If $1/2 \notin \mathcal{J}$, then $l(g(\mathcal{J})) = 2l(\mathcal{J})$, where $l(\cdot)$ is the length of the interval; if $1/2 \in \mathcal{J}$, then there exists $\varepsilon > 0$ such that $g(\mathcal{J}) \supset [0, \varepsilon]$ and $g^m([0, \varepsilon]) = [0, \varepsilon \cdot 2^m]$ for $\varepsilon \cdot 2^m < 1$ and $g^m([0, \varepsilon]) = I$, otherwise.

A similar assertion can be established for any other map topologically equivalent to (3). In particular, it holds for map (2).

Lemma 1.1. *For any open (in I) interval $\mathcal{J} \subset I$, there exists a number m such that $f^m(\mathcal{J}) = I$.*

This lemma does not seem to be obvious because map (2) strongly contracts intervals in the vicinity of $x = 1/2$ ($f'(1/2) = 0$). Nevertheless, in view of the fact that $h \circ f^n = g^n \circ h$ for any $n \ge 0$, where $h(x) = \frac{2}{\pi} \arcsin \sqrt{x}$, we conclude that, under the action of the map f, the interval $\mathcal{J}$ will also cover the interval I after about

$$m = \frac{\log 1/l(h(\mathcal{J}))}{\log 2}$$

steps (because $h(\mathcal{J})$ is an interval).

This lemma enables us to establish many important properties of the dynamical system generated by map (2).

Proposition 1.1. *Periodic points are dense in I. Moreover, any open interval contains periodic points with arbitrarily large periods.*

Proposition 1.2. *There exists a trajectory everywhere dense in I. Moreover, almost all trajectories are everywhere dense in I (these trajectories form a set of the second Baire category in I).*

We prove Proposition 1.1. Let $\mathcal{J}$ be an arbitrary open interval and let m be such that $f^m(\mathcal{J}) = I$. Then there are points $x', x'' \in \mathcal{J}$ such that $f^m(x') = 0$ and $f^m(x'') = 1$. Due to the continuity of f (and, consequently, of f^m), one can find a point x_0 lying between x' and x'' such that $f^m(x_0) = x_0$. The point x_0 is periodic and its period is a divisor of m. In order to prove that the interval $\mathcal{J}$ contains periodic points whose periods are greater than m_0, it suffices to consider the map f^m on $\mathcal{J}$ with $m = m_0!$ There is an open interval $\mathcal{J}' \subset \mathcal{J}$ such that $f^m(x) \neq x$ for any $x \in \mathcal{J}'$. Therefore, $\mathcal{J}'$ does not contain periodic points with periods $1, 2, 3, \ldots, m_0$. At the same time, according to what has been proved above, $\mathcal{J}'$ contains periodic points and, hence, their periods are greater than m_0.

To prove Proposition 1.2, we take an arbitrary countable base on I, e.g., the base formed on I by open intervals $\mathcal{J}_1, \mathcal{J}_2, \ldots, \mathcal{J}_s, \ldots$. The fact that the family of $\mathcal{J}_s$ forms a base means that, for any point $x \in I$, one can indicate a sequence of intervals $\mathcal{J}_{s_1} \supset \mathcal{J}_{s_2} \supset \ldots$ such that

$$\bigcap_{i=1}^{\infty} \mathcal{J}_{s_i} = \{x\}.$$

Thus, one can choose a basis in the form of the family of intervals whose ends are binary rational points on I. It is clear that a trajectory that visits all intervals $\mathcal{J}_s$, $s = 1, 2, \ldots$, is dense in I. Let us show that one can find a point $x_0 \in \mathcal{J}_1$ such that $\{f^i(x_0)\}_{i=0}^{\infty} \cap \mathcal{J}_s \neq \varnothing$ for any $s = 1, 2, \ldots$. By virtue of the lemma, there are positive numbers $m_1, m_2, \ldots$ such that $f^{m_s}(\mathcal{J}_s) = I$ for $s = 1, 2, \ldots$. Since $f^{m_1}(\mathcal{J}_1) = I \supset \mathcal{J}_2$, one can find an open (in I) interval $\mathcal{J}^{(1)} \subset \mathcal{J}_1$ such that $f^{m_1}(\mathcal{J}^{(1)}) = \mathcal{J}_2$. In view of the fact that $f^{m_2}(\mathcal{J}_2) = I \supset \mathcal{J}_3$ and $f^{m_1}(\mathcal{J}^{(1)}) = \mathcal{J}_2$, one can find an open interval $\mathcal{J}^{(2)} \subset \mathcal{J}^{(1)}$ such that $f^{m_1+m_2}(\mathcal{J}^{(2)}) = \mathcal{J}_3$. Since $f^{m_3}(\mathcal{J}_3) = I \supset \mathcal{J}_4$, there exists an open interval

$\mathcal{J}^{(3)} \subset \mathcal{J}^{(2)}$ such that $f^{m_1+m_2+m_3}(\mathcal{J}^{(3)}) = \mathcal{J}_4$, and so on. We arrive at a sequence of enclosed open intervals $\mathcal{J}_1 \supset \mathcal{J}^{(1)} \supset \mathcal{J}^{(2)} \supset \mathcal{J}^{(3)} \supset \ldots \supset \mathcal{J}^{(s)} \supset \ldots$ such that

$$f^{\sum_{i=1}^{s} m_i}(\mathcal{J}^{(s)}) = \mathcal{J}_{s+1}.$$

It is clear that, for each point of the set $\bigcap_{s=1}^{\infty} \overline{\mathcal{J}^{(s)}}$, one can indicate a trajectory which passes through this point and is dense in $\mathcal{J}$.

The second part of Proposition 1.2 holds for dynamical systems in a general (Baire) space X:

If X contains a dense trajectory, then the points of the trajectories dense in X form a set of the second Baire category in X.

This is a consequence of the fact that the set of these points is a G_δ-set, i.e., it can be represented as an intersection of countably many open sets (Birkhoff [1]). This G_δ-set is dense in X (because it contains a trajectory dense in X). Therefore, it is a set of the second class. Thus, almost all points of the space X (and, in particular, I) generate trajectories dense in X. Here, the notion "almost all" is understood in the topological sense. For map (2), the Lebesgue measure of these points is equal to $\operatorname{mes} I = 1$, but one can find C^1-mappings of I onto itself which have trajectories dense in I and are such that the Lebesgue measure of all trajectories dense in I is less than 1.

For map (2), the Lebesgue measure of the set of points generating dense trajectories in I is equal to $\operatorname{mes} I$. Nevertheless, for general continuous maps on I that have trajectories everywhere dense in I, this condition may be not satisfied (generally speaking, it is often quite difficult to verify this fact (see, e.g., Lyubich and Milnor [1], Keller and Nowicki [1])).

All stated above for map (2) is true for the equivalent map (3). Consider the following important property of maps (2) and (3):

A measure μ defined in the space X is called *invariant* under a map $f: X \to X$ if, for any μ-measurable set $A \subset X$, we have $\mu(f^{-1}(A)) = \mu(A)$.

The Lebesgue measure is invariant under map (3). Map (2) possesses an invariant measure which is absolutely continuous with respect to the Lebesgue measure, namely,

$$\mu(dx) = dh(x) = \frac{1}{\pi} \frac{dx}{\sqrt{x(1-x)}}.$$

The existence of a finite invariant measure whose support has a positive Lebesgue measure means that, in order to characterize the properties of dynamical systems after a long period of time, one should use the language of probability theory.

In particular, for maps (2) and (3), even the statement of the problem concerning the construction of a trajectory that passes through a point $x_0 \in I$ must be made more precise. Thus, it is possible to determine 5 or 10 points of the trajectory of x_0 by using a computer: $x_1 = f(x_0), \ldots, x_n = f^n(x_0)$. At the same time, the exact computation of a sufficiently large segment of the trajectory, e.g., up to $n = 100$, is impossible for standard precision of computers used for this purpose and, hence, the problem of constructing large segments of trajectories is incorrect. To explain this idea, we note that, for map (3), any interval of length ε covers $[0, 1]$ after $m \approx (\log(1/\varepsilon))/\log 2$ steps. If our computer is capable of discerning $\varepsilon = 10^{-20}$, then it makes no sense to ask at which point of $[0, 1]$ the trajectory under investigation is located for $m > 20 \log_2 10$ ($\approx$ 70). Maps (2) and (3) "forget" initial conditions (x_0) very quickly and, for large m, one should ask: With what probability can the trajectory be found in a set $I' \subset I$? For example, if $I' = (\alpha, \beta)$, then this probability is equal to $\beta - \alpha$ for map (3) and, for map (2), we can write

$$\frac{1}{\pi}\int_{\alpha}^{\beta}\frac{dx}{\sqrt{x(1-x)}} = \frac{2}{\pi}\left(\arcsin\sqrt{\beta} - \arcsin\sqrt{\alpha}\right) = h(\beta) - h(\alpha).$$

Sometimes, it is used to say (see Blokh [2] and Guckenheimer [2]) that maps (2) and (3) are characterized by highly sensitive dependence on the initial conditions (on I). For such maps, every trajectory is unstable in Lyapunov's sense, for any $x \in I$ and $\varepsilon > 0$, there exist x' such that $|x - x'| < \varepsilon$ and $n > 0$ for which

$$\rho(x, x') = \max_{0 \le i \le n} |f^i(x) - f^i(x')| > 1/2$$

(this is a consequence of the lemma on expansion). Any two trajectories with distinct but close initial points diverge, and the rate of divergence is characterized by the *Lyapunov exponent* equal to

$$\lim_{n \to \infty} \ln \left| \frac{d}{dx} f^n(x) \Big|_{x = x_0} \right|^{1/n}$$

at the point x_0 (if this limit exists). Hence, the Lyapunov exponent is the parameter that enables one to estimate the maximum length of a segment of the trajectory the consideration of which makes sense.

The divergence of close trajectories in the bounded interval I leads to the situation where the number of trajectories with different asymptotic behavior becomes too large.

As a quantitative measure of the variety in the behavior of trajectories, we can take topological entropy defined as follows:

Let X be a compact topological space and let $f: X \to X$. If $\mathcal{A}$ is a family of subsets of X, then

$$\mathcal{A}^n = \mathcal{A}^n_f = \left\{ \bigcap_{i=0}^{n-1} f^{-i}(A_i) \,\middle|\, A_i \in \mathcal{A} \text{ for } i = 0, \ldots, n-1 \text{ and } \bigcap_{i=0}^{n-1} f^{-i}(A_i) \ne \varnothing \right\}.$$

If $\mathcal{A}$ is an open covering of X, we denote by $\mathcal{N}(\mathcal{A})$ the minimal possible cardinality of a subcovering extracted from $\mathcal{A}$. Then

$$h(f, \mathcal{A}) = \lim_{n\to\infty} \frac{1}{n} \log \mathcal{N}(\mathcal{A}_f^n)$$

is the *topological entropy* of f *on the covering* $\mathcal{A}$. The *topological entropy* of f is then defined by (Adler, Konheim, and McAndrew [1])

$$h(f) = \sup \{h(f, \mathcal{A}) \mid \mathcal{A} \text{ is an open covering of } X\}.$$

Let us also present the Bowen's definition of topological entropy (see Bowen [2] and Dinaburg [1]), which is equivalent to that given above. Let (X, ρ) be a compact metric space. A subset E of X is called *(n, ε)-separated* if, for every two different points x, $y \in E$, there exists $0 \le j < n$ with $\rho(f^j(x), f^j(y)) > \varepsilon$. A set $F \subset X$ *(n, ε)-spans* another set $S \subset X$ provided that, for each $x \in S$, one can indicate $y \in F$ such that $\rho(f^j(x), f^j(y)) \le \varepsilon$ for all $0 \le j < n$.

For a compact set $S \subset X$, let $r_n(\varepsilon, S)$ be the minimal possible cardinality of a set F which (n, ε)-spans S and let $s_n(\varepsilon, S)$ be the maximal possible cardinality of an (n, ε)-separated set E contained in S (we write $r_n(\varepsilon, S, f)$ and $s_n(\varepsilon, S, f)$ to stress that the relevant quantities depend on f). Finally, we define

$$r(\varepsilon, S, f) = \limsup_{n\to\infty} \frac{1}{n} \log r_n(\varepsilon, S, f)$$

and

$$s(\varepsilon, S, f) = \limsup_{n\to\infty} \frac{1}{n} \log s_n(\varepsilon, S, f).$$

Then we set

$$h_\rho(f, X) = \lim_{\varepsilon\to\infty} s(\varepsilon, S, f) = \lim_{\varepsilon\to\infty} r(\varepsilon, S, f)$$

and (see Bowen [3] and Dinaburg [1])

$$h(f) = h_\rho(f, X).$$

For $\lambda \in (0, \lambda^*)$, every trajectory of the map $x \to \lambda x(1-x)$ is a periodic trajectory or its ω-limit set is a cycle. It is not difficult to check that, in this case, $h(f) = 0$. The following statement was proved by Bowen and Franks [1] and Misiurewicz [1]:

The topological entropy of a continuous mapping $f: I \to I$ is equal to zero if and only if the period of every cycle is a power of two.

For piecewise monotone maps (in particular, for maps with a single extremum), there exists a simple formula for topological entropy (Misiurewicz and Szlenk [1])

$$h(f) = \lim_{n \to \infty} \frac{1}{n} \log m_n,$$

where m_n is the number of intervals of monotonicity of f^n. Consequently, for $\lambda = 4$, the topological entropy of the map $x \to \lambda x(1-x)$ is equal to $\log 2$.

10. $\lambda > 4$. Finally, we consider the map f_λ for $\lambda > 4$ and $x \in \mathbb{R}$. In this case, we have $f_\lambda(1/2) = \lambda/4 > 1$ and, consequently, $f_\lambda(I) \not\subset I$ (Fig. 10). In particular, $f_\lambda(1/2) \notin I$ and $f_\lambda^n(1/2) \to -\infty$ as $n \to \infty$. The same behavior is exhibited by all trajectories starting at the points of the interval $\mathcal{J} = \{x \in \mathbb{R} : f_\lambda(x) > 1\}$ (the ends of the interval $\mathcal{J}$ are the roots of the equation $\lambda x(1-x) = 1$ and, consequently, are given by the expression $\frac{1}{2}(1 \pm \sqrt{1 - 4/\lambda})$). The interval I contains two intervals $\mathcal{J}_0$ and $\mathcal{J}_1$ which are preimages of the interval $\mathcal{J}$ (i.e., $f(\mathcal{J}_0) = f(\mathcal{J}_1) = \mathcal{J}$). Thus, the interval I also contains two preimages $\mathcal{J}_{00}$ and $\mathcal{J}_{10}$ of $\mathcal{J}_0$ and two preimages $\mathcal{J}_{01}$ and $\mathcal{J}_{11}$ of $\mathcal{J}_1$, and so on. Obviously, all trajectories starting from the set

$$\mathcal{J}^* = \bigcup_{i=0}^{\infty} f^{-i}(\mathcal{J})$$

(in particular, from the intervals $\mathcal{J}_0$ and $\mathcal{J}_1$, $\mathcal{J}_{00}$, $\mathcal{J}_{01}$, $\mathcal{J}_{10}$, and $\mathcal{J}_{11}$) eventually leave the interval I and approach $-\infty$ as $n \to \infty$.

The set $\mathcal{J}^*$ is open. Moreover, one can show that it is dense on I and its complement $K = I \backslash \mathcal{J}^*$ is a perfect nowhere dense set. Consequently, it is homeomorphic to the Cantor set. Furthermore, $\operatorname{mes} K = 0$ (see Theorem 6.3 and Henry [1]).

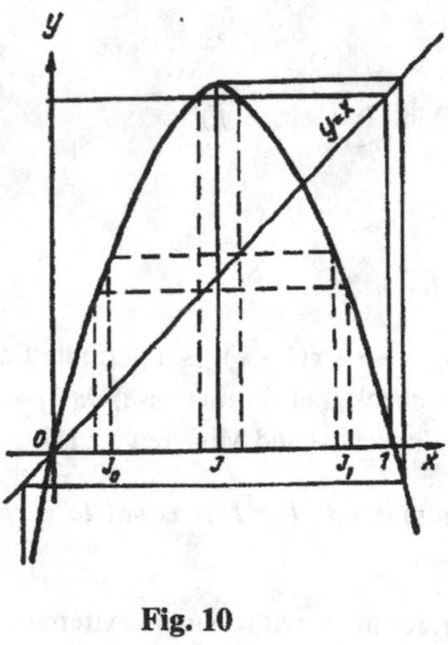

Fig. 10

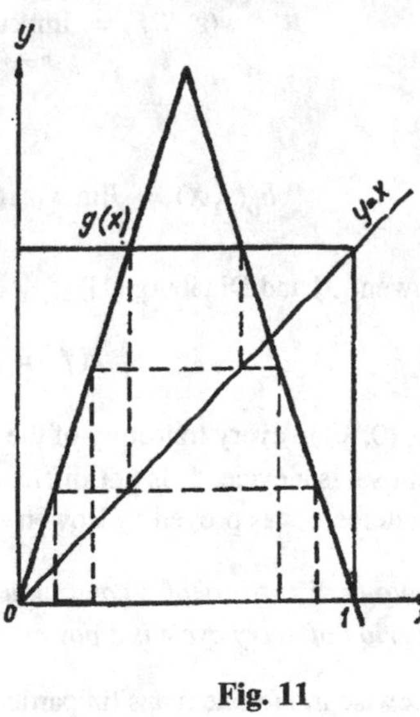

Fig. 11

The dynamical system defined on the set K possesses the same properties as the dynamical system generated by the map f_λ with $\lambda = 4$ on I. Namely, the periodic points are dense in K. Moreover, in any neighborhood of any point K, one can find periodic points with arbitrarily large periods. The set K contains an everywhere dense trajectory.

For any $\lambda > 4$, the map f_λ is conjugate (on $\mathbb{R}$) to the map

$$g\colon x \mapsto g(x) = \begin{cases} 3x, & x \le \dfrac{1}{2}, \\ 3(1-x), & x > \dfrac{1}{2} \end{cases}$$

(there exists a homeomorphism $h_\lambda : \mathbb{R} \to \mathbb{R}$ such that $f_\lambda = h_\lambda^{-1} \circ g \circ h_\lambda$) (see Fig. 11). The points that do not leave the interval $[0, 1]$ under the action of g form a set

$$\{x \in [0, 1] \mid g^n(x) \in [0, 1],\ n \ge 0\}\ \ (= h_\lambda K).$$

This is actually the standard Cantor set. Indeed, let us use the ternary representation of the points $x \in [0, 1]$, i.e., we set $x = 0.\alpha_1\alpha_2\alpha_3 \ldots \alpha_i \ldots$, where $\alpha_i \in \{0, 1, 2\}$. For ternary rational points, $x = 0.\alpha_1 \ldots \alpha_m 000\ldots = 0.\alpha_1 \ldots \alpha'_m 222\ldots$ ($\alpha_m \ne 0$, $\alpha'_m = \alpha_m - 1$), we use the first representation if $\alpha_m = 2$ and the second representation if $\alpha_m = 1$. Then

$$g(x) = \begin{cases} 0, \alpha_2\alpha_3 \ldots \alpha_i \ldots, & \text{if } \alpha_1 = 0, \\ > 1, & \text{if } \alpha_1 = 1, \\ 0, \overline{\alpha}_2\overline{\alpha}_3 \ldots \overline{\alpha}_i \ldots, & \text{if } \alpha_1 = 2, \end{cases}$$

where $\overline{\alpha}_i = 2 - \alpha_i$ (cf. the similar representation for map (3)). Therefore, the points that contain 1 in their ternary representation leave the interval $[0, 1]$ under the action of g, while points of the Cantor set

$$\{x \in [0, 1] \mid x = 0.\alpha_1\alpha_2\alpha_3 \ldots \alpha_i \ldots,\ \alpha_i \in \{0, 2\}\}$$

do not leave the interval $[0, 1]$.

These examples demonstrate how sets homeomorphic to the Cantor set appear in the theory of dynamical systems, and it becomes clear why these sets play an important role in the dynamics of systems.

2. ω-Limit and Statistically Limit Sets. Attractors and Quasiattractors

As indicated above, the asymptotic behavior of trajectories is described by ω-limit sets. The examples considered in Section 1 demonstrate that, in the most simple cases, the ω-limit sets are fixed points and cycles. In more complicated cases, they can be Cantor sets (as in the case where $\lambda > 4$ or $\lambda = \lambda^* \approx 3.57$ for map (1)) or intervals (as for $\lambda = 4$, i.e, for map (2)).

What other types of ω-limit sets can be discovered for one-dimensional dynamical systems? Is it possible for an ω-limit set to consist of finitely many points but not to be a cycle, for example, to consist of two different cycles?

For a dynamical system on an arbitrary locally compact space X, the following statement is true (Sharkovsky [5]):

If an ω-limit set consists of finitely many points, then these points form a cycle.

This is true due to the following property of a dynamical system on the ω-limit set of any compact trajectory (Sharkovsky [5]):

(*) *If F is an ω-limit set, then $f(\overline{U}) \not\subset U$ for any set $U \subset F$ $(U \neq F)$ open with respect to F.*

In this case, we say that the dynamical system possesses the property of *weak incompressibility.* If we assume that F consists of finitely many points and contains a cycle $\tilde{F}$ that does not coincide with F, then $\tilde{F}$ should be a closed invariant set and, at the same time, it should be open with respect to F, i.e., we arrive at the inclusion $f(\overline{F}) \subset F$ that contradicts the property of incompressibility.

By the same reason, we have the following assertion (Sharkovsky [5]):

Each cycle that lies in an ω-limit set but does not coincide with this set is not isolated in this ω-limit set; more precisely, each point of this cycle is not isolated.

This means that each point of this cycle is limiting for the points of the ω-limit set.

This situation can be encountered in the case where an ω-limit set consists of infinitely many points. This ω-limit set can be either countable or continual. It is worth noting that, in the first case, the ω-limit set F necessarily contains at least one cycle. Indeed, a sequence of closed sets $F_1 \supset F_2 \supset \ldots \supset F_\alpha \supset \ldots$, where $F_1 = F$, $F_{\alpha+1} = \omega(x_\alpha)$ (x_α is an arbitrary point from F_α) and $F_{\alpha'} = \bigcap_{\alpha<\alpha'} F_\alpha$ whenever α' is a limiting ordinal number, is always stabilized, i.e., there exists a finite or countable ordinal number α^* such that $F_{\alpha*} = F_{\alpha*+1}$ (Aleksandrov [1]). If F_α is countable, then $F_{\alpha+1} \neq F_\alpha$ because isolated points of F_α do not belong to $F_{\alpha+1}$ unless these points are periodic and belong to the trajectory $\{f^n(x_\alpha)\}$. Therefore, F_α consists of finitely many points forming a cycle.

Is it possible for an ω-limit set of a one-dimensional dynamical system to be countable? The answer is positive. Let us present a simple example. For this purpose, we recall the definition of homoclinic trajectories. If time in a dynamical system is reversible, i.e., if the dynamical system under consideration is a group (but not a semigroup) of maps, a trajectory is called *homoclinic* if it approaches the same periodic trajectory both as time infinitely increases and infinitely decreases. In our case, this definition is not correct because, generally speaking, time is not reversible. One of the possibilities to preserve this notion for semigroups is to consider bilateral trajectories $\{x_i\}_{i=-\infty}^{i=+\infty}$, where $x_{i+1} = f(x_i)$. However, one may arrive both at the situation where there are many negative trajectories $\{x_i\}_{i=-\infty}^{i=-1}$ for the point x_0 (if f^{-1} is an ambiguous function) and at the situation where there are no negative trajectories at all (if $f(I) \neq I$).

We can now apply the definition of homoclinic trajectories presented above to the trajectory $\{x_i\}_{i=-\infty}^{i=+\infty}$. A bilateral trajectory is homoclinic to some periodic trajectory γ if $\omega(x_0) = \gamma$ and the set of limiting points of the sequence $\{x_{-i}\}_{i=0}^{\infty}$ coincides with γ.

To present an example of a countable ω-limit set, we consider map (3) once again. Assume that the point $x_0 \in I$ is such that $\omega(x_0) = \{0\}$. Since $x = 0$ is a repelling fixed point, this is possible only in the case where $g^m(x_0) = 0$ for some $m > 0$. The set of points $\{g^i(x_0),\ i = 0, 1, \ldots, m,\ x_0/2^i,\ i = 1, 2, \ldots\}$ forms a homoclinic trajectory. It is easy to show that this homoclinic trajectory is the ω-limit set for other trajectories (there are many trajectories of this sort; the points of these trajectories form a set of the third Baire class (Sharkovsky [7]).

Property (∗) implies the following statement, which is also valid in the general case (Sharkovsky [6]):

If an ω-limit set F is different from a cycle, then any its open (with respect to F) zero-dimensional subset (if it exists) contains at least one nonperiodic point.

The requirement that an open set be zero-dimensional is essential. For example, for maps on the plane, an ω-limit set may be an interval consisting only of fixed points.

The abovementioned property also implies that, for one-dimensional maps, the following stronger statement (Sharkovsky [6]) is true:

If $X = I$, then on any ω-limit set that is not a cycle, nonperiodic points are dense (i.e., nonperiodic points form a dense subset on any ω-limit set of this type).

An ω-limit set may contain a trajectory for which it is the ω-limit set (this means that the trajectory is dense in this set). Then, similarly to the reasoning in Section 1 (see Proposition 1.2), we conclude that almost every trajectory is dense in the ω-limit set (such trajectories form a G_δ-set in it). If not all trajectories are dense in this set (i.e., if the set is not minimal), then the set of points that generate nondense trajectories is also relatively large: such points are dense in this ω-limit set (Sharkovsky [6]).

If an ω-limit set contains a trajectory that is dense in this set, then, on this set, the dynamical system possesses the following "mixing" property stronger than property (*):

(**) *For any two open (with respect to* F*) subsets* $U_1, U_2 \subset F$*, there exists* $m > 0$ *such that* $f^m U_1 \cap U_2 \neq \varnothing$.

Properties (*) and (**) completely describe the behavior of a dynamical system on ω-limit sets (Sharkovsky [5], [9]) in the following sense:

Suppose that a continuous map f *given on a closed set* $F \subset X$ *satisfies the condition* $fF = F$*. Then*

– if property (**) holds, there exists a point $x \in F$ *such that* $\omega(x) = F$*;*

– if property (*) holds, then, provided that F *is nowhere dense in* X *(i.e., it does not contain open subsets of* X*), the map* f *can be extended to a closed set* X'*,* $F \subset X' \subseteq X$*, such that the map* f *on the set* X' *is continuous and there exists a point* $x \in X'$ *for which* $\omega(x) = F$*.*

Thus, the question about the admissible topological structure of ω-limit sets can be reduced to the following one: What topological structure should the closed set F have in order that one can define a continuous map on it that possesses either property (*) or the stronger property (**) ?

Since, on any connected set, the identity map (all trajectories of which are fixed points) possesses property (*), *any closed connected set can be an* ω*-limit set of a dynamical system.*

On the other hand, one can easily give examples of closed sets that cannot be ω-limit sets. For example, it is not possible to define a continuous map with property (*) if the set F consists of

– finitely many connected components, at least one of which is a point and another one differs from a point;

– infinitely many connected components, only finitely many of which are not one-point sets and at least one of the components is isolated from the others.

These statements are simple consequences of the fact that the components which are not one-point sets must form an invariant set. It follows from (Kolyada, Snoha [1]) that there are no exceptions for sets that can be imbedded into the real line: in this case, one can find a continuous map possessing property (*) if and only if the set is not a set of the form indicated above, i.e., if it is not the union of finitely many intervals and finitely or infinitely many points the closure of which has no common points with at least one of these intervals.

This means, that in the case of continuous maps on an interval, one can define a continuous map with property (*) on a closed subset of the real line only if the set

(i) does not contain intervals (i.e., is nowhere dense on the real line);

(ii) consists of finitely many intervals;

(iii) consists of finitely many intervals and a countable set of points the closure of which intersects each of these intervals;

(iv) consists of finitely many intervals and an uncountable nowhere dense set;

(v) consists of countably many intervals.

As shown by Kolyada and Snoha [1], any set with the structure described above can be an ω-limit set for continuous maps on the plane.

However, for continuous maps on an interval, sets that contain intervals and separated points cannot be ω-limit sets. Thus,

for continuous maps on an interval, a closed set F can be an ω-limit set only in the following cases:

– F is an arbitrary nowhere dense set;

– F consists of finitely many intervals.

The second possibility can easily be realized, e.g., by maps similar to (2) and (3). The realizability of the first possibility was proved by Agronsky, Bruckner, Ceder, and Pearson [1].

For the ω-limit set of each trajectory, one can select its smallest closed subset such that, for any neighborhood of this subset, the trajectory stays in it almost all time. This is especially important in connection with the fact that it is often impossible to get a precise mathematical description of dynamical systems and just this subset (but not the entire ω-limit set) is, as a rule, observed in experiments.

For a map $f: I \to I$, we define

$$p(x, U) = \lim_{n \to \infty} \sup \frac{1}{n} \sum_{k=0}^{n-1} \chi_U(f^k(x)),$$

where $x \in I$, U is an arbitrary subset of the interval I, and χ_U is the indicator of U. The trajectory of a point x is called *statistically asymptotic* with respect to the set M if the equality $p(x, U) = 1$ holds for any neighborhood U of M (Krylov and Bogolyubov [1]). It is clear that each trajectory is statistically asymptotic with respect to its own ω-limit set. The smallest closed set for which the trajectory of a given point $x \in I$ is statistically asymptotic is called the *statistical limit set* or the σ-limit set of the trajectory of a point x; it is denoted by $\sigma_f(x)$ or simply by $\sigma(x)$. As indicated above, we have the inclusion $\sigma(x) \subset \omega(x)$ but the situation where $\sigma(x) \neq \omega(x)$ is also possible. Indeed, if $\omega(x)$ is the closure of a trajectory homoclinic to a certain cycle, then $\sigma(x)$ coincides with this cycle and, hence, does not coincide with $\omega(x)$. Unlike $\omega(x)$, the set $\sigma(x)$ may consist of finitely many points being not a cycle, e.g., it may consist of a pair of cycles. This situation is observed when $\omega(x)$ is a pair of cycles joined by heteroclinic trajectories (a trajectory is called *heteroclinic* to cycles B and B' if it approaches B as time increases and B' as time decreases).

The dynamical system generated by the map f_λ for $\lambda \geq 4$ (see Section 1) possesses an important property, which is known as mixing of trajectories. For $\lambda = 4$, mixing takes place on $I = [0, 1]$, while for $\lambda > 4$, it takes place on an invariant Cantor subset of I. The definition of mixing can be formulated as follows:

If $\{X, f\}$ is a dynamical system and $\Lambda \subset X$ is a compact invariant set but not a cycle, then we say that this dynamical system is *mixing* on Λ or, for the sake of brevity, that Λ is a mixing set if, for any open (in Λ) set V and any open finite covering $\Sigma = \{\sigma_j\}$ of the set Λ, there exists $m = m(V, \Sigma)$ and $r \geq 1$ depending only on Λ and such that

$$f^m\left(\bigcup_{i=0}^{r-1} f^i(V)\right) \cap \sigma_j \neq \varnothing$$

for all j. This property can be characterized by the following physical analogy: Imagine a "drop" (a set V open in Λ) that gets into Λ and, after a certain period of time, fills the entire set Λ.

The property of mixing on Λ implies *transitivity*. Indeed, for any two open (in Λ) sets V_1 and V_2 $(\subset \Lambda)$, there exists a number m such that $f^m(V_1) \cap V_2 \neq \varnothing$. Transitivity is equivalent (if $f\Lambda = \Lambda$) to the existence of a dense trajectory in Λ and, in this sense, any transitive set is indecomposable. We have already noted that, for the map f_λ, both for $\lambda = \lambda^*$ and $\lambda > 4$, there exists a Cantor-type subset of the interval $I = [0, 1]$ that contains dense trajectories. For $\lambda = \lambda^*$, it is easy to show that mixing is absent. At the same time, for $\lambda > 4$, on this invariant set (denoted by K), the dynamical system possesses not only the property of mixing but also a stronger property of expansion, i.e., for any set $V \subset K$ open in K, there exists a number m depending on V and such that $f^m(V) = K$.

The map f_λ with $\lambda = 4$ possesses the same property in the entire I (this is just the assertion of Lemma 1.1) by virtue of the fact that the map f_λ is expanding on K. Hence, both for $\lambda = 4$ and $\lambda > 4$, we observe mixing; moreover, the number m in the definition of this property can be chosen independently of the covering Σ and r can be chosen to be equal to 1.

If $\{X, f\}$ is a dynamical system and Λ is a compact invariant set, then we say that Λ is a *strange* or *mixing attractor* (Sharkovsky, Maistrenko, and Romanenko [2]) whenever

(a) Λ is an attractor, i.e., there exists a neighborhood U of Λ such that $U \supset f(U) \supset f^2(U) \supset \ldots$, $U \neq \Lambda$, and $\bigcap_{i \geq 0} f^i(U) = \Lambda$;

(b) Λ is a mixing set.

For the map f_λ, a closed interval I and a Cantor-type subset $K \subset I$ are mixing sets but not attractors for $\lambda = 4$ and $\lambda > 4$, respectively. Indeed, for $\lambda = 4$, we have $f^n(x) \to -\infty$ as $n \to \infty$ for any point $x \in \mathbb{R} \setminus I$. For $\lambda > 4$, the set $K = I \setminus \mathcal{J}^*$, where

$$\mathcal{J}^* = \bigcup_{n=0}^{\infty} f^{-n}(\mathcal{J}), \quad \mathcal{J} = \{x \in I \mid f(x) > 1\},$$

and $f^n(x) \to -\infty$ as $n \to \infty$ for any point $x \in \mathcal{J}^*$, is mixing.

It is not difficult to "improve" the map f_λ with $\lambda = 4$ on the set $\mathbb{R} \setminus I$ to transform the mixing set I into an attractor. Thus, one can set

$$\bar{f}_\lambda(x) = \begin{cases} \lambda x(1-x), & x \geq 0, \\ 0, & x < 0. \end{cases}$$

In this case, the interval $[0, 1]$ is a mixing attractor of the map $\bar{f}_\lambda$ with $\lambda = 4$.

This proves that one-dimensional maps may have mixing attractors either in the form of intervals containing a dense trajectory or in the form of a collection of intervals cyclically mapped into each other. The following statement is true:

If $f \in C^0(I, I)$ and I is an interval, then a mixing attractor consists of one or finitely many intervals cyclically mapped into each other.

Thus, a Cantor-type set cannot be a mixing attractor. In particular, the mixing Cantor-type set K mentioned above for the map $\bar{f}_\lambda$ with $\lambda > 4$ does not possess property (a) despite the fact that $\omega(x) \subset K$ for any $x \in \mathbb{R}$.

Let us explain this in brief. Since any mixing set contains a trajectory dense in this set, it is a perfect set and if it is not dense at least at one point of I, then it is nowhere dense on this interval. Hence, the mixing sets are either homeomorphic to the Cantor set or consist of finitely many intervals. Any neighborhood of a nowhere dense set which is dense in itself always contains points that are not attracted to this set. This result is due to Sharkovsky [2, 8]. In [2], Sharkovsky established the following fact:

Every nonisolated point of an arbitrary ω-limit set is a limiting point of the set of periodic points.

Therefore, in order that a set be an attractor, it is necessary that the periodic points be dense in it (for this reason, the minimal set that exists for $\lambda = \lambda^*$ and is not a cycle cannot be an attractor). At the same time, if the ω-limit set contains periodic points, then the dynamical system possesses on this set the property expansion of (relative) neighborhoods (Sharkovsky [8]). As a result, any sufficiently small neighborhood of this set contains points leaving this neighborhood after a certain period of time.

For some values of the parameter λ, the mapping f_λ may have mixing attractors, e.g., for $\lambda = 3.678\ldots$ when the point $x = 1/2$ hits the fixed point $x = 1 - 1/\lambda$ after 3 steps (Fig. 12). In this case, the interval

$$\mathcal{J} = [f^2(1/2), f(1/2)],$$

where $f(1/2) = \lambda/4 \approx 0.92$ and $f^2(1/2) = \lambda^2(1-\lambda/4)/4 \approx 0.27$ is an attractor. Indeed, for any closed interval I' such that $I' \subset (0, 1)$, one can indicate m such that $f^m(I') \subset \mathcal{J}$. In the interval $\mathcal{J}$, the mapping is mixing and, in particular, possesses all properties exhibited by the map f_λ with $\lambda = 4$ on the interval I (the set of periodic points is dense, there are everywhere dense trajectories, there is an invariant measure absolutely continuous with respect to the Lebesgue measure). In the interval $\mathcal{J}$, the map f_λ is conjugate to the piecewise linear map

$$x \mapsto g(x) = \begin{cases} (2/3)(1+x), & x \le 1/2, \\ 2(1-x), & x \ge 1/2, \end{cases}$$

defined on the interval $[0, 1]$. The interval $[0, 1]$ is a mixing attractor of the map g (Fig. 13).

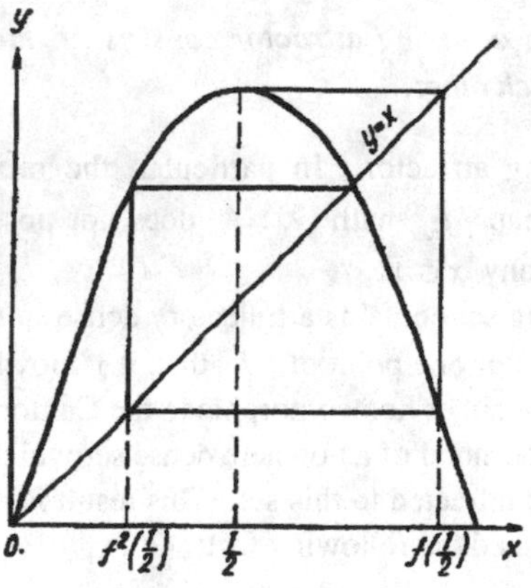

Fig. 12

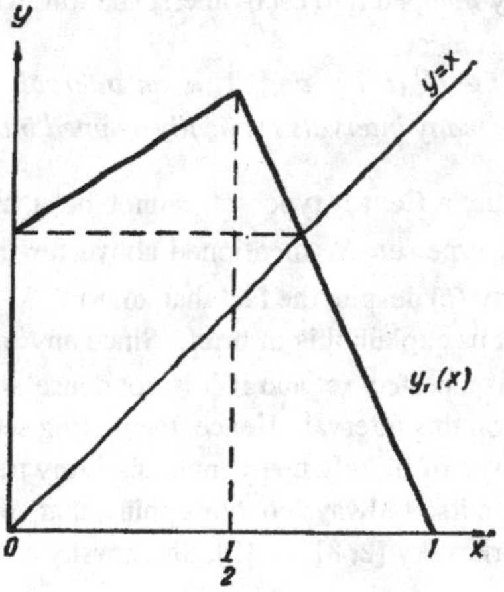

Fig. 13

It should be noted that the mapping f_λ possesses a mixing set whenever the value of the parameter λ is chosen so that the point $x = 1/2$ (the point of extremum) hits some repelling periodic point of period m for finitely many steps. This set is an attractor and consists of m intervals provided that the periodic point does not coincide with the ends of one of the intervals (as for $\lambda = 4$, where $x = 1/2$ hits the fixed point $x = 0$ which is one of the ends of the interval $[0, 1]$). In particular, if the point $x = 1/2$ hits the repelling cycle with period 2 (as already mentioned, it is formed by the points

$$\frac{\lambda + 1 \pm \sqrt{\lambda^2 - 2\lambda - 3}}{2\lambda})$$

and the parameter λ takes the least possible value ($\lambda \approx 3.593$), then the mixing attractor consists of 2 intervals.

A mixing set which is not an attractor and, in addition, does not belong to any larger ω-limit set is sometimes called a *mixing repeller*. We have already encountered such sets in our presentation. The map f_λ possesses a mixing repeller for $\lambda = 4$ (the interval $I = [0, 1]$) and for $\lambda > 4$ (a Cantor-type set on I). Repellers and attractors play an important role in the theory of difference equations and, especially, in the theory of equations with continuous argument.

As already mentioned, the minimal set K which exists for the map $x \to \lambda x(1-x)$ with $\lambda = \lambda^*$ and differs from a cycle is not an attractor. However, the set K is, in a certain sense, a quasiattractor. (Moreover, the ω-limit sets of almost all points in I coincide with K.)

A set $A \subset I$ is called a *quasiattractor* if

(i) for any neighborhood U of the set A, there exists a neighborhood $V \subset U$ such that $f^i(V) \subset U$ for all $i \geq 0$;

(ii) there exists a neighborhood U of the set A such that the ω-limit sets of almost all its points belong to A.

3. Return of Points and Sets

As already mentioned, the asymptotic behavior of the trajectories of a dynamical system may be fairly diverse. In order to understand a dynamical system as a whole, it is convenient to select in its phase space the sets which attract all or almost all trajectories. One of the most important properties of trajectories belonging to such sets is the property of *return*.

In the theory of dynamical systems, it is customary to distinguish between several types of return. The simplest type is connected with the return of points to their initial location after a certain period of time. Points with this property are called periodic (in the previous sections, they have been studied in detail). The set of periodic points of a map f is usually denoted by $\mathrm{Per}(f)$.

A more general type of return is connected with the return of a point into its own neighborhood (even after an arbitrarily large period of time): A point $x \in X$ is called *recurrent* if $x \in \omega(x)$, i.e., for any neighborhood U of x, there exists an integer $m > 0$ such that $f^m(x) \in U$ and, consequently, one can find an infinite sequence of return times $m_1 < m_2 < \ldots$ such that $f^{m_i}(x) \in U$ for $i = 1, 2, \ldots$. Recurrent points can be, in turn, classified depending on the properties of the sequence $\{m_i\}$. For example, if $\{m_i\}$ is a relatively dense sequence, then x is called a *regularly recurrent point;* if, in addition, $m_i = mi$ (m depends on U), then x is called an *almost periodic point,* and so on.

The set of recurrent points of a map f is denoted by $R(f)$, the set of regularly recurrent points is denoted by $RR(f)$, and the set of almost periodic points by $AP(f)$. (It should be noted that some authors use the terms "Poisson stable", "almost periodic", and "isochronous" points instead of "recurrent", "regularly recurrent", and "almost periodic" points, respectively).

It follows from the definitions introduced above that $AP(f) \subseteq RR(f) \subseteq R(f)$. Note that there exist maps such that $R(f) \setminus RR(f) \neq \varnothing$ (for example, it follows from Proposition 1.2 that map (2) has a trajectory everywhere dense in I whose points belong to $R(f) \setminus RR(f)$) and maps such that $RR(f) \setminus AP(f) \neq \varnothing$ (e.g., the piecewise linear map f in Fig. 14, where $f|_{[a,b]}$ is topologically conjugate to $f^2|_{[f(a),\, b]}$. For this map, the point b belongs to $AP(f)$ while its preimage b' belongs to $RR(f) \setminus AP(f)$. For the proof of this property, see Section 4 in Chapter 4).

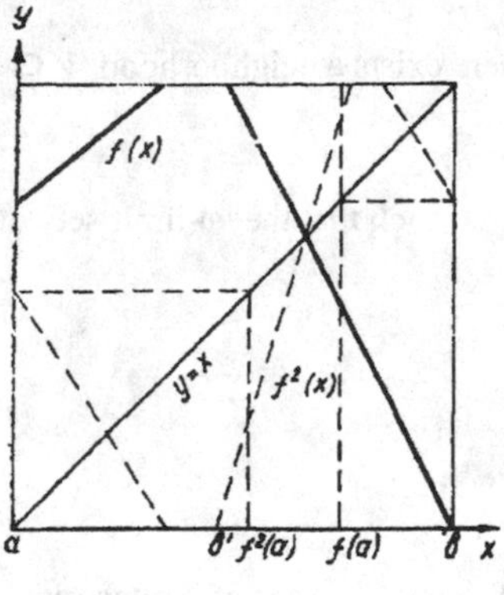

Fig. 14

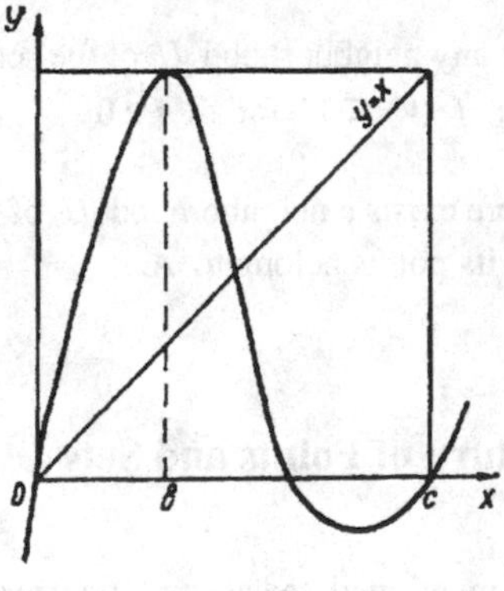

Fig. 15

A weaker type of return is exhibited by the so-called nonwandering points. A point $x \in X$ is called *nonwandering* if, for any its neighborhood U, there exists an integer $m > 0$ such that $f^m(U) \cap U \neq \varnothing$, i.e., a subset of U returns into U after m steps. It is clear that the points exhibiting all types of return described above are nonwandering as well as the ω-limit points of the trajectories. The set of all nonwandering points of a dynamical system generated by a map f is denoted by $NW(f)$.

It follows from the definition of $NW(f)$ that $NW(f)$ is always a closed set and if the dynamical system is a group of maps, then NW(f) is invariant (i.e., $f(\mathrm{NW}(f)) = \mathrm{NW}(f)$).

The following assertion is well known (the Birkhoff theorem):

Consider a dynamical system defined in a space X. Assume that the space X is compact. Then, for any neighborhood U of NW(f), there exists an integer m (depending on U) such that the time of stay of any trajectory outside U does not exceed m, i.e., the following inequality is true for any $x \in X$:

$$\tau(x, U) = \sum_i \chi_{X \setminus U}\left(f^i(x)\right) \leq m;$$

here, χ_A is the indicator of a set A.

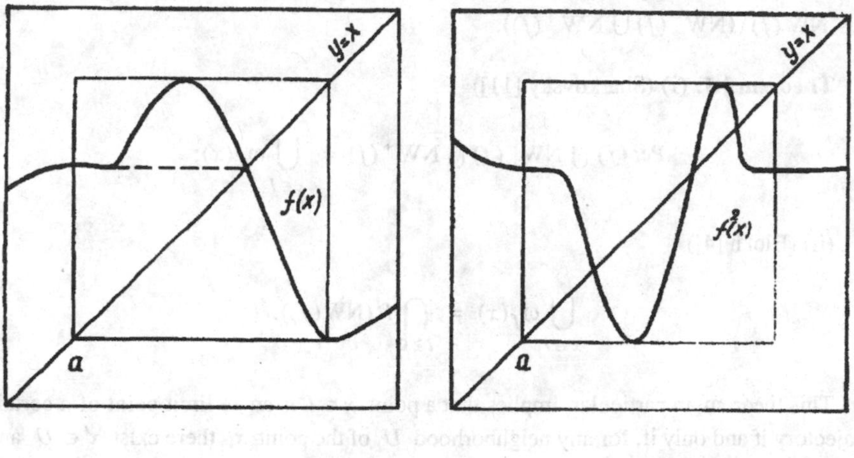

a b

Fig. 16

If a dynamical system is generated by a continuous map (and is nothing more than a semigroup of maps), then it is possible that $f(\mathrm{NW}(f)) \neq \mathrm{NW}(f)$, although it is obvious that the inclusion $f(\mathrm{NW}(f)) \subset \mathrm{NW}(f)$ is always true. As an example (Sharkovsky [2]), we consider the map represented in Fig. 15. For this map, the point $x = c$ is nonwandering but one can indicate no points $x \in \mathrm{NW}(f)$ such that $f(x) = c$. It is easy to see that the point $x = b$ does not belong to NW(f). Note that $c \notin \overline{\mathrm{Per}(f)}$ and c is not an ω-limit point for any trajectory. Hence, for this map, we have $\mathrm{NW}(f) \neq \overline{\mathrm{Per}(f)}$ and $\mathrm{NW}(f) \neq \bigcup_x \omega_f(x)$. It is not difficult to verify that, in this case, $\overline{\mathrm{Per}(f)} = \bigcup_x \omega_f(x)$ and

$$\mathrm{NW}(f) = \overline{\mathrm{Per}(f)} \cup \{c\}.$$

Note that, in any neighborhood of the point $x = c$, there exists a point x' such that $f^{m'}(x') = c$ for some $m' > 0$. It turns out that any nonwandering point possesses this property, provided that $X = I$.

Since the point $x = c$ is nonwandering, one of the images of its neighborhood $(c - \varepsilon, c + \varepsilon)$ necessarily intersects this neighborhood in the course of time. However, for sufficiently small ε, the images of the left and right unilateral neighborhoods never intersect the corresponding unilateral neighborhoods. This type of behavior exhibited by dynamical systems indicates the necessity of distinguishing between the subsets of unilateral nonwandering points $\mathrm{NW}^-(f)$ and $\mathrm{NW}^+(f)$ in the set of nonwandering points.

Namely, a point x belongs to $\mathrm{NW}^-(f)$ $(\mathrm{NW}^+(f))$ if, for any open (in I) interval U whose right (left) end coincides with the point x, we have $f^m(U) \cap U \neq \varnothing$ for some $m > 0$ which depends on U. Thus, for the map displayed in Fig. 15, we can write $c \in \mathrm{NW}(f) \setminus (\mathrm{NW}^-(f) \cup \mathrm{NW}^+(f))$.

Theorem 1.1. (i) (Sharkovsky [11])

$$\mathrm{Per}(f) \cup \mathrm{NW}^-(f) \cup \mathrm{NW}^+(f) = \bigcup_{x \in I} \omega_f(x);$$

(ii) (Blokh [4])

$$\bigcup_{x \in I} \omega_f(x) = \bigcap_{i \geq 0} f^i(\mathrm{NW}(f)).$$

This theorem, in particular, implies that a point $x \in I$ is an ω-limit point of a certain trajectory if and only if, for any neighborhood U of the point x, there exist $x' \in U$ and integer numbers $0 < m_1 < m_2$ such that $f^{m_i}(x') \in U$, $i = 1, 2$.

Denote the set $\bigcup_x \omega_f(x)$ by $\Omega(f)$. This set is smaller than $\mathrm{NW}(f)$ but satisfies the following analog of the Birkhoff theorem:

Theorem 1.2 (Sharkovsky [11]). *For any neighborhood U of the set $\Omega(f)$, there exists an integer $m = m(U)$ such that the time of stay of the trajectory of any point from I outside U does not exceed m.*

The set $\Omega(f)$ in Theorem 1.2 cannot be replaced by a smaller closed subset: Indeed, for any point $x' \in \Omega(f)$, there exists a point x'' such that $\omega_f(x) \ni x'$ and, hence, the trajectory of the point x'' hits any neighborhood of the point x' infinitely many times.

Note that $\mathrm{Per}(f^m) = \mathrm{Per}(f)$ for any m. Generally speaking, the set $\mathrm{NW}(f)$ does not possess this property. The example given in Fig. 16 (Coven and Nitecki [1]) is characterized by the property $NW(f^2) \neq NW(f)$ (note that this example is a modification of the previous one). In this case, $x = a$ is a nonwandering point of the map f but, for the map f^2, this point is not nonwandering as can easily be seen from its graph. Nevertheless, the equality $NW(f^m) = NW(f)$ always holds for odd m (Coven and Nitecki [1]).

By definition, the set $NW(f)$ consists of points at which one observes the return of domains of the space X. At the same time, the situation where relative regions (i.e., subsets of $NW(f)$ open with respect to $NW(f)$) do not return is possible. Therefore, in the theory of dynamical systems, parallel with $NW(f)$, it is reasonable to consider a smaller set $C(f)$ called the *center* of a dynamical system and characterized by the return of relative domains.

If $f \in C^0(X, X)$ and X is an arbitrary compact space, then we can define $C(f)$ as follows: Let $C_1 = NW(f)$ and let, for $\alpha \geq 1$, $C_{\alpha+1}$ be a set of the nonwandering points of the space C_α, i.e., $NW(f|_{C_\alpha})$. If α is the limiting ordinal number, then we set

$$C_\alpha = \bigcap_{\beta<\alpha} C_\beta .$$

According to the Baire–Hausdorff theorem, we have $C_r = C_{r+1} = \ldots$ for some finite or countable ordinal number r. Then $C(f) = C_r$. This r is called the *depth of the center*, provided that it is the least possible ordinal number of this sort.

The center of a dynamical system can also be defined as follows: $C(f)$ is the largest closed invariant set characterized by the property of incompressibility of the regions, i.e., for any subset $U \subset C(f)$ open in $C(f)$, we have either $f(U) = U$ or $f(U) \not\subset U$.

It is well known (see, e.g., Birkhoff [1], Nemytsky and Stepanov [1]) that $C(f)$ is the closure of the set of recurrent points. For any trajectory, the probability of its stay in any neighborhood of the center is equal to one, i.e., for any set $U \supset C(f)$ open in X, we have

$$\lim_{m \to \infty} \frac{1}{m} \sum_{i=0}^{m-1} \chi_U\left(f^i(x)\right) = 1$$

for any point $x \in X$.

In the case where $X = I$, some statements can be made more precise. Thus, in the general case, the depth of the center can be equal to any finite or countable ordinal number but, for $X = I$, the depth of the center is not greater than 2.

Theorem 1.3 (Sharkovsky [2]). $C(f) = NW(f|_{NW(f)})$.

For the map whose graph is depicted in Fig. 16, the depth of the center is equal to 2.

As mentioned above, in the general case, the recurrent points are dense in $C(f)$. This does not mean that periodic points are also dense in $C(f)$. Thus, for the circle S^1 and f defined as a rotation of S^1 about an irrational angle, we have $\mathrm{Per}(f) = \varnothing$ but $C(f) = S^1$. At the same time, periodic points are everywhere dense in $C(f)$ for $X = I$.

Theorem 1.4 (Sharkovsky [2]). $C(f) = \overline{\mathrm{Per}(f)}$.

Note that there exist (nonsmooth!) mappings $f \in C^0(I, I)$ with ω-limit points that are not limiting points for the set of periodic points (see Chapter 4). For these mappings, we have $C(f) \neq \Omega(f)$.

The weakest property of return that may take place for some points of dynamical systems is *chain recurrence.* A point $x \in I$ is called *chain recurrent* if, for any $\varepsilon > 0$, there exists a sequence $\{x_i\}_{i=0}^{n}$ such that $x_0 = x = x_n$ and $|f(x_i) - x_{i+1}| < \varepsilon$ for any $i < n$ (the points $\{x_i\}_{i=0}^{n}$ are called *ε-trajectories* of the point x_0).

The concept of chain recurrence is closely related to the notion of weak incompressibility (Vereikina and Sharkovsky [2]). We recall that a closed invariant set F exhibits the property of *weak incompressibility* if, for any subset $U \subset F$ open with respect to F and not equal to F, one can write $f(\overline{U}) \not\subset U$ (Sharkovsky [15]).

We have already mentioned in the previous section that the property of weak incompressibility is observed for any ω-limit set; it is also typical of cycles, the closures of homoclinic trajectories, etc. On the other hand, the set that consists of two cycles does not exhibit the property of weak incompressibility. In general, this property is not observed for any set that consists of two disjoint closed invariant subsets (one can choose U in the definition of incompressibility in the form of one of these invariant subsets and, in this case, $f(\overline{U}) = U$).

A point that belongs to a set with the property of weak incompressibility can be called an *almost returning point.* One can easily show that every point of this sort can be made periodic by arbitrarily small perturbations of the dynamical system, i.e., this point will have the strongest property of return. This fact enables us to say that the points of sets with the property of weak incompressibility are *almost returning points.*

The set of almost returning points coincides with the set of chain recurrent points for any map f (Vereikina and Sharkovsky [2]).

This set is denoted by $CR(f)$, i.e., $x \in CR(f)$ if there exists a set $F \ni x$ with the property of weak incompressibility.

As a rule, there is no weak incompressibility in the entire set $CR(f)$. For example, this is true if $CR(f)$ consists of two fixed points—a sink and a source—as in the case of mapping (1) with $1 < \lambda \leq 3$ (see Section 1).

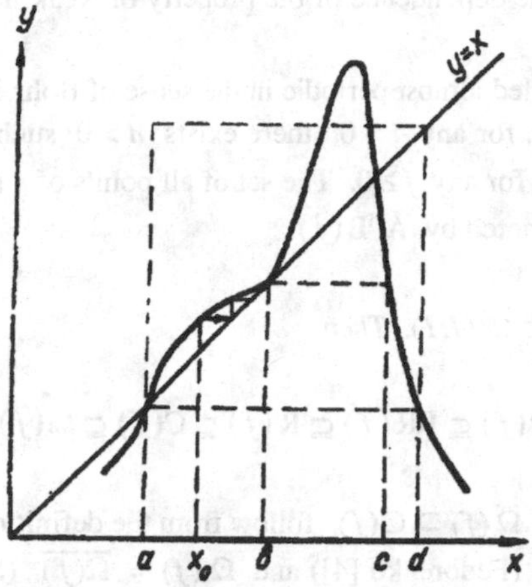

Fig. 17

We now recall some properties of the set $CR(f)$, which can easily be derived from the definition. The set $CR(f)$ is closed and invariant. Every nonwandering point is chain recurrent, i.e., $NW(f) \subset CR(f)$ but it is possible that $NW(f) \neq CR(f)$. For the map depicted in Fig. 17,

$$CR(f) \setminus NW(f) = (a, b) \cup \bigcup_{i=0}^{\infty} f^{-i}([c, d]),$$

while for the map presented in Fig. 15, we have

$$CR(f) \setminus NW(f) = \bigcup_{i=1}^{\infty} f^{-i}(c).$$

The definition of $CR(f)$ immediately implies that $CR(f|_{CR(f)}) = CR(f)$. Unlike the map $f \mapsto NW(f)$ of the space $C^0(X, X)$ into 2^X, which admits Ω-explosions (i.e., is neither upper nor lower semicontinuous), the map $f \mapsto CR(f)$ is upper semicontinuous.

Finally, in many cases (in particular, for $X = I$), the set $CR(f)$ coincides with the set of points that return as a result of infinitesimal perturbations of dynamical systems themselves, i.e., it coincides with the set of weakly nonwandering points.

A point $x \in X$ is called a *weakly nonwandering point* of the map $f \in C^r(X, X)$ if, for any neighborhood $U(x)$ of the point x and any neighborhood $\mathfrak{A}(f)$ of the map f (in $C^r(X, X)$), one can find $\tilde{f} \in \mathfrak{A}(f)$ and integer $m > 0$ such that $\tilde{f}^m(U(x)) \cap U(x) \neq \varnothing$.

As for as we know, the dependence of the property of weak nonwandering on r has not been studied yet.

The point $x \in I$ is called almost periodic in the sense of Bohr if, for any $\varepsilon > 0$, one can find $N > 0$ such that, for any $i > 0$, there exists $n > 0$ such that $i + 1 \le n \le i + N$ and $|f^{j+n}(x) - f^j(x)| < \varepsilon$ for any $j \ge 0$. The set of all points of a map f almost periodic in the sense of Bohr is denoted by $\mathrm{APB}(f)$.

Theorem 1.5. *Let* $f \in C^0(I, I)$. *Then*

$$\mathrm{Per}(f) \subseteq \mathrm{APB}(f) \subseteq \mathrm{AP}(f) \subseteq \mathrm{RR}(f) \subseteq \mathrm{R}(f) \subseteq \mathrm{C}(f) \subseteq \Omega(f) \subseteq \mathrm{NW}(f) \subseteq \mathrm{CR}(f).$$

All inclusions, except $\Omega(f) \supset \mathrm{C}(f)$, follow from the definitions. For $C^0(I, I)$, we have $\mathrm{APB}(f) \subseteq \mathrm{AP}(f)$ (Fedorenko [4]) and $\Omega(f) = \overline{\Omega(f)}$ (Sharkovsky [2]). This enables us to conclude that $\Omega(f) \supset \mathrm{C}(f)$.

Sometimes, it is possible to study the structure of sets indicated in Theorem 1.5 and represent these sets in the form of a finite (or countable) union of subsets which are, in a certain sense, dynamically indecomposable (e.g., contain a dense trajectory). Representations of this sort are usually called spectral decompositions. The spectral decomposition of the set of nonwandering points is the most popular object of investigations. As a rule, in terms of this decomposition, one can easily describe the typical behavior of the trajectories of the corresponding dynamical system.

To explain this in detail, we consider quadratic mappings from the family f_λ described in Section 1. For these mappings, the sets $\overline{\mathrm{Per}(f)}$, $\Omega(f)$, and $\mathrm{NW}(f)$ always coincide as follows from the results of Chapter 5.

The examples presented in Section 1 demonstrate that, for $0 < \lambda \le 1$, the set $\mathrm{NW}(f)$ consists of a single fixed point $x = 0$. For $1 < \lambda \le 3$, it consists of two fixed points, namely, the repelling point $x = 0$ and the attracting point $x = 1 - 1/\lambda$. For $3 < \lambda \le 1 + \sqrt{6}$, we observe the appearance of an attracting cycle of period 2 and $\mathrm{NW}(f)$ consists of three dynamically indecomposable components, namely, the repelling fixed points $x = 0$ and $x = 1 - 1/\lambda$ and the attracting cycle of period 2. One can check that the trajectories of all points on the interval $I = [0, 1]$ (except countably many) are attracted by the cycle of period 2 (see Chapter 5 for detailed explanation). Further, if the value of λ increases to $\lambda = \lambda^*$, then the number of elements in the spectral decomposition of the set $\mathrm{NW}(f)$ increases to infinity. Thus, for $\lambda = \lambda^*$, the set $\mathrm{NW}(f)$ is a union of two repelling fixed points $x = 0$ and $x = 1 - 1/\lambda$, infinitely many cycles of periods 2^i, $i = 1, 2, \ldots$ (with one cycle of each period), and the minimal Cantor set K. Note that, for $\lambda < \lambda^*$, the generic behavior of trajectories on the interval I (i.e., the behavior of trajectories of almost all (with respect to the Lebesgue measure) points) can be described as the asymptotic convergence to an attracting cycle. For $\lambda = \lambda^*$, a typical trajectory on I is asymptotically approaching the set K, i.e., it is asymptotically almost periodic.

For $\lambda = 4$, the set $\mathrm{NW}(f)$ coincides with $I = [0, 1]$ and can be regarded as dynamically indecomposable because I contains a dense trajectory. It has already been indicated that, in this case, the trajectories of almost all points from I are dense in I. We recall once again that the structure of the set $\mathrm{NW}(f)$ is investigated in more details in Chapter 5.

From the practical point of view, it seems reasonable to select the properties typical of the trajectories not of all points of the phase space but of almost all points of this space. In this case, the term "almost all points" may denote either a collection of points forming a set of the second Baire category (i.e., almost all points in the topological sense) or almost all points with respect to a certain measure in the phase space (i.e., almost all points in the metric sense).

This point of view, in particular, leads to the notion of probabilistic limit sets (or *Milnor attractors,* see Arnold, Afraimovich, Il'yashenko, and Shilnikov [1] and Milnor [2]), i.e., to the notion of the smallest closed set that contains the ω-limit sets of trajectories of almost all points in the phase space (this set is denoted by $\mathcal{M}(f)$).

In a similar way, the notion of statistical limit set introduced in the previous section leads to the notion of the *minimal center of attraction* of almost all trajectories of a dynamical system (or to the notion of statistical limit set, as it is defined in Arnold, Afraimovich, Il'yashenko, and Shilnikov [1], i.e., to the smallest closed set that contains statistical limit sets of the trajectories of almost all points of the phase space; this set is denoted by $\mathcal{A}(f)$). It follows from the definition that, as a rule, this is just the set observed in the experimental investigation of dynamical systems.

It is worth noting that if we replace the words "almost all" by "all" in this definition, then we arrive at the notion of the minimal center of attraction (of all trajectories), which is well known in the theory of dynamical systems since thirties; this set is defined as the smallest set in any neighborhood of which all trajectories stay almost always. As already mentioned, the trajectories stay almost always in the neighborhood of the center of the dynamical system. Therefore, the minimal center of attraction is a subset of $C(f)$.

It follows from the definition that $\mathcal{A}(f) \subset \mathcal{M}(f)$. There exist maps for which these sets do not coincide. An example of this sort is presented in Chapter 6 (Fig. 44); for this map, the set $\mathcal{M}(f)$ is an interval with a dense trajectory and $\mathcal{A}(f)$ consists of a single repelling fixed point.

Consider a mapping almost all trajectories of which are attracted by a repelling fixed point. For $x \in [0, 1]$, we define

$$g(x) = \begin{cases} 3x, & 0 \le x < 1/3, \\ 1, & 1/3 \le x \le 2/3, \\ 3(1-x), & 2/3 < x \le 1. \end{cases}$$

By using the reasoning applied in Section 1 to the investigation of the family f_λ for $\lambda > 4$, one can show that the trajectories of all points, except the points of the Cantor set $K \subset [0, 1]$, hit the repelling fixed point $x = 0$ after finitely many steps. Moreover, K contains a dense trajectory. Therefore, both $\Omega(g)$ and the minimal center of attraction of the map g coincide with the set $K \cup \{0\} = K$. At the same time, the sets $\mathcal{M}(f)$ and $\mathcal{A}(f)$ consist of a single repelling fixed point $x = 0$.

The fact that there exists a mapping for which its generic trajectory "is attracted" by a repelling cycle seems to be unexpected. However, the map g may be untypical or even, in a certain sense, exclusive. As an argument for this assertion, one can recall, e.g., the following fact: The repelling fixed point may lose its property to attract almost all trajectories as a result of infinitesimally small perturbations of the map g.

It is also interesting to study a more general question: What properties of a dynamical system generated by a map from a certain space $\mathfrak{M}$ of maps can be regarded as typical? Any property can be regarded as generic (typical) if a collection of maps characterized by this property forms a set of the second Baire category in $\mathfrak{M}$ Clearly, the answer to the posed question depends on the space $\mathfrak{M}$ under consideration. Thus, as shown in Chapter 6, for a sufficiently broad class of smooth mappings, almost all trajectories are attracted either by an attracting cycle, or by a Cantor-type set, or by a set that consists of finitely many intervals cyclically permutable by the map and contains an everywhere dense trajectory. At the same time, none of the indicated types of behavior is observed for typical mappings in $C^0(I, I)$. In particular, for these mappings, no cycle is attracting and no trajectory is dense on any interval.

Let us now formulate an assertion about the typical behavior of the trajectories of C^0-typical dynamical systems recently proved by Agronsky, Bruckner, and Laczkovich [1].

The space $C^0(I, I)$ contains a set $C^\#$ of the second category such that, for any map $f \in C^\#$, there are continuum many minimal Cantor-type sets F_α on each of which f is a homeomorphism and, moreover,

(a) $P(F_\alpha) = \{x \in I \mid \omega_f(x) = F_\alpha\}$ is nowhere dense in I;

(b) $\bigcup_\alpha P(F_\alpha)$ is a set of the second category.

This means that almost all (on I) trajectories of a dynamical system are asymptotically regularly recurrent almost always in $C^0(I, I)$.

This result is, to a certain extent, unexpected. Actually, almost all mappings (in particular, in $C^\#$) possess cycles with periods $\neq 2^i$ and, consequently, Cantor-type quasiminimal sets that contain cycles on which almost all trajectories are recurrent but not regularly recurrent or asymptotically regularly recurrent. Although each quasiminimal set of this sort contains continuum many Cantor-type minimal sets, in the typical case, they attract not too many trajectories (which form a set of the first category in I). Therefore, in the case of smooth mappings, almost all trajectories almost always are either recurrent or asymptotically approach recurrent trajectories. The above-mentioned result by Agronsky, Bruckner, and Laczkovich [1] states that, for C^0-typical mappings, the situation is absolutely different: Due to the very complicated structure of typical continuous C^0-maps, one observes the appearance of (continuum) many Cantor-type minimal sets, which "seize" almost all trajectories.

2. ELEMENTS OF SYMBOLIC DYNAMICS

1. Concepts of Symbolic Dynamics

Symbolic dynamics is a part of the general theory of dynamical systems dealing with cascades generated by shifts in various spaces of sequences

$$\Theta = (\ldots \theta_{-2}, \theta_{-1}, \theta_0, \theta_1, \theta_2, \ldots) \quad \text{or} \quad \Theta = (\theta_0, \theta_1, \theta_2, \ldots),$$

where θ_n are letters of an alphabet $\mathcal{A} = \{\theta^1, \theta^2, \ldots, \theta^m\}$. The methods of symbolic dynamics are now widely applied to the investigation of various types of dynamical systems.

Let Π be the space of all unilateral sequences $\Theta = (\theta_0, \theta_1, \theta_2, \ldots)$ (or infinite words, if it is reasonable to omit commas) with the metric

$$\rho(\Theta', \Theta'') = \sum_{n \geq 0} \frac{r(\theta_n', \theta_n'')}{m^n},$$

where

$$r(\theta_n', \theta_n'') = \begin{cases} 0, & \text{if } \theta_n' = \theta_n'', \\ 1, & \text{if } \theta_n' \neq \theta_n''. \end{cases}$$

We define a shift $\sigma : \Pi \to \Pi$ as follows: If $\Theta = (\theta_0, \theta_1, \theta_2, \ldots)$, then $\sigma\Theta = (\theta_1, \theta_2, \ldots)$. For the dynamical system (Π, σ), many standard problems of the theory of dynamical systems, in particular, those concerning periodic trajectories can be solved almost trivially.

Thus, for the dynamical system (Π, σ), every point Θ corresponding to the periodic sequence $\theta_1 \ldots \theta_k \theta_1 \ldots \theta_k \theta_1 \ldots$ with the least period k generates a k-periodic trajectory in the space Π (since $\sigma^k\Theta = \Theta$ and $\sigma^i\Theta \neq \Theta$ for $1 \leq i < k$). Hence, this dynamical system possesses periodic trajectories of all periods and these periodic trajectories are everywhere dense in Π. The last property follows from the fact that, for any $\Theta = (\theta_0, \theta_1, \ldots, \theta_{k-1}, \theta_k, \ldots) \in \Pi$ and $\varepsilon > 0$, the point $\Theta' = (\theta_0, \ldots, \theta_{k-1}, \theta_0, \ldots, \theta_{k-1},$

$\theta_0, \ldots)$ belongs to the ε-neighborhood of the point Θ and is periodic whenever k satisfies the inequality $1/2^k < \varepsilon$.

The space Π also contains dense trajectories. Thus, the trajectory that passes through the point

$$\Theta^* = (\theta^1\theta^2 \ldots \theta^m \Gamma^{11}\Gamma^{12} \ldots \Gamma^{mm}\Gamma^{111} \ldots \Gamma^{mmm}\Gamma^{1111} \ldots),$$

where $\Gamma^{i_1 i_2 \ldots i_{s-1} i_s} = \Gamma^{i_1 \ldots i_{s-1}} \theta^{i_s}$, $s = 2, 3, \ldots$, $\Gamma^{i_1} = \theta^{i_1}$, $i_1, i_2, \ldots, i_s = 1, \ldots, m$, is a trajectory of this sort. The sequence Θ^* consists of all possible words written in the following succession: First, we write the words of length 1, then the words of length 2, and so on. The trajectory $\Theta^*, \sigma\Theta^*, \sigma^2\Theta^*, \ldots$ is dense in Π because the "cylinders" $D^{i_1 \ldots i_s} = \{\Theta \in \Pi \mid \Theta = (\Gamma^{i_1 \ldots i_s}\theta_{s+1}\theta_{s+2} \ldots)\}$, $1 \le i_1, \ldots, i_s \le m$, $s = 1, 2, \ldots$, form a base of the space Π and, for any $i'_1, \ldots, i'_s$, one can indicate an integer k such that $\sigma^k\Theta^* \in D^{i'_1 \ldots i'_s}$.

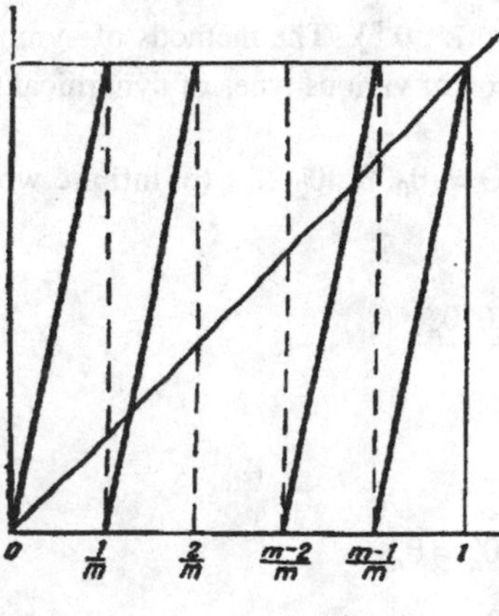

Fig. 18

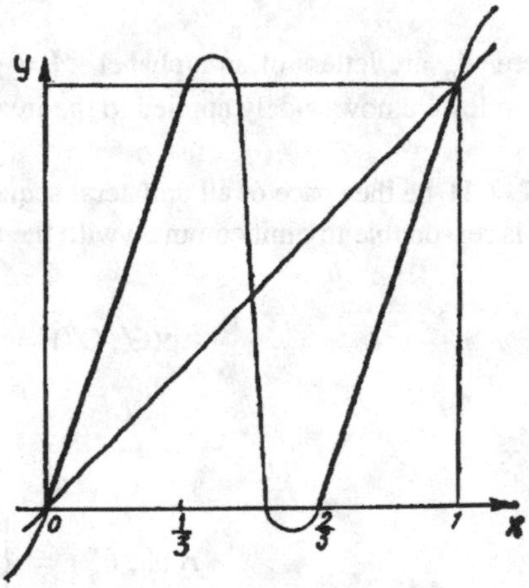

Fig. 19

Let us now analyze the possibility of application of symbolic dynamics to the investigation of individual dynamical systems, e.g., on the real axis $\mathbb{R}$.

The dynamical system on $[0, 1]$ generated by the map (see Fig. 18)

$$f: x \to mx \,(\text{mod } 1) \tag{1}$$

is isomorphic to the dynamical system of shifts with alphabet $\theta^1, \ldots, \theta^m$. If we use the m-digit representation of the points $x \in [0, 1]$, then, clearly, $\theta^i \sim i - 1$, where $i = 1, \ldots, m$. Hence, the dynamical system generated by (1) possesses the properties of the dynamical system of shifts, i.e., there are periodic trajectories of arbitrarily large periods, periodic trajectories form an everywhere dense set in $[0, 1]$, and there are trajectories dense in $[0, 1]$, e.g.,

$$\Theta^* \sim x^* = 0.\,0\,1 \ldots m-1\;00\,01 \ldots\; m-1\;m-1\;000\,001 \ldots = 1/m^2 + 2/m^3 + \ldots .$$

The dynamical system generated by (1) does not belong to the class of one-dimensional dynamical systems considered in the book because map (1) is not continuous. At the same time, the methods of symbolic dynamics can be successfully applied to the analysis of continuous maps. To illustrate this assertion, we present several examples.

Let f be an arbitrary continuous function $\mathbb{R} \to \mathbb{R}$ satisfying the condition

$$f(x) = 3x \pmod 1 \quad \text{for} \quad x \in \mathcal{J} = [0, 1/3] \cup [2/3, 1] \tag{2}$$

(Fig. 19). Clearly, one can use symbolic dynamics to investigate the dynamical system generated by (2): If we use the ternary representation of points in $\mathcal{J}$, i.e., if we have $x = 0.\,\theta_1\theta_2\theta_3\theta_4 \ldots$, where $\theta_i \in \{0, 1, 2\}$ and $\theta_1 \neq 1$ for $x \in \mathcal{J}$, then $f(x) = 0.\,\theta_2\theta_3 \ldots$. Some points of $\mathcal{J}$ eventually leave $\mathcal{J}$ under the action of f. The points leave $\mathcal{J}$ if and only if their ternary representations $0.\,\theta_1\theta_2\theta_3 \ldots$ contain at least one θ_i equal to 1 (the point Θ leaves $\mathcal{J}$ after $l = l(\Theta)$ steps, where $l(\Theta) = \min\{i \mid \theta_i = 1\}$). The points of the set $K = \{0.\,\theta_1\theta_2 \ldots \theta_i \ldots \mid \theta_i \in \{0, 2\}\}$ (K is the standard Cantor set) and only these points do not leave $\mathcal{J}$ and $fK = K$. The map f acts on the set K as the dynamical system of shifts with alphabet $\{0, 2\}$. Hence, the map $f\colon \mathbb{R} \to \mathbb{R}$ has periodic points of all periods on $\mathcal{J}$ (thus, the point $0.\,20202\ldots = 3/4$ has period 2, the point $0.\,2002002\ldots = 9/13$ has period 3, the point $0.\,\underbrace{20\ldots0}_{m}\,\underbrace{20\ldots0}_{m}\,2\ldots = 2/3\,\frac{3^m}{3^m-1}, \ldots$ has period m, etc.).

The possibility of application of the methods of symbolic dynamics (with two-letter alphabets) to the investigation of map (2) is certainly explained not by the special form of the map f on $\mathcal{J}$ but by the fact that $\mathcal{J}$ is the union of two intervals $\mathcal{J}_1$ and $\mathcal{J}_2$ such that $f\mathcal{J}_1 \supset \mathcal{J}_1 \cup \mathcal{J}_2$ and $f\mathcal{J}_2 \supset \mathcal{J}_1 \cup \mathcal{J}_2$. This means that any word $\theta_{i_1}\theta_{i_2} \ldots \theta_{i_k}$ generated by the two-letter alphabet θ^1, θ^2 can be associated with a sequence of intervals $\mathcal{J}_{i_1}, \mathcal{J}_{i_2}, \ldots, \mathcal{J}_{i_k}$, where $i_s = 1$ or 2 and $s = 1, \ldots, k$, which contains trajectories of the map f that pass through the intervals $\mathcal{J}_1$ and $\mathcal{J}_2$ in the indicated order. There are many trajectories of this sort. Moreover, these are the only trajectories passing through the points of the set $\mathcal{J}^{i_1} \subset \mathcal{J}_{i_1}$ successively constructed from the intervals $\mathcal{J}_{i_k}, \mathcal{J}_{i_{k-1}}, \ldots, \mathcal{J}_{i_1}$ as follows: $\mathcal{J}^{i_s} = \mathcal{J}_{i_s} \cap f^{-1}\,\mathcal{J}^{i_{s+1}}$, $s = k-1, k-2, \ldots, 1$; $\mathcal{J}^{i_k} = \mathcal{J}_{i_k}$. The set $\mathcal{J}^{i_1}$ always contains a nondegenerate interval.

In the next chapter, a similar approach is used to study periodic trajectories of arbitrary continuous maps $\mathbb{R} \to \mathbb{R}$. More precisely, we consider the problem of coexistence of periodic trajectories of various periods and types.

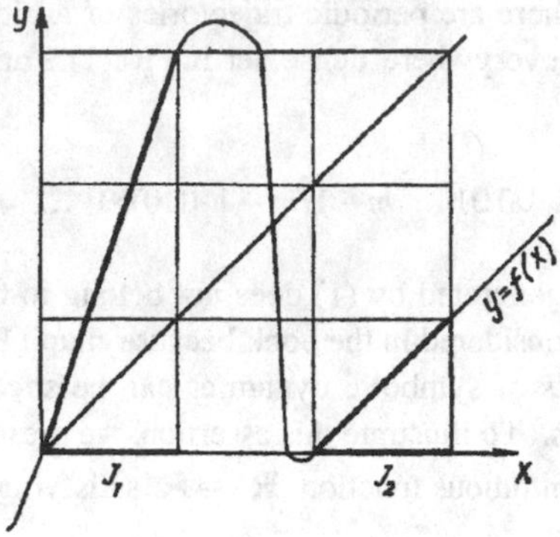

Fig. 20

It is important to mention the following fact: We have always assumed that the shift map is defined in the entire space Π. At the same time, in analyzing individual dynamical systems by using symbolic dynamics, we most often encounter the situation where the shift map σ is defined not in the entire Π but in a certain subspace Π' (of "admissible" sequences $\theta_1\theta_2\ldots$). Thus, for the map displayed in Fig. 20, there are intervals $\mathcal{J}_1$ and $\mathcal{J}_2$ such that $f\mathcal{J}_1 \supset \mathcal{J}_1 \cup \mathcal{J}_2$ and $f\mathcal{J}_2 \supset \mathcal{J}_1$. Therefore, if we pass to a symbolic dynamical system with alphabet $\{\theta^1, \theta^2\}$ to study the map on the intervals $\mathcal{J}_1$ and $\mathcal{J}_2$, then we arrive at the shift map defined only on the sequences $\theta_1\theta_2\ldots\theta_n\ldots$ satisfying the following condition: If $\theta_n = \theta^2$, then $\theta_{n+1} = \theta^1$. It is clear that restrictions of this sort may significantly complicate the investigation of symbolic dynamical systems.

Note that, in simple cases, the subspace Π' of admissible sequences may be determined by a matrix of admissible transitions (of the mth order). In the last example, this is the matrix $\begin{pmatrix} 1 & 1 \\ 1 & 0 \end{pmatrix}$ (the only forbidden transition is $\mathcal{J}_2 \to \mathcal{J}_2$ because $f\mathcal{J}_2 \not\supset \mathcal{J}_2$).

In the set Π of sequences with alphabet $(\theta^1, \ldots, \theta^m)$ we introduce the following natural lexicographic ordering: $\Theta' = (\theta_1'\theta_2'\ldots) < \Theta'' = (\theta_1''\theta_2''\ldots)$ if, for some n, we have $\theta_i' = \theta_i''$ for $i < n$ and $\theta_n' = \theta^{s'}$, $\theta_n'' = \theta^{s''}$ for some $s' < s''$.

In the examples presented above, the map f is monotonically increasing on the intervals $\mathcal{J}_1$ and $\mathcal{J}_2$. Thus, when we pass to symbolic dynamics, the correspondence between the points $x \in \bigcup_r \mathcal{J}_r$ and the sequences $\Theta(x) = (\theta_1\theta_2\ldots)$ is monotone, i.e., if x', $x'' \in \bigcup_r \mathcal{J}_r$ and $x' < x''$, then $\Theta(x') < \Theta(x'')$ in the lexicographic order.

In studying one-dimensional dynamical systems, we mostly deal with piecewise monotone maps and, hence, both with intervals of increase and intervals of decrease of the function f. At the same time, if the intervals of decrease of the function f are involved in the construction of symbolic dynamics, the monotonicity of the correspondence between x and $\Theta(x)$ is violated. Therefore, it is necessary to modify the method used to construct symbolic sequences.

In this chapter, we analyze the possibilities of the method of symbolic dynamics in more details for fairly simple piecewise monotone maps, namely, for unimodal maps.

Let $f: I \to I$ be a unimodal map, let $I = \mathcal{J}_1 \cup \mathcal{J}_2$, let f be a function monotonically increasing on $\mathcal{J}_1$ and monotonically decreasing on $\mathcal{J}_2$, and let c be the point of extremum.

Let us define the *address* of a point $x \in I$:

$$A(x) = \begin{cases} \mathcal{J}_s, & \text{if } x \in \mathcal{J}_s,\ x \neq c, \\ c, & \text{if } x = c. \end{cases}$$

The *route* is defined as a sequence of addresses

$$A_f(x) = (A(x), A(f(x)), A(f^2(x)) \ldots) = (A_0, A_1, A_2, \ldots).$$

The operation of shift σ on the space of unilateral sequences $(A_0, A_1, A_2, \ldots)$ is defined, as usual, by the equality $\sigma(A_0, A_1, A_2, \ldots) = (A_1, A_2, \ldots)$. The map f and the shift σ are connected by the equality $\sigma(A_f(x)) = A_f(f(x))$.

In constructing symbolic dynamics, it turns out to be useful to take into account not only the changes in the addresses $A(f^n(x))$ but also the changes in orientation. This idea was applied by Milnor and Thurston [1] to the theory of kneading invariants (see also Guckenheimer [1]).

We associate the intervals $\mathcal{J}_s$ with the signs

$$\varepsilon(\mathcal{J}_s) = \begin{cases} +1, & \text{if } f / \mathcal{J}_s \text{ increases}, \\ -1, & \text{if } f / \mathcal{J}_s \text{ desreases}, \end{cases}$$

and set $\varepsilon(c) = 0$.

Parallel with a route $A_f(x) = (A_0, A_1, A_2, \ldots)$, we consider a sequence $\Theta_f(x) = (\theta_0, \theta_1, \theta_2, \ldots)$, where $\theta_0 = \varepsilon_0$, $\theta_1 = \varepsilon_0 \varepsilon_1, \ldots,$ $\theta_n = \varepsilon_0 \varepsilon_1 \ldots \varepsilon_n \ldots,$ $\varepsilon_i = \varepsilon(A_i)$. The sign of $\varepsilon_0 \cdot \varepsilon_1 \cdot \ldots \cdot \varepsilon_n$ corresponds to the local behavior of f^n in the vicinity of the point x, i.e., it is equal to $+1$, -1, or 0, respectively, if f^n increases, decreases, or has an extremum at the point x. Due to the fact that the phase space is one-dimensional, the lexicographic ordering is connected with the natural ordering of real numbers by the following assertion:

Lemma 2.1 (on monotonicity). *The map $x \to \Theta_f(x)$ is monotone.*

Proof. Note that the map $x \to \Theta_f(x)$ is either nonincreasing or nondecreasing depending on the type of extremum (minimum or maximum) attained at the point c by the function $f: I \to I$. Assume that c is the maximum point. If $x' < x''$, then let n be the least integer for which $\theta_n(x') \neq \theta_n(x'')$. For $n = 0$, we have $x' \leq c \leq x''$ and $\theta_0(x') \geq$

$\theta_0(x'')$. For $n \geq 1$, f^n is a homeomorphic map of the interval $\langle x', x''\rangle$ onto the interval $\langle f^n(x'), f^n(x'')\rangle$ and the interval $\langle f^n(x'), f^n(x'')\rangle$ contains the point c (here and below, $\langle a, b\rangle$ denotes the closed interval bounded by the points a and b). Assume that $\theta_{n-1}(x') = \theta_{n-1}(x'') = -1$. Then the homeomorphism f^n: $\langle x', x''\rangle \to \langle f^n(x'), f^n(x'')\rangle$ changes orientation. Therefore, $f^n(x') \in \mathcal{J}_2$, $f^n(x'') \in \mathcal{J}_1$, and $\theta_n(x'') \leq \theta_n(x')$. The cases where $\theta_n(x') = \theta_n(x'') = 1$ and $\theta_{n-1}(x') = \theta_{n-1}(x'') = 0$ are analyzed similarly.

Let $C = \bigcup_{i \geq 0} f^{-i}(c)$. The equality $\theta_f(x') = \theta_f(x'')$ holds for $x' \neq x''$ if and only if the points x' and x'' belong to the same component of the set $I \setminus C$. In this case, f^n: $\langle x', x''\rangle \to \langle f^n(x'), f^n(x'')\rangle$ is a homeomorphism for all $n > 0$. Note that $\Theta_f(x) = (\theta_0, \theta_1, \theta_2, \ldots)$ with $\Theta_i \in \{-1, +1\}$ for $x \in I \setminus C$.

The topology of coordinatewise convergence generates a metric ρ in the set $\{\Theta_f(x)\}$. Due to the lemma on monotonicity, the limits

$$\Theta_f(x^+) = \lim_{y \searrow x} \Theta_f(y) \quad \text{and} \quad \Theta_f(x^-) = \lim_{y \nearrow x} \Theta_f(y)$$

exist for all $x \in I$. Moreover, for any $x \in I$, all elements of the sequences $\Theta_f(x^\pm)$ are nonzero.

Let σ be a shift in the set Σ of unilateral sequences with alphabet $\{-1, +1\}$.

For $\alpha = (\alpha_0, \alpha_1, \alpha_2, \ldots) \in \Sigma$ and $\beta_0 \in \{-1, 1\}$, we set $\beta_0\alpha = (\beta_0\alpha_0, \beta_0\alpha_1, \beta_0\alpha_2, \ldots)$ and define $|\alpha|$ by the equality $|\alpha| = \alpha_0\alpha$. Then $\sigma(\Theta_f(x)) = \theta_0(x)\Theta_f(f(x))$.

Let $\Sigma' = \{\Theta_f(x) : x \in I \setminus C\}$. The lemma on monotonicity implies that the map $x \to \Theta_f(x)$ is continuous at the points of the set Σ'. Consequently, the set $\overline{\Sigma'} \setminus \Sigma'$ consists of at most countably many sequences of the form $\Theta_f(z^\pm)$, where $z \in C$. Moreover, the set Σ' is invariant under the transformation σ': $(\theta_0, \theta_1, \theta_2, \ldots) \to \theta_0(\theta_1, \theta_2, \theta_3, \ldots)$, which coincides with the shift map multiplied by sign θ_0.

Thus, by neglecting a countable set of points and identifying points x and y such that $f^n|_{\langle x, y\rangle}$ is a homeomorphism for all $n \geq 0$, we reduce the investigation of the dynamics of unimodal mappings to the investigation of the symbolic system (Σ', σ').

2. Dynamical Coordinates and the Kneading Invariant

In this section, we give a description of the theory of kneading invariants (Milnor and Thurston [1]) for unimodal maps. It suffices to consider unimodal maps f: $[-1, 1] \to [-1, 1]$ such that $f(-1) = f(1) = -1$.

Let $x \to \Theta_f(x)$ be the correspondence constructed in the previous section. A sequence $\Theta_f(x) = (\theta_0, \theta_1, \theta_2, \ldots)$ can be associated with the formal power series

$$\Theta(x) = \sum_{i=0}^{\infty} \Theta_i(x) t^i.$$

This power series is called the dynamical coordinate of the point x. The lexicographic ordering and topology of coordinatewise convergence on $\{\Theta_f(x)\}$ induce the lexicographic ordering and topology of coordinatewise convergence on the set $\{\Theta(x)\}$. Moreover, the correspondence $x \to \Theta(x)$ remains monotone and, for any $x \in I$, there exist

$$\Theta(x^+) = \lim_{y \searrow x} \Theta(y) \quad \text{and} \quad \Theta(x^-) = \lim_{y \nearrow x} \Theta(y).$$

The series $\nu_f = \Theta(c^-)$ is called the *kneading invariant* of the map f.

We have chosen the series $\Theta(c^-)$ but not $\Theta(c)$ because $\Theta(c) = 0$. However, if $c \notin \operatorname{Per} f$, then the map $x \to \Theta(x)$ is continuous at the point $f(c)$ and we have $\Theta(c^-) = 1 + t\Theta(f(c))$. If c is a periodic point with period n, then the sequence $\Theta_f(c^-)$ is also periodic with period n or $2n$ (depending on the side on which $f^n(c^-)$ approaches the point c). In both cases, $\Theta(c^-) = 1 + t\Theta(f(c)) (\operatorname{mod} t^n)$. Hence, the series $\Theta(f(c))$ contains the same amount of information as the series $\Theta(c^-)$ and, therefore, $\Theta(f(c))$ can also be chosen as kneading invariant. We also note that $\Theta(c^-) = -\Theta(c^+)$.

The lemma presented below demonstrates that kneading invariants contain almost complete information about the behavior of the orbits of maps.

The formal power series Θ is called ν_f-admissible if, for any $n \geq 0$, we have either $|\sigma^n(\Theta)| \geq \nu_f$ or $\sigma^n(\Theta) = 0$, where σ corresponds to the operation of shift. By virtue of the lemma on monotonicity, the dynamical coordinate of any point is a ν_f-admissible power series.

Lemma 2.2. *For any ν_f-admissible formal power series Θ, there exists a point $x \in I$ such that Θ is equal either to $\Theta(x)$, or to $\Theta(x^-)$, or to $\Theta(x^+)$.*

Proof. Let $x = \inf\{y \mid \Theta(y) \leq \Theta\}$. Then $\Theta(x^-) \geq \Theta \geq \Theta(x^+)$. If $\Theta(y)$ is continuous at the point x, we have $\Theta(x) = \Theta$. If $\Theta(y)$ has a jump at the point x, then $f^n(x) = c$ for some $n \geq 0$ and, consequently, $\sigma^n(\Theta(x^-)) = -\sigma^n(\Theta(x^+))$ and is equal to $\pm\nu_f$. The series $\sigma^n(\Theta)$ is ν_f-admissible and lies between $\sigma^n(\Theta(x^-))$ and $\sigma^n(\Theta(x))$. Therefore, we have either $\sigma^n(\Theta) = \pm\nu_f$ or $\sigma^n(\Theta) = 0$, and this implies the required assertion.

Corollary 2.1. *Let f and g be unimodal maps and let c_f and c_g be their maximum points. If $\nu_f = \nu_g$, then there exists an orientation preserving map*

$$h \colon \bigcup_{i \geq 0} f^{-i}(c_f) \to \bigcup_{i \geq 0} g^{-i}(c_g)$$

such that $h \circ f = g \circ h$.

Proof. For $x \in \bigcup_{i\geq 0} f^{-i}(c_f)$, we set $h(x) = \inf\{y \mid \Theta_g(y) \leq \Theta_f(x)\}$. As in the proof of the previous lemma, we show that $h(x) \in \bigcup_{i\geq 0} g^{-i}(c_g)$ and $h \circ f = g \circ h$.

The assertions established above demonstrate that ν_f contains all information on the behavior of trajectories except the answer to the following question: Is the map $x \to \Theta_f(x)$ constant on some intervals?

Lemma 2.3. *If the map $x \to \Theta_f(x)$ is constant on an interval $\mathcal{J}$, then one of the following possibilities is realized:*

(i) there exists an integer $n < \infty$ for which $f^n(\mathcal{J})$ consists of a single point;

(ii) there exist $n \geq 0$, $k > 0$, and an interval L such that $f^n(\mathcal{J}) \subset L$, $f^k(L) \subset L$, and $f^k|_L$ is a homeomorphism;

(iii) $\mathcal{J}$ is a wandering interval of f, i.e., $\mathcal{J}, f(\mathcal{J}), f^2(\mathcal{J}), \ldots$ are mutually disjoint intervals.

Proof. If (i) is excluded, then, for any $n \geq 0$, $f(\mathcal{J})$ is an interval and $f^n(x) \neq c$ for $x \in \operatorname{int}(\mathcal{J})$. In particular, $f^n|_{\mathcal{J}}$ is a homeomorphism for any $n \geq 0$. Assume that (iii) is also not true. Then there are $n \geq 0$ and $k > 0$ such that $f^{n+k}(\mathcal{J}) \cap f^n(\mathcal{J}) \neq \varnothing$ and f^k is a homeomorphic mapping of the interval $L = \bigcup_{i\geq 0} f^{n+ik}(\mathcal{J})$ into itself.

Let us now return to the concept of ν_f-admissible series. Let Θ_f be the sequence that corresponds to the series ν_f. Then all elements of the sequence Θ_f differ from zero and the inequality $|\sigma^n\Theta_f| \geq \Theta_f$ is true for all $n \geq 0$ because the series ν_f is also ν_f-admissible.

Any sequence $\alpha = (\alpha_0, \alpha_1, \alpha_2, \ldots)$ is called *admissible* if $\alpha_0 = +1$, $\alpha_i = \pm 1$ for $i \in \mathbb{N}$, and $|\sigma^n \alpha| \geq \alpha$ for any $n \geq 0$.

Thus, for any unimodal map f, the sequence Θ_f is admissible. On the other hand, by the intermediate-value theorem (see Theorem 2.6 below), for any admissible sequence α, there exists a map f such that $\Theta_f = \alpha$.

The structure of the set of admissible sequences was investigated by Jonker and Rand [1].

For a given periodic sequence $\beta = (\beta_1, \beta_2, \ldots, \beta_m, \beta_1, \beta_2, \ldots, \beta_m, \ldots)$ with minimal period m, we set

$$\beta^{(1)} = (\beta_1, \ldots, \beta_m, -\beta_1, \ldots, -\beta_m, \beta_1, \ldots, \beta_m, -\beta_1, \ldots, -\beta_m, \ldots).$$

The sequence $\beta^{(1)}$ is called an antiperiodic sequence of period m. For $n > 1$, we successively define the sequence $\beta^{(n)} = (\beta^{(n-1)})^{(1)}$ with period $m2^n$.

Any periodic sequence α is admissible if and only if $\alpha^{(1)}$ is admissible. If, for some periodic sequence α, we have $\alpha < \beta < \alpha^{(1)}$, then β cannot be admissible.

Let Π' be the set of all admissible sequences, let $P \subset \Pi'$ be its subset of periodic sequences, and let $P' \subset P$ be the subset of periodic sequences that are not antiperiodic. The structure of the set Π' is described by the following theorem:

Theorem 2.1. (Jonker and Rand [1]). *Every sequence $\nu \in \Pi' \backslash P$ is limiting both for $\alpha > \nu$ and $\alpha < \nu$. Everysequence $\nu \in P'$ is limiting for $\alpha > \nu$ and isolated for $\alpha < \nu$. Every sequence $\nu \in P \backslash P'$ is isolated in Π'. Moreover, any antiperiodic sequence is equal to $\nu^{(k)}$ for some $\nu \in P'$ and belongs to the sequence $\nu > \nu^{(1)} > \nu^{(2)} > \ldots > \nu^{(\infty)}$ generated by ν.*

Consider two arbitrary sequences $\alpha = (\alpha_0, \alpha_1, \alpha_2, \ldots)$ and $\beta = (\beta_0, \beta_1, \beta_2, \ldots)$. We say that α is β-admissible if, for any $n \geq 0$, either $|\sigma^n \alpha| \geq \beta$ or $\sigma^n \alpha = (0, 0, 0, \ldots)$. Due to the existence of one-to-one correspondence between unilateral sequences and formal power series, one can apply the notation and notions introduced for sequences to power series. Furthermore, by Lemma 2.2, if β is a formal power series and, for the map f, one can find a point x such that either $\Theta(x)$, or $\Theta(x^-)$, or $\Theta(x^+)$ is equal to β, then, for any β-admissible series α, there exists a point y such that either $\Theta(y)$, or $\Theta(y^-)$, or $\Theta(y^+)$ is equal to α.

Lemma 2.4. *If, for some point x, $\Theta(x)$, $\Theta(x^-)$, or $\Theta(x^+)$ is either an admissible periodic or an admissible antiperiodic series of period $n \geq 1$, then f possesses a periodic point β of period n. If f possesses a periodic orbit of period n, then there is a point x such that one of its series $\Theta(x)$, $\Theta(x^-)$, or $\Theta(x^+)$ is either an admissible periodic or an admissible antiperiodic series of period n.*

Proof. First, we prove the second statement. Let β be a periodic point of period n and let B be the cycle that contains β. If the point of extremum c belongs to B, then we set $x = c$. If $c \notin B$, we consider a point $\beta_0 \in B$ such that $f(\beta_0) \geq \beta_1$ for any $\beta_1 \in B$ and assume that x coincides with a (unique) point β_0' such that $\beta_0' < c$ and $f(\beta_0') = f(\beta_0)$. In both cases, x is the required point.

Now assume that, for some point x, one of the series $\Theta(x)$, $\Theta(x^-)$, or $\Theta(x^+)$ is either admissible periodic or admissible antiperiodic with period $n \geq 1$. Denote this series by α. If $x \notin \bigcup_{i \geq 0} f^{-i}(c)$, then $\alpha = \Theta(x)$ and one can easily show that $\omega(x)$ is a periodic trajectory of period n. If $x \in \bigcup_{i \geq 0} f^{-i}(c)$, then the admissibility of α implies that $x = c$, $\alpha = \Theta(x^-)$, and c is a periodic point of period n.

Let $\mu(k)$ be the maximal admissible series in the lexicographically ordered collection of admissible periodic (but not antiperiodic) series of period k.

Thus, in particular, the following admissible sequences occupy the first positions in the indicated lexicographically ordered collection (here, we write only the relevant signs instead of ± 1):

$$\mu(2^0) = (+++\ldots),$$

$$\mu(2^1) = (\mu(2^0))^{(1)} = (+-+-+-\ldots),$$

$$\mu(2^2) = (\mu(2^0))^{(2)} = (\mu(2^1))^{(1)} = (+--++--++--\ldots),$$

$$\cdots\cdots\cdots\cdots\cdots\cdots\cdots\cdots\cdots\cdots\cdots\cdots$$

$$\mu(2^\infty) = (\mu(2^0))^{(\infty)} = (+--+-++--++-+--+-+\ldots).$$

Lemma 2.5. *The following sequence is lexicographically ordered:*

$$\mu(1) > \mu(2) > \mu(4) > \ldots > \mu(2^n) > \ldots > \mu(2^n\cdot 7) > \mu(2^n\cdot 5) > \mu(2^n\cdot 3)$$
$$> \ldots > \mu(2\cdot 7) > \mu(2\cdot 5) > \mu(2\cdot 3) > \ldots > \mu(7) > \mu(5) > \mu(3).$$

To prove this assertion, it is necessary to determine the sequences $\mu(2^n\cdot k)$. Let $k = 2i+3,\ i \geq 0$. Then $\mu(2^n\cdot k)$ is generated by the periodic replication of the finite chain $\alpha(2^n\cdot k)$ of length $2^n\cdot k$. For $n = 0$, we have $\alpha(k) = \alpha(2i+3) = (+--+-+-\ldots+-)$, where the pair $(+-)$ is repeated i times. For $n \geq 1$, we get $\alpha(2^n\cdot k) = \alpha(2^{n-1}k)\cdot(+-)$, where the chain on the right-hand side is obtained from the chain $\alpha(2^{n-1}k)$ by replacing every sign with $(+-)$ or $(-+)$ depending on the sign to be replaced.

The statements established above imply the following assertion about the coexistence of periods of cycles for unimodal mappings:

Corollary 2.2. *Assume that the natural numbers are arranged in the following order:* $1 \lhd 2 \lhd 4 \lhd \ldots \lhd 2\cdot 7 \lhd 2\cdot 5 \lhd 2\cdot 3 \lhd \ldots \lhd 7 \lhd 5 \lhd 3$. *If a unimodal map* f *has a cycle of period* n *and* $k \lhd n$, *then it has a cycle of period* k.

In the next chapter, we prove this assertion for general continuous maps.

3. Periodic Points, ζ-Function, and Topological Entropy

In this section, we establish the relationship between the kneading invariant and the number of extrema γ_n, the number of the intervals of monotonicity l_n, and the number η_n of fixed points of the map f^n.

We set γ_n to be equal to $\operatorname{card}\{f^{-n}(c)\}$ and consider the power series $\gamma_f = \sum_{n=0}^{\infty} \gamma_n t^n$. Similarly, let l_n be the number of the intervals of monotonicity of f^n and let $l_f = \sum_{n=0}^{\infty} l_n t^n$. Since $l_n = 1 + \sum_{k=0}^{n-1} \gamma_k$, we have $l_f = (1 + t\gamma_f)/(1-t)$.

If f^k has finitely many fixed points for any $k \geq 1$, then the function ζ can be defined as the formal series

$$\zeta_f = \exp\left\{\sum_{k\geq 1} \frac{\eta_k}{k} t^k\right\},$$

where $\eta_k = \operatorname{card}\{x \mid f^k(x) = x\}$.

The contribution of every orbit of period p to ζ_f is $\sum_{i=0}^{\infty} t^{i\cdot p} = (1-t^p)^{-1}$. Therefore, $\zeta_f^{-1} = \prod_\beta (1 - t^{p(\beta)})$, where the product is taken over all periodic orbits.

The following definitions enable us to establish the relationship between the ζ-function and the kneading invariant: Points x and y are called monotonically equivalent with respect to f^k whenever f^k is a homeomorphic mapping of $\langle x, y\rangle$ onto $\langle x, y\rangle$. Let $\hat{\eta}_k$ be the number of equivalence classes of points with period k and let

$$\hat{\zeta}_f = \exp\left\{\sum_{k\geq 1} \frac{\hat{\eta}_k}{k} t^k\right\}.$$

Lemma 2.6. *The following equalities hold:*

$$\nu_f \gamma_f = (1-t)^{-1} \quad \textit{and} \quad -\nu_f \hat{\zeta}_f = (Q(t)(1-t))^{-1},$$

where $Q(t) \equiv 1$ *whenever* ν_f *is nonperiodic, and* $Q(t) = 1 - t^p$ *if* ν_f *is periodic with period* p.

Proof. We prove only the first equality. Consider

$$\nu_f \gamma_f = \sum_{n\geq 0} \left(\sum_{0\leq i\leq n} \gamma_{n-i} \nu_i \right) t^n.$$

Let x be a point of the set $f^{-(n-i)}(c)$. Then $f^{(n-i)}(x) = c$ and f^{n+1} possesses a local minimum (maximum) at the point x if and only if $\nu_i = 1$ (or $\nu_i = -1$). This implies that $\sum_{0\leq i\leq n} \gamma_{n-i} \nu_i$ is equal to the difference between the numbers of maxima and minima of f^{n+1}. Therefore, this sum can be equal to $+1$, 0, or -1. Thus, it is equal to zero if and only if f^{n+1} either simultaneously increases or simultaneously decreases at the ends of the interval.

Let $\mathcal{J} = (a, b)$ and let $f^{n+1} : \mathcal{J} \to \mathbb{R}$. Then the difference between the numbers of maxima and minima of $f^{n+1} \mid \mathcal{J}$ is equal to $\frac{1}{2}(\theta_n(a^+) - \theta_n(b^+))$. Therefore, $\nu_f \gamma_f(\mathcal{J}) = \frac{1}{2}(\theta(a^+) - \theta(b^-))$. For the map $f: [-1, 1] \to [-1, 1]$, we have $f(-1) = f(1) = -1$,

$$\theta(1^-) = -1 - t - t^2 - \ldots, \qquad \theta(-1^+) = 1 + t + t^2 + \ldots,$$

and $\nu_f \gamma_f = 1 + t + t^2 + \ldots = (1-t)^{-1}$.

Consider some examples of finding ν_f, γ_f, and ζ_f for maps from the family $f_\lambda(x) = \lambda x(1-x)$, $x \in [0, 1]$, $\lambda \in (0, 4]$, corresponding to examples presented in Section 1 of Chapter 1.

1. $0 < \lambda \le 1$ (see Fig. 3). In this case, we have

$$\theta_{f_\lambda}(c^-) = (+++\ldots),$$

$$\nu_{f_\lambda} = 1 + t + t^2 + \ldots = (1-t)^{-1},$$

$$\gamma_{f_\lambda} = 1, \quad (\hat{\zeta}_{f_\lambda})^{-1} = 1 - t, \quad \zeta_{f_\lambda} = \hat{\zeta}_{f_\lambda}.$$

2. $1 < \lambda \le 3$ (see Fig. 4). It is necessary to consider the following two subcases:

(i) $1 < \lambda \le 2$. The map f_λ has two equivalent fixed points. Here, $\theta_{f_\lambda}(c^-)$, ν_{f_λ}, γ_{f_λ}, and $\hat{\zeta}_{f_\lambda}$ are as in the previous example but $(\zeta_{f_\lambda})^{-1} = (1-t)^2 \ne (\hat{\zeta}_{f_\lambda})^{-1}$.

(ii) $2 < \lambda \le 3$. As λ passes through the point 2, the sign of the multiplier of one of the fixed point changes and fixed points become nonequivalent. Here,

$$\theta_{f_\lambda}(c^-) = (+-+-\ldots),$$

$$\nu_{f_\lambda} = (1+t)^{-1},$$

$$\gamma_{f_\lambda} = \frac{1+t}{1-t}, \quad (\hat{\zeta}_{f_\lambda})^{-1} = (1-t)^2, \quad \zeta_{f_\lambda} = \hat{\zeta}_{f_\lambda}.$$

3. $3 < \lambda \le 1 + \sqrt{6}$ (see Fig. 5):

(i) $3 < \lambda \le 1 + \sqrt{5}$. The map f_λ has two nonequivalent fixed points and the cycle of period 2. Both points of this cycle are monotonically equivalent to one of the fixed points. Here, $\theta_{f_\lambda}(c^-)$, ν_{f_λ} γ_{f_λ}, and $\hat{\zeta}_{f_\lambda}$ are the same as in case 2(ii) but

$$(\zeta_{f_\lambda})^{-1} = (1-t)^2(1-t^2) \neq (\hat{\zeta}_{f_\lambda})^{-1}.$$

(ii) $1+\sqrt{5} < \lambda \leq 1+\sqrt{6}$. As λ passes through the value $1+\sqrt{5}$, the multiplier of the 2-periodic cycle becomes negative. This results in the doubling of the period of kneading invariant and the points of 2-periodic cycle become non-equivalent to the fixed point. In this case,

$$\theta_{f_\lambda}(c^-) = (+--++--+\ldots),$$

$$\nu_{f_\lambda} = \frac{1-t}{1+t^2},$$

$$\gamma_{f_\lambda} = \frac{1+t^2}{(1-t)^2}, \quad (\hat{\zeta}_{f_\lambda})^{-1} = (1-t)^2(1-t^2) = (\zeta_{f_\lambda})^{-1}.$$

4. $\lambda = 3.83$ (see Fig. 8). For this value of λ, the map f_λ has two nonequivalent cycles of period 3 and the kneading invariant is antiperiodic with period 3. Moreover,

$$\theta_{f_\lambda}(c^-) = (+---+++--\ldots),$$

$$\nu_{f_\lambda} = 1-t-t^2-t^3+t^4+t^5+\ldots = \frac{1-t-t^2}{1+t^3} = \frac{1}{1+t},$$

$$\gamma_{f_\lambda} = \frac{1+t^2}{(1-t)^2}, \quad (\zeta_{f_\lambda})^{-1} = (\hat{\zeta})^{-1} = (1-t)(1-t-t^2)(1-t^3).$$

5. $\lambda = 4$. In this case (see Fig. 9),

$$\theta_{f_\lambda}(c^-) = (+---\ldots),$$

$$\nu_{f_\lambda} = 1-t-t^2+\ldots = \frac{1-2t}{1-t},$$

$$\gamma_{f_\lambda} = \frac{1}{1-2t}, \quad (\hat{\zeta}_{f_\lambda})^{-1} = 1-2t, \quad \zeta_{f_\lambda} = \hat{\zeta}_{f_\lambda}.$$

The importance of Lemma 2.6 is corroborated by the following assertion:

Corollary 2.3. *The power series* ν_f *is convergent for all* t. *The power series* γ_f *and* l_f *are meromorphic and convergent inside the circle* $|t| < r(f)$, *where* $r(f)$ *is the smallest positive real zero of the function* $(1-t)\nu_f(t)$.

Proof. Since $\nu_f = \sum_{i\geq 0} \nu_i t^i$, where $\nu_i = \pm 1$, the series ν_f is convergent for all t with $|t| < 1$. The equalities $\nu_f\gamma_f = (1-t)^{-1}$ and $l_f = (1 + t\gamma_f)/(1-t)$ imply that l_f and γ_f are meromorphic and their poles in the circle $|t| < 1$ coincide. It follows from the positivity of the terms of the series γ_f and Abel's theorem that the radius of convergence $r(f)$ of γ_f is a pole of the function $\gamma_f(t)$ if $r(f) > 0$. If $r(f) < 1$, then $r(f)$ is the smallest positive (real) zero of the function $\nu_f(t)$.

Corollary 2.3 and the inequality $l(f \circ g) \leq l(f)l(g)$ imply that

$$\lim_{n\to\infty} (l_n)^{1/n} = \lim_{n\to\infty} (\gamma_n)^{1/n} = \frac{1}{r(f)}.$$

The number $s(f) = \frac{1}{r(f)}$ is called the *growth exponent* of the map f. It is obvious that, for unimodal maps, we have $1 \leq s(f) \leq 2$.

Let $h(f)$ be the topological entropy of the map f. Misiurewicz and Szlenk [1] proved that $h(f) = \lim_{n\to\infty} \frac{1}{n} \log \gamma_n$. Hence, $s(f) = \exp(h(f))$ and, consequently, $h(f) \in [0, \log 2]$ for unimodal maps.

Theorem 2.2 (Milnor and Thurston [1]). *The mapping* $h : f \to h(f)$ *is continuous in the space of unimodal* C^1*-maps endowed with* C^1*-topology.*

Proof. First, we assume that the critical point of f is not periodic. Then the dependence of the series $\nu_f = \sum_{i\geq 0} \nu_i t^i = \theta(c^-)$ on f is continuous. Moreover, the Cauchy integral theorem implies that the smallest zero of ν_f continuously depends on f, but this means that entropy is continuous.

Now let $c \in \operatorname{Per} f$ and let the period of c be equal to n. In this case, any unimodal map g sufficiently close to f in C^1 possesses an attracting periodic orbit that contains a (unique) critical point in its domain of immediate attraction. Let $\nu_g = \sum_{i\geq 0} \nu_i t^i$. Then the sequence ν_g is either periodic or antiperiodic with period n. Thus, ν_g can be written in the form

$$\left(\sum_{0\leq i<n} \nu_i t^i\right)(1-t^n)^{-1} \quad \text{or} \quad \left(\sum_{0\leq i<n} \nu_i t^i\right)(1+t^n)^{-1}.$$

Inside the circle $|t| < 1$, the poles of these two functions coincide. Therefore, the entropy of the map g is equal to $h(f)$. Theorem 2.2 is proved.

Consider the space of unimodal maps with C^0-topology. To define $F: [0, 1] \to [0, 1]$, we set $F(x) \equiv x$ for $x \in [0, \frac{3}{4}]$ and $F(x) = 3(1-x)$ for $x \in [\frac{3}{4}, 1]$. Let $f_\lambda(x) = \lambda F(x)$, $\lambda \geq 1$.

By using the equality $h(f_\lambda) = \lim_{n\to\infty} \frac{1}{n} \log \gamma_n(\lambda)$, we can show that $h(f_\lambda) = 0$ for $\lambda = 1$. For $\lambda > 1$, we consider the map f_λ^2 on the interval $[\beta', \beta]$, where $\beta = \beta(\lambda)$ is the nonzero fixed point of the map f_λ, $\beta' \neq \beta$, and $f_\lambda(\beta') = f_\lambda(\beta)$. In this interval, the map $g = f_\lambda^2$ is unimodal and expanding (moreover, $|(f_\lambda^2)'(x)| > 3$ for all $x \in [\beta', \beta]$). Therefore, $l(g^n) \geq 2^n$ and $h(g) \geq \log 2$. Consequently, $h(f_\lambda) \geq \frac{1}{2} \log 2$ for any $\lambda > 1$.

At the same time, for unimodal maps with positive topological entropy, continuous changes in the map induce continuous changes in entropy. More precisely, Misiurewicz [5] proved the following statement:

Let U^0 be the space of unimodal maps endowed with C^0-topology. Then the map $h: f \to h(f)$ is continuous at a point f_0 of the space U^0 whenever $h(f_0) > 0$.

4. Kneading Invariant and Dynamics of Maps

For mappings $f: I \to I$ whose exponents of growth $s(f)$ are greater than 1, one can construct piecewise linear models. Consider a function $l_f(\mathcal{J}) / l_f$, where

$$l_f(\mathcal{J}) = \sum_{n+1}^{\infty} l(f^n|_{\mathcal{J}})\, t^{n-1}, \; l_f = l_f(I).$$

It follows from the results of the previous section (see Corollary 2.3) that this function is meromorphic in the circle $|t| < 1$ and satisfies the condition $l_f(\mathcal{J}) / l_f \leq 1$ for $t > 0$. Hence, $l_f(\mathcal{J}) / l_f$ possesses a removable singularity at $t = r(f)$. We define

$$\Lambda(\mathcal{J}) = \lim_{t \to r(f)} \frac{l_f(\mathcal{J})}{l_f}.$$

It is easy to show that

(i) if $\mathcal{J}_1$ and $\mathcal{J}_2$ have a common end, then

$$\Lambda(\mathcal{J}_1 \cup \mathcal{J}_2) = \Lambda(\mathcal{J}_1) + \Lambda(\mathcal{J}_2);$$

(ii) if $\mathcal{J}$ does not contain points of extrema, then

$$\Lambda(f\mathcal{J}) = s(f)\Lambda(\mathcal{J});$$

(iii)
$$\lim_{|\mathcal{J}|\to 0} \Lambda(\mathcal{J}) = 0.$$

Let $I = [0, 1]$. We set $\lambda(x) = \Lambda([0, x])$. Then there exists a unique map $F: [0, 1] \to [0, 1]$ that satisfies the condition $F \circ \lambda = \lambda \circ f$. The map F is piecewise linear, its derivative is equal to $\pm s(f)$, and the number of the intervals of monotonicity of F does not exceed the number of the intervals of monotonicity of f.

The function $\lambda(x)$ may be not strictly monotone. Thus, if $\Theta(x) = \Theta(y)$ and $y \neq x$, then λ maps the interval $\langle x, y\rangle$ into a point. The map $x \to \lambda(x)$ is called the semi-conjugation of f to F.

The kneading invariants of f and the piecewise linear model

$$F_s = \begin{cases} sx, & x \in \left[0, \dfrac{1}{2}\right], \\ s - sx, & x \in \left[\dfrac{1}{2}, 1\right], \end{cases}$$

where $s = s(f)$, are connected as follows: If $\lambda^{-1}(\{\frac{1}{2}\})$ is a point, then $\nu_f = \nu_{F_s}$. If $\lambda^{-1}(\{\frac{1}{2}\}) = \mathcal{J}$ is an interval, then ν_{F_s} is periodic with period n. In this case, $f^n(\mathcal{J}) \subset \mathcal{J}$ and $g = f^n|_{\mathcal{J}}$ is a unimodal map for which it is possible to introduce the kneading invariant ν_g.

Let $\Theta_{F_s} = (\theta_0\theta_1\theta_2 \ldots \theta_{n-1}, \theta_0\theta_1 \ldots \theta_{n-1}, \ldots)$ and $\Theta_g = \alpha = (\alpha_0, \alpha_1, \alpha_2, \ldots)$. Then Θ_f satisfies the equality

$$\Theta_f = \Theta_{F_s} * \Theta_g \overset{\text{def}}{=} ((\theta_0\theta_1 \ldots \theta_{n-1}), \alpha_1(\theta_0\theta_1 \ldots \theta_{n-1}), \alpha_\alpha(\theta_0\theta_1 \ldots \theta_{n-1}), \ldots).$$

In terms of kneading invariants, we can write

$$\nu_{F_s}(t) = P_{n-1}(t)(1 - t^n)^{-1}, \qquad \nu_g(t) = \alpha_0 + \alpha_1 t + \alpha_2 t^2 + \ldots,$$

$$\nu_f(t) = P_{n-1}(t)\nu_g(t^n) = (1 - t^n)\nu_{F_s}(t)\nu_g(t^n).$$

In the general case where $\nu(t)$ is an admissible periodic series of period n and $\alpha(t)$ is admissible, we define $\nu * \alpha(t)$ as the following admissible series:

$$\nu * \alpha(t) = (1 - t^n)\nu(t)\alpha(t^n).$$

For $\nu \in \Pi'$, we set $s(\nu) = s(f)$, where f is an arbitrary map such that $\nu_f = \nu$. We have the following theorem on decomposition of ν into a product of irreducible factors:

Theorem 2.3. *Let $\nu \in \Pi'$ and $s_1 = s(\nu)$. If $s_1 = 1$, then $\nu = \mu(2^i)$, $0 \le i \le \infty$. If $s_1 > 1$, then ν admits one of the following decompositions:*

(i) there exist $s_2, s_3, \ldots, s_{m*} \in (1, 2]$ and $\alpha \in \Pi'$, $s(\alpha) = 1$, such that

$$\nu = \nu_{F_{s_1}} * \nu_{F_{s_2}} * \ldots * \nu_{F_{s_m}} * \alpha,$$

$\nu_{F_{s_m}}$ are periodic for all $1 \le m \le m^*$ and $s_1 > s_2^{1/p_2} > \ldots > s_{m*}^{1/p_{m*}}$ (here,

$$p_m = \prod_{k=1}^{m-1} n_k$$

and n_k is the period of $\nu_{F_{s_k}}$), and $\alpha = \mu(2^i)$, $i \le \infty$;

(ii) $\nu = \nu_{F_{s_1}} * \ldots * \nu_{F_{s_{m*}}}$, where the series $\nu_{F_{s_m}}$ with $1 \le m \le m^* - 1$ have the same properties as in (i) and $\nu_{F_{s_{m*}}}$ is a periodic series;

(iii) $\nu = \lim_{m^* \to \infty} \nu_{F_{s_1}} * \ldots * \nu_{F_{s_{m*}}}$, where all $\nu_{F_{s_m}}$ have the same properties as in (i) and $\lim_{m^* \to \infty} s_{m*}^{1/p_{m*}} = 1$.

For a unimodal map f, the decomposition of ν_f in Theorem 2.3 corresponds to the spectral decomposition of the set $NW(f)$ (see Theorem 2.5).

The theory of kneading invariants takes the most perfect form for maps with negative Schwarzian. For $f \in C^3(I, I)$ with $f'(x) \ne 0$, the Schwarzian (or Schwarzian derivative) Sf is defined by the equality

$$Sf(x) = \frac{f'''(x)}{f'(x)} - \frac{3}{2}\left(\frac{f''(x)}{f'(x)}\right)^2.$$

Let SU denote the set of unimodal maps f such that $Sf(x) < 0$ for all x except c (in what follows, maps of this sort are studied in detail).

Theorem 2.4 (Jonker and Rand [1]). *Let $\nu_f = \nu_g$ for $f, g \in$ SU.*

(i) If the series ν_f is not periodic, then f and g are topologically conjugate.

(ii) If ν_f is periodic with period n, then both f and g possess an attracting or neutral periodic trajectory of period n or $n/2$; moreover, f and g are topo-

logically conjugate whenever these trajectories are of the same type, i.e., either both attracting or both neutral, and the corresponding points of these trajectories have the same dynamical coordinates.

By applying Theorem 2.4 to maps with negative Schwarzian, we arrive at the following decomposition of the set $NW(f)$.

Theorem 2.5 (Jonker and Rand [1]). *For* $f \in SU$, *the set* $NW(f)$ *admits the following decomposition*

$$NW(f) = \bigcup_{1 \le m \le m^*} \Omega_m, \qquad m^* \le \infty,$$

where Ω_m *are closed mutually disjoint invariant sets such that*

(i) for $1 \le m < m^*$, *the sets* Ω_m *are representable in the form* $\Omega_m = P_m \cup C_m$, *where* P_m *consists of finitely many periodic orbits and* $f|_{C_m}$ *is topologically conjugate to a transitive topological Markov chain; moreover,* Ω_m *are hyperbolic sets;*

(ii) for $m = m^*$, *one of the following cases is realized: if* $m^* < \infty$, *then*

(a) Ω_{m^*} *is the same as* Ω_m *with* $m < m^*$ *but* C_{m^*} *contains a periodic trajectory whose derivative on the period is equal to* $+1$ *(this corresponds to case (i) of Theorem 2.3,* $\alpha = (+++\ldots)$*);*

(b) Ω_{m^*} *is the union of finitely many repelling periodic orbits of periods* $p_{m^*}2^i$, $0 \le i < n$, *and an attracting periodic orbit of period* $p_{m^*}2^n$ *(case (i) of Theorem 2.3,* $\alpha = \mu(2^n)$, $n \le 0$*);*

(c) Ω_{m^*} *is the union of repelling periodic orbits of periods* $p_{m^*}2^i$, $0 \le i < \infty$, *and a minimal invariant set equal to* $\overline{\text{orb}\,(c)}$ *(case (i) of Theorem 2.3,* $\alpha = \mu(2^\infty)$*);*

(d) Ω_{m^*} *is the union of finitely many repelling periodic orbits and finitely many intervals (cyclically permuted by the map* f*) such that* $f^{p_{m^*}}$ *is topologically conjugate on these intervals to a piecewise linear map (case (ii) of Theorem 2.3).*

If $m^* = \infty$, *then* $\Omega^\infty = \overline{\text{orb}\,(c)}$ *is a minimal invariant set (case (iii) of Theorem 2.3).*

In all cases, the domain of attraction of the set Ω_{m^*}, i.e., $P(\Omega_{m^*}, f) = \{x \in I \mid \omega_f(x) \subset \Omega_{m^*}\}$ is a set of the second category and $\text{mes}\, P(\Omega_{m^*}, f) = \text{mes}\, I$, where mes denotes the Lebesgue measure of the corresponding sets.

For the exponent of growth of the family of piecewise linear maps F_s, $s \in [1, 2]$, we have $s(F_s) = s$. The kneading invariant monotonically changes from $1 + t + t^2 + \ldots = (1 - t)^{-1}$ for $s = 1$ to $1 - t - t^2 - \ldots = (1 - 2t)(1 - t)^{-1}$ for $s = 2$. This is a consequence of the following result of Jonker and Rand [1]:

If $\nu > \tilde{\nu}$ *in the sense of lexicographic ordering, then* $s(\nu) \leq s(\tilde{\nu})$. *Moreover,* ν_{F_s} *does not take all admissible values, e.g.,* $\nu_{F_s} \neq \mu(2^i)$ *for any* $s, i \in \mathbb{N}$.

At the same time, this is impossible for the families of smooth maps.

In the case of smooth maps, we have the following intermediate-value theorem for kneading invariants:

Theorem 2.6 (Milnor and Thurston [1]). *Let* $\{f_s\}_{s \in [0,1]}$ *be a family of* C^1*-class maps which continuously depend on* s *in the* C^1*-topology, let* $\nu_{f_0} > \nu_{f_1}$, *and let* $\alpha \in \Pi$ *satisfy the inequalities* $\nu_{f_0} > \alpha > \nu_{f_1}$. *Then there exists* $s_0 \in [0, 1]$ *such that* $\alpha = \nu_{f_{s_0}}$.

3. COEXISTENCE OF PERIODIC TRAJECTORIES

Dynamical systems generated by continuous maps of an interval into itself are characterized by the following important property: The data on the relative location of points of a single trajectory on the interval I may contain much information about the dynamical system as a whole. Clearly, this is explained by the fact that the phase space (the interval I) is one-dimensional. The points of a trajectory define a decomposition of the phase space, and information on the mutual location of these points often enables one to apply the methods of symbolic dynamics. These ideas are especially useful for the investigation of periodic trajectories.

As already shown, the existence of cycles of some periods implies the existence of cycles of other periods. At present, the problem of coexistence of cycles is fairly well studied and there are numerous papers dealing with this problem. Many important results on the coexistence of cycles were established for continuous maps of a circle, of one-dimensional branched manifolds, and some other classes of topological spaces. In this chapter, we present the most important facts established for continuous maps of an interval into itself.

1. Coexistence of Periods of Periodic Trajectories

First, we present several simple assertions.

If a map $f \in C^0(I, I)$ has a cycle of period $m > 1$, then it also possesses a fixed point. Indeed, if β' and β'' are the smallest and the largest points of this cycle, respectively, then $f(\beta') > \beta'$ and $f(\beta'') < \beta''$ and it follows from the continuity of the function f on $[\beta', \beta'']$ that $f(\beta) = \beta$ for a point $\beta \in [\beta', \beta'']$.

In what follows, $\langle a, b \rangle$ denotes a closed interval with ends at $a, b \in \mathbb{R}$. This notation is convenient in the case where the relative location of the points a and b is unknown or inessential.

Lemma 3.1. *A map f has a cycle of period 2 $\Leftrightarrow$ there exists a point $a \in I$ such that $a \neq f(a)$ and $a \in f(\langle a, f(a) \rangle)$.*

Proof. For any periodic point a of period 2, we have $a \neq f(a)$ and $a \in f(\langle a, f(a)\rangle)$. Therefore, it remains to prove the converse assertion.

For definiteness, we assume that $f(a) > a$. Then there exists a point $a' \in (a, f(a)]$ such that $f'(a) = a$. If $a' = f(a)$, then a is a periodic point with period 2. For $a' < f(a)$, we have only two possible cases:

(i) there are fixed points for $x > a'$;

(ii) there are no fixed points for $x > a'$.

We consider each of these possibilities separately.

(i) Let b be the smallest fixed point in the interval $\{x > a'\}$. Since $f([a, a']) \supset [a, a']$, there are fixed points in the interval $[a, a']$. Let b' be one of these points. Since $f([a', b]) \supset [a, a']$, one can find a point $c \in [a', b]$ such that $f(c) = b'$. Hence, $f^2(a') > a'$ and $f^2(c) = b' < c$ and, consequently, there are periodic points of period 2 in the interval $[a', c]$.

(ii) Consider f^2. Since the interval I is mapped into itself, there exists a point $d \geq a'$ such that $f^2(d) \leq d$. Moreover, $f^2(a') > a'$ and there are no fixed points of f in the interval $[a', d]$. Therefore, the interval $[a', d]$ must contain a periodic point of period 2.

Lemma 3.2. *If a map has a cycle of period $m > 2$, then this map has a cycle of period 2.*

Proof. Let B be a cycle of the map f of period m and let $\beta_0 = \max\{\beta \in B \mid f(\beta) > \beta\}$. It is clear that $\beta_0 \in f([\beta_0, f(\beta_0)])$ and it remains to apply Lemma 3.1.

Corollary 3.1. *If a map f has a cycle of period 2^l for $l \geq 0$, then f has cycles of periods 2^i, $i = 0, 1, \ldots, l-1$.*

Corollary 3.2. *If a map f has a cycle of period $\neq 2^i$, $i = 0, 1, 2, \ldots$, then f also has cycles of periods 2^i, $i = 0, 1, 2, \ldots$.*

In order to prove that f has a cycle of period 2^n, it suffices to apply Lemma 3.2 to the map $g = f^{2^{n-1}}$. Thus, in the case of Corollary 3.2, the map f has a periodic point of period $2^l m$ with odd m and $l \geq 1$. For the map g, the period of this periodic point is greater than 2 (namely, it is equal to $2^{l-n+1} m$ for $n \leq l$ and to m whenever $n > l$). According to Lemma 3.2, the map g possesses a periodic point of period 2 which is obviously a periodic point of period 2^n for f.

Actually, we have the following theorem (Sharkovsky [1]):

Theorem 3.1 (on coexistence of cycles). *If a continuous map of the interval onto itself has a cycle of period* m, *then it also has cycles of any period* m' *such that* $m' \lhd m$, *where*

$$1 \lhd 2 \lhd 2^2 \lhd 2^3 \lhd \ldots \lhd 2^2 \cdot 7 \lhd 2^2 \cdot 5 \lhd 2^2 \cdot 3 \lhd \ldots$$

$$\lhd 2 \cdot 7 \lhd 2 \cdot 5 \lhd 2 \cdot 3 \lhd \ldots \lhd 9 \lhd 7 \lhd 5 \lhd 3.$$

Moreover, for any m, *there exists a map with cycle of period* m *and no cycles of periods* m' *if* $m \lhd m'$.

(Maps of this sort are studied in Section 2 of this chapter).

A part of the statement of Theorem 3.1 (concerning the existence of cycles of periods 2^i, $i = 0, 1, 2, \ldots$) is contained in Corollaries 3.1 and 3.2.

At present, there are several known versions of the proof of Theorem 3.1 (Arneodo, Ferrero, and Tresser [1], Block, Guckenheimer, Misiurewicz, and Young [1], Burkart [1], Guckenheimer [1], Ho and Morris [1], Jonker [1] Shapiro and Luppov [1], Sharkovsky [1]). Here, we present a proof based on the use of symbolic dynamics and properties of cyclic permutations (a cyclic permutation of length m is defined as a map of the set $\{1, 2, \ldots, m\}$ onto itself which has no invariant subsets other than $\{1, 2, \ldots, m\}$).

Every cycle can be associated with a cyclic permutation π, a transition matrix, and (or) an oriented transition graph. The investigation of these objects gives vast information on properties of a dynamical system as a whole.

If a cycle B consists of the points $\beta_1 < \beta_2 < \ldots < \beta_m$ and $f(\beta_i) = \beta_{s_i}$, $1 \le s_i \le m$, $i = 1, 2, \ldots, m$, then

$$\pi = \begin{pmatrix} 1 & 2 & \ldots & m \\ s_1 & s_2 & \ldots & s_m \end{pmatrix}.$$

Due to the continuity of the map f on the intervals $\mathcal{J}_i = [\beta_i, \beta_{i+1}]$, $i = 1, \ldots, m-1$, we can write

$$f(\mathcal{J}_i) \supseteq \begin{cases} \mathcal{J}_{s_i} \cup \ldots \cup \mathcal{J}_{s_{i+1}-1}, & \text{if } s_i < s_{i+1}, \\ \mathcal{J}_{s_{i+1}} \cup \ldots \cup \mathcal{J}_{s_i-1}, & \text{if } s_i > s_{i+1}. \end{cases}$$

In this case, we say that $\mathcal{J}_i$ covers (or f-covers) the corresponding intervals $\mathcal{J}_{s_i}$. This cycle can be associated with a matrix $\{\mu_{is}\}$ of admissible transitions (of points of the intervals $\mathcal{J}_i$), where

$$\mu_{is} = \begin{cases} 0, & \text{if } f(\mathcal{J}_i) \not\supset \mathcal{J}_s, \\ 1, & \text{if } f(\mathcal{J}_i) \supset \mathcal{J}_s, \end{cases}$$

and with an oriented transition graph with vertices $\mathcal{J}_1, \ldots, \mathcal{J}_{m-1}$ and oriented edges that connect $\mathcal{J}_i$ and $\mathcal{J}_s$ if $f(\mathcal{J}_i) \supset \mathcal{J}_s$. For convenience, we write $\mathcal{J}_i \to \mathcal{J}_s$ if $f(\mathcal{J}_i) \supset \mathcal{J}_s$ (i.e., in the case where $\mathcal{J}_i$ f-covers $\mathcal{J}_s$). In what follows, the transition graph is called the *B*-graph of a cycle. Thus, the map displayed in Fig. 21 has a 3-periodic cycle formed by the points β_1, β_2, and β_3. For this cycle, we have $\pi = \begin{pmatrix} 1 & 2 & 3 \\ 2 & 3 & 1 \end{pmatrix}$, $f(\mathcal{J}_1) \supset \mathcal{J}_2$, and $f(\mathcal{J}_2) \supset \mathcal{J}_1 \cup \mathcal{J}_2$. Hence, the transition matrix has the form $\begin{pmatrix} 0 & 1 \\ 1 & 1 \end{pmatrix}$ and the *B*-graph is depicted in Fig. 22.

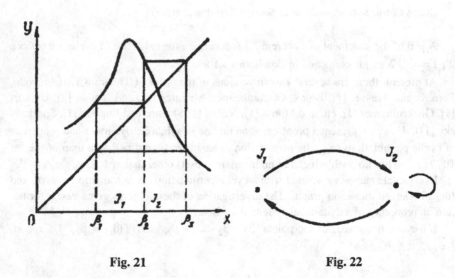

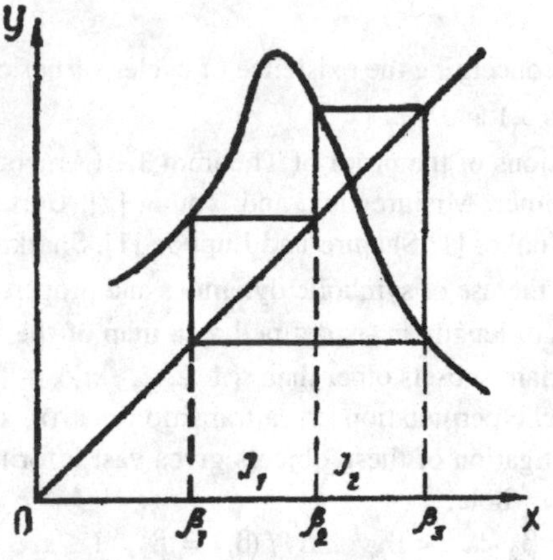

Fig. 21

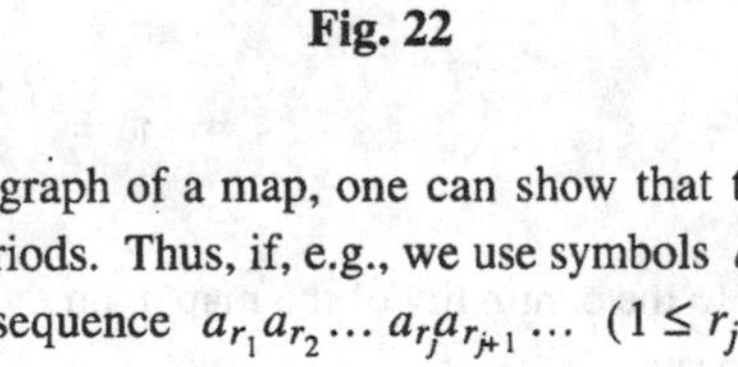

Fig. 22

By analyzing the transition matrix or the *B*-graph of a map, one can show that the map possesses periodic trajectories of various periods. Thus, if, e.g., we use symbols a_1, $a_2, \ldots, a_{m-1}$ as an alphabet, then any symbolic sequence $a_{r_1} a_{r_2} \ldots a_{r_j} a_{r_{j+1}} \ldots$ $(1 \le r_j \le m-1)$ admitted by the transition matrix ($\mu_{r_j r_{j+1}} = 1$ for all $j = 1, 2, \ldots$) corresponds to (at least one) trajectory of the system which passes through the intervals $\mathcal{J}_1, \ldots, \mathcal{J}_{m-1}$ in the following order: $\mathcal{J}_{r_1} \to \mathcal{J}_{r_2} \to \ldots \to \mathcal{J}_{r_j} \to \mathcal{J}_{r_{j+1}} \to \ldots$. In particular, if the symbolic sequence is periodic with the smallest period n, then the system has at least one periodic trajectory of period m (which passes through the intervals $\mathcal{J}_1, \mathcal{J}_2, \ldots, \mathcal{J}_{m-1}$ in the indicated order).

This fact is a consequence of the following simple geometric lemma, which allows us to pass from the intervals covering each other to periodic points:

Lemma 3.3. *1. If there exists a closed path* $\mathcal{J}_{r_0} \to \mathcal{J}_{r_1} \to \ldots \to \mathcal{J}_{r_{n-1}} \to \mathcal{J}_{r_0}$ $(1 \le r_i \le m-1)$, *then there exists a periodic point* β *such that*

$$f^i(\beta) \in \mathcal{J}_{r_i}, \quad i = 0, 1, \ldots, n-1, \quad f^n(\beta) = \beta.$$

2. Furthermore, if n *is the smallest period of the sequence* $r_0, r_1, \ldots, r_{n-1}, r_0, r_1, \ldots$ *and* $\beta \notin B$, *then the period of* β *is equal to* n.

Proof. There exists a closed interval $I' \subset \mathcal{J}_{r_0}$ such that $f^i(I') \subset \mathcal{J}_{r_i}$, $i = 0, 1, \ldots, n-1$, and $f^n(I') = \mathcal{J}_{r_0}$. Therefore, there exists a point $\beta \in I'$ for which $f^n(\beta) = \beta$.

We prove the second assertion. Assume that the period of β is n' and $n' < n$ (clearly, in this case, n' is a divisor of n). The condition $\beta \notin B$ means that the points β, $f(\beta), f^2(\beta), \ldots$ lie in the interior of the intervals $\mathcal{J}_1, \mathcal{J}_2, \ldots, \mathcal{J}_{m-1}$. Since the interiors of these intervals are mutually disjoint and $f^i(\beta) = f^{i+h}(\beta)$ for $i = 0, 1, 2, \ldots$, the intervals $\mathcal{J}_{r_i}$ and $\mathcal{J}_{r_{i'}}$ coincide whenever $|i' - i| = n'$. Hence, the sequence $r_0, r_1, \ldots, r_{n-1}, \ldots$ is periodic with period $n' < n$ but this contradicts the assumption of the lemma.

Remark 1. It is obvious that the condition $\beta \notin B$ is not necessary in the case where m is not a divisor of n. If m is a divisor of n, this condition is essential. For any $m > 2$ and any n divisible by m, which is regarded as the (smallest) period of a sequence $r_0, r_1, \ldots, r_{n-1}, r_0, r_1, \ldots,$ one can always indicate a map such that the periodic point b from the condition of Lemma 3.3 belongs to the cycle B (and, hence, is m-periodic).

Lemmas 3.4 and 3.5 presented below deal with the properties of cyclic permutations.

Let $|i_1, i_2|$ denote the segment of the sequence of natural numbers lying between i_1 and i_2, i.e., the set $\{i \in \mathbb{N} \mid i_1 \le i \le i_2\}$. Segments $|i_1, i_2|$ with $i_2 = i_1 + 1$ are denoted by $|i_1, *|$. Any permutation π of length n generates an operator A_π acting on the segments $|i_1, i_2| \subset |1, n|$ as follows:

$$A_\pi |i_1, i_2| = \Big| \min_{i \in |i_1, i_2|} \pi(i), \ \max_{i \in |i_1, i_2|} \pi(i) \Big|.$$

Thus, for a map $f \in C^0(I, I)$ with a periodic trajectory $\beta_1 < \beta_2 < \ldots < \beta_n$ of type π, the action of the operator A_π has the following sense: If $A_\pi |i_1, i_2| = |i_1', i_2'|$, then $f([\beta_{i_1}, \beta_{i_2}]) \supset [\beta_{i_1'}, \beta_{i_2'}]$. The operator A_π possesses the following obvious properties:

1. $\operatorname{card} A_\pi |i_1, i_2| \ge \operatorname{card} |i_1, i_2|$
2. $A_\pi |i_1, i_2| \subseteq |i_1, i_2|$ if and only if $|i_1, i_2| = |1, n|$.

Property 1 follows from the fact that π is a one-to-one map. Property (ii) is a consequence of the absence of proper π-invariant subsets of the segment $|1, n|$.

In what follows, we use the notation $|i_1, i_2| \to |i_1', i_2'|$, which means that $A_\pi |i_1, i_2| \supset |i_1', i_2'|$. Also let A_π^k be the operator A_π applied k times, i.e., $\underbrace{A_\pi \circ \ldots \circ A_\pi}_{k \text{ times}}$.

Lemma 3.4. *Let* π *be a cyclic permutation of length* $n > 2$. *Then*

(i) one can indicate $i_0 \in |1, n-1|$ *and* $k \in \{1, \ldots, n-2\}$ *such that* $|i_0, *| \subset A_\pi |i_0, *| \subset \ldots \subset A_n^k |i_0, *| = |1, n|$;

(ii) for any $i_1 \in |1, n-1|$ *other than* i_0, *there exists a set of distinct elements* i_j, $2 \le j \le r \le k$, *of the set* $|1, n-1|$ *such that* $|i_0, *| \to |i_r, *| \to |i_{r-1}, *| \to \ldots \to |i_j, *| \to \ldots \to |i_1, *|$.

Proof. (i). Let $i_0 \in \max\{i \in |1, n| \mid \pi(i) > i\}$. It is clear that $i_0 \in |1, n-1|$ because $\pi(1) > 1$ and $\pi(n) < n$. Since $\pi(i_0 + 1) < i_0 + 1 \le \pi(i_0)$, we have $|i_0, *| \subseteq A_\pi |i_0, *|$ and, consequently, $|i_0, *| \subseteq A_\pi |i_0, *| \subseteq \ldots \subseteq A_\pi^j |i_0, *| \subseteq \ldots$, $j = 1, 2, \ldots$. Let $k = \min\{j \mid A_\pi^j |i_0, *| = A_\pi^{j+1} |i_0, *|\}$. Property (ii) of the operator A_π implies that $A_\pi^k |i_0, *| = |1, n|$. Since $|i_0, *| \ne |1, n|$, we have $k \ge 1$, $\operatorname{card} |i_0, *| = 2$, and, therefore, $\operatorname{card} A_\pi^j |i_0, *| \ge 2 + j$, $0 \le j \le k$. Hence, $k \le n - 2$.

(ii) Since $i_0 \ne i_1$ and $|i_1, *| \subset |1, n|$, one can find j_1 such that $A_\pi^{j_1 - 1} |i_0, *| \not\supset |i_1, *|$ and $A_\pi^{j_1} |i_0, *| \supset |i_1, *|$. This implies the existence of an element i_2 such that $|i_2, *| \subset A_\pi^{j_1 - 1} |i_0, *|$ and $|i_2, *| \to |i_1, *|$. Given i_2, by applying the same procedure, we choose an element i_3. Since $A_\pi^{j+1} |i_0, *| \supset A_\pi^j |i_0, *|$, by repeating the same arguments r times, $r < k$, we arrive at the element i_0. The fact that the elements i_j, $0 \le j \le r$, are distinct is a consequence of the fact that, in each step, we choose an element of the set $A_\pi^{j+1} |i_0, *| \setminus A_\pi^j |i_0, *|$.

Let $\mathfrak{A}$ be the set of cyclic permutations π with the following property: The lengths n of all π are larger than 2 and one can indicate an element $i^* \in |1, n|$ such that i^* and $\pi(i^*)$ simultaneously belong either to the segment $|1, i_0|$ or to the segment $|i_0 + 1, n|$, where $i_0 = \max\{i \in |1, n| \mid \pi(i) > i\}$.

Lemma 3.5. *Let* π *be a permutation of length* n *from the set* $\mathfrak{A}$. *Then*

(i) there exists a collection of distinct elements i_j *of the set* $|1, n-1|$, $0 \le j \le r$, $1 \le r \le k$, *such that the diagram depicted in Fig. 23 is realized;*

(ii) there exist elements $i_1 < i_2 < i_3$ *of the set* $|1, n|$ *such that the diagram depicted in Fig. 24 holds for the operator* A_π^2.

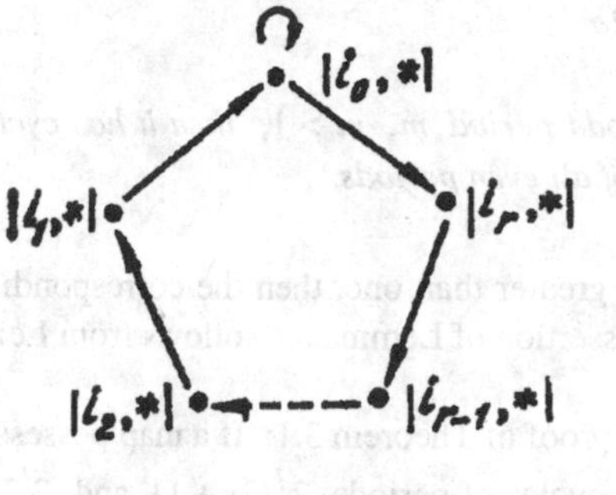

Fig. 23

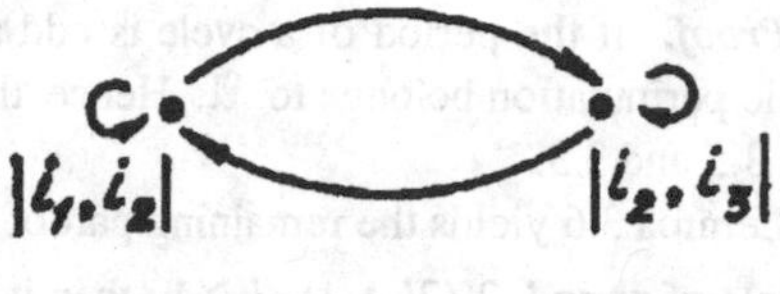

Fig. 24

Proof. (i). Let $i_0 = \max\{i \in |1, n| \mid \pi(i) > i\}$. As follows from statement (i) of Lemma 3.4, $|i_0, *| \to |i_0, *|$. It is clear that at least one of the sets $\{i \in |1, i_0| \mid \pi(i) \in |1, i_0|\}$ or $\{i \in |i_0 + 1, n| \mid \pi(i) \in |i_0 + 1, n|\}$ is nonempty. For definiteness, we assume that this is true for the first set. We set $i_1 = \max\{i \in |1, i_0| \mid \pi(i) \in |1, i_0|\}$. Then $\pi(i_1) \leq i_0 < i_0 + 1 \leq \pi(i_1 + 1)$, i.e., $|i_1, *| \to |i_0, *|$. To complete the proof, it suffices to apply statement (ii) of Lemma 3.4.

(ii). Let $\{i \in |1, i_0| \mid \pi(i) \in |1, i_0|\} \neq \varnothing$ (the case where this set is empty is investigated analogously). As in (i), we assume that i_1 is the maximal element of the indicated set and i_2 is such that

$$\pi(i_2) = \max_{i \in \{i_1, i_2\}} \pi(i),$$

and $i_3 = i_0 + 1$. Obviously, $i_1 < i_2 < i_3$.

There exists an element $\tilde{i} \in |i_0 + 1, \pi(i_2)|$ such that $\pi(i) \leq i_1$. Indeed, due to the choice of i_0, i_1, and i_2, we have $A_\pi |i_1 + 1, i_0| \subset |i_0 + 1, \pi(i_2)|$ and $\pi(i) < i$ for all $i \in |i_0 + 1, \pi(i_2)|$. Therefore, if there are no such $\tilde{i}$, then $A_\pi |i_1 + 1, \pi(i_2)| \subseteq |i_1 + 1, \pi(i_2)|$. But this contradicts Property 2 of the operator A_π.

Let us now check the required inclusions. It follows from the inequalities $\pi(i_1) \leq i_0 < \tilde{i} \leq \pi(i_2)$ that $A_\pi |i_1, i_2| \supset \{i_0, \tilde{i}\}$. Since $\pi(\tilde{i}) \leq i_1 < i_3 \leq \pi(i_2)$, we have $A_\pi^2 |i_1, i_2| \supset |i_1, i_3|$. Similarly, the inequalities $\pi(i_3) \leq i_2 < \tilde{i} \leq \pi(i_2)$ imply the inclusion $A_\pi^2 |i_2, i_3| \supset |i_1, i_3|$.

Remark 2. The set $\mathfrak{A}$ contains permutations such that it is impossible to add any edge to the graph depicted in Fig. 23 (e.g.,

$$\pi = \begin{pmatrix} 1 & 2 & 3 & 4 & 5 & 6 & 7 & 8 & 9 \\ 4 & 6 & 9 & 8 & 7 & 5 & 3 & 2 & 1 \end{pmatrix},$$

where $i_0 = 5,\ i_1 = 1,\ i_2 = 8,\ i_3 = 3,$ and $i_4 = 6$).

Lemma 3.6. *If a map possesses a cycle of odd period* $m,\ m > 1,$ *then it has cycles of all odd periods greater than* m *and cycles of all even periods.*

Proof. If the period of a cycle is odd and greater than one, then the corresponding cyclic permutation belongs to $\mathfrak{A}$. Hence, the assertion of Lemma 3.6 follows from Lemmas 3.5 and 3.3.

Lemma 3.6 yields the remaining part of the proof of Theorem 3.1: If a map possesses a cycle of period $2^l(2k+1),\ k \geq 1,$ then it has cycles of periods $2^l(2r+1)$ and $2^{l+1}s$ with $r > k$ and $s \geq 1$. Indeed, if the map f has a cycle of period $2^l(2k+1)$, then the map f^{2^l} has a cycle of period $2k+1$. It follows from Lemma 3.6 that f^{2^l} has a cycle of period $2s$ for any $s \geq 1$ and, consequently, f has a cycle of period $2^{l+1}s$. Moreover, Lemma 3.6 implies that the map f^{2^l} has a cycle of period $2r+1$ whose points are periodic with period $2^l(2r+1)$ or $2^{l'}(2r+1),\ l' < l,$ for f. In the latter case, the existence of a cycle with period $2^l(2r+1)$ follows from the already proved part of the theorem. This completes the proof of Theorem 3.1.

The theorem on coexistence of cycles guarantees the existence of cycles of any period $m' \lhd m$ when the map has a cycle of period m, but this theorem contains no information about the number of these cycles. For $m = 2^l$, there are maps (e.g., $\lambda^* x(1-x)$) which have a single cycle of period $m' \lhd m$. However, this is not true for $m \neq 2^l,\ l = 0, 1, 2, \ldots$. Numerous papers (e.g., Bowen and Franks [1] and Du [1]) are devoted to the estimation of the lower bound of the number of cycles of period $m' \lhd m$.

We now present another formulation of Theorem 3.1. Let $\mathcal{P}_m = \{f \in C^0(I, I) \mid f$ has a cycle of period $m\}$. Obviously, $C^0(I, I) = \mathcal{P}_1$.

Theorem 3.2 (on the stratification of the space of maps). *If* $m' \lhd m$, *then* $\mathcal{P}_m \supset \mathcal{P}_{m'}$, *i.e.,* $\mathcal{P}_1 \supset \mathcal{P}_2 \supset \mathcal{P}_4 \supset \ldots \supset \mathcal{P}_{10} \supset \mathcal{P}_6 \supset \ldots \supset \mathcal{P}_9 \supset \mathcal{P}_7 \supset \mathcal{P}_5 \supset \mathcal{P}_3$ *and all inclusions are strict.*

Note that there are maps $f \in \mathcal{P}_m \setminus \mathcal{P}_{m'}$ of arbitrarily high smoothness (including analytic maps). The corresponding examples can be constructed by analogy with maps from the class $C^0(I, I)$ indicated above.

The theorem on coexistence of cycles determines the periods of cycles of the map f in the case where f has a cycle of period m. Is it possible to say anything about cycles of maps that are close to f? Is it possible to establish any (lower and upper) bounds for the periods of cycles of maps that are C^r-close to $f,\ r \geq 0$?

If a map f has a cycle of period m, then it is possible that a map arbitrarily C^r-close to f (for any $r \geq 0$) has no cycles of period m (for any m). This may happen, e.g., in the case where the existence of a cycle of period m is guaranteed, say, by the tangency of the curves $y = f^n(x)$ and $y = x$ (the corresponding examples can be constructed quite easily). Nevertheless, the following theorem establishes a "lower" bound for the periods of cycles of maps close to f (Block [3]).

Theorem 3.3. *If a map f has a cycle of period m, then there exists a neighborhood $\mathfrak{U} \subset C^0(I, I)$ of the map f such that $\mathfrak{U} \subset \mathcal{P}_{m'}$ for any $m' \triangleleft m$.*

At the same time, it is impossible to establish an "upper" bound of the periods of cycles in $C^0(I, I)$. Actually, the maps with cycles of all periods are dense in $C^0(I, I)$. Indeed, if I is a bounded closed interval, then any map $f: I \to I$ has a fixed point. If β is a fixed point, then, for any $\varepsilon > 0$, one can find $\delta > 0$ such that $|f(x) - \beta| < \varepsilon$ for $|x - \beta| < \delta$. Hence, it remains to replace $f(x)$ on the interval $(\beta - \delta, \beta + \delta)$ by any ε-close map with cycles of all periods, preserving continuity. Thus, one can choose $\delta' < \min\{\varepsilon, \delta\}$ and set $\tilde{f}(x) = \beta + \delta' - 2|x - \beta|$ for $|x - \beta| < \delta'$.

For any map $\tilde{f}$, there exists a neighborhood in $C^0(I, I)$ that consists of maps with cycles of all periods. Therefore, the collection of maps that have cycles of all periods (i.e., $\mathcal{P}_3$) has an open dense subset contained in $C^0(I, I)$.

Hence, for any m, the set $\mathcal{P}_m$ contains an open dense subset of $C^0(I, I)$.

It follows from Theorem 3.3 that the set of maps with cycles of periods $\neq 2^i$, $i = 0, 1, 2, \ldots$ (i.e., $\bigcup_{m \neq 2^i} \mathcal{P}_m$) is open in $C^0(I, I)$. Therefore, this set is open and dense in $C^0(I, I)$.

It should be noted that maps with cycles of all periods cannot be constructed in a similar way by using C^1-perturbations because these maps are not dense in $C^1(I, I)$.

Theorem 3.4 (Sharkovsky [17]). *If $f \notin C^1(I, I)$ and $f \in \mathcal{P}_m$, then there exists a neighborhood $\mathfrak{U} \subset C^1(I, I)$ of the map f such that*

(i) $\mathfrak{U} \cap \mathcal{P}_m = \varnothing$ *if* $m \neq 2^i$, $i = 0, 1, 2, \ldots$,

(ii) $\mathfrak{U} \cap \mathcal{P}_{2m} = \varnothing$ *if* $m = 2^i$, $i \geq 0$.

In particular, it follows from Theorem 3.4 that, for any $m \neq 2^i$, the set $\mathcal{P}_m$ is closed in $C^1(I, I)$.

2. Types of Periodic Trajectories

The cyclic permutation associated with a cycle is called the *type* of this cycle. Since the type of a cycle depends not only on its period (this is the length of the permutation) but also on relative positions of the points of this cycle, the classification of cycles by types is more comprehensive than the classification by periods. Thus, cycles of period 3 may have only one type, namely,

$$\pi_3 = \begin{pmatrix} 1 & 2 & 3 \\ 2 & 3 & 1 \end{pmatrix}$$

(to within the inverse permutation). At the same time, cycles of period 4 may have several different types, e.g.,

$$\pi_4^{(1)} = \begin{pmatrix} 1 & 2 & 3 & 4 \\ 2 & 3 & 4 & 1 \end{pmatrix}$$

and

$$\pi_4^{(2)} = \begin{pmatrix} 1 & 2 & 3 & 4 \\ 3 & 4 & 2 & 1 \end{pmatrix}.$$

If a map f has a cycle of type $\pi_4^{(1)}$, then it is easy to show that f has a cycle of type π_3 and, consequently, it has cycles of all periods (in this case, Theorem 3.1 implies only the existence of cycles of periods 2 and 1).

Let us study the problem of "unimprovability" of Theorem 3.1. This means that we indicate maps which have a cycle of period m $(m > 1)$ and have no cycles of period m' if $m' \lhd m$.

Any cyclic permutation

$$\pi = \begin{pmatrix} 1 & \ldots & m \\ s_1 & \ldots & s_m \end{pmatrix}, \quad m > 1,$$

and an arbitrary collection of points $\beta_1 < \ldots < \beta_m$ are associated with a continuous piecewise linear map $f_\pi\colon [\beta_1, \beta_m] \to [\beta_1, \beta_m]$ linear in the intervals $I_i = [\beta_i, \beta_{i+1}]$ and such that $f_\pi(\beta_i) = \beta_{s_i}$, $i = 1, \ldots, m$. The map f_π does not depend on the choice of the points β_i in the sense that the maps f_π and $\tilde{f}_\pi$ constructed for two collections of points $\{\beta_i\}$ and $\{\tilde{\beta}_i\}$ are topologically equivalent; the corresponding conjugating homeomorphism h is an arbitrary homeomorphism such that $h(\beta_i) = \tilde{\beta}_i$.

We say that a permutation π of length m, $m > 1$, is minimal if the map f_π introduced above has no cycles of period m' for $m' \lhd m$ (the permutation $\begin{pmatrix} 1 \\ 1 \end{pmatrix}$ is also called minimal). A cycle is called a *cycle of minimal type* if the corresponding permutation is minimal. Minimal permutations can be described in the following way:

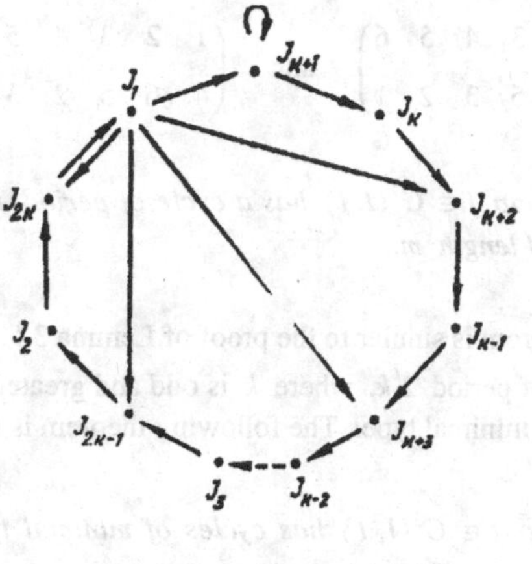

Fig. 25

1. If $m = 2k + 1$, $k \geq 1$, then the following permutations are minimal:

$$\pi_3 = \begin{pmatrix} 1 & 2 & 3 \\ 2 & 3 & 1 \end{pmatrix}$$

for $k = 1$ and

$$\pi_{2k+1} = \begin{pmatrix} 1 & 2 & 3 & \ldots & k+1 & k+2 & \ldots & 2k & 2k+1 \\ k+1 & 2k+1 & 2k & \ldots & k+2 & k & \ldots & 2 & 1 \end{pmatrix}$$

for $k > 1$ (and the inverse permutations). The B-graph of a π_{2k+1}-type cycle for the map $f_{\pi_{2k+1}}$ is displayed in Fig. 25.

2. For $m = 2k$, $k \geq 1$, permutations π are minimal if they possess the following property: The sets $\{1, \ldots, k\}$ and $\{k + 1, \ldots, 2k\}$ are invariant under π^2 and the restriction of π^2 to each of these sets is a minimal permutation (the same is true for the inverse permutations).

In order to prove that all these permutations π of length m are minimal, one must directly check that the B-graph of f_π has no closed loops of length m' for $m' \lhd m$. For the first time, cycles of minimal type were described by Sharkovsky [1]. For more details concerning this problem, see Alseda, Llibre, and Misiurewicz [1], Alseda, Llibre, and Serra [1], Block [2], Block and Coppel [2], Coppel [1], Snoha [1], and Stefan [1].

Note that if m is odd, then the minimal permutation of length m is unique (to within the inverse permutation). At the same time, for even m, $m \geq 6$, there are several minimal permutations of length m. Thus, there are two minimal permutations for $m = 6$:

$$\begin{pmatrix} 1 & 2 & 3 & 4 & 5 & 6 \\ 4 & 6 & 5 & 3 & 2 & 1 \end{pmatrix} \quad \text{and} \quad \begin{pmatrix} 1 & 2 & 3 & 4 & 5 & 6 \\ 4 & 6 & 5 & 2 & 3 & 1 \end{pmatrix}.$$

Theorem 3.5. *If a map $f \in C^0(I, I)$ has a cycle of period m, then f also has a cycle of minimal type of length m.*

The proof of this theorem is similar to the proof of Lemma 3.5.

If a map has a cycle of period $2^l k$, where k is odd and greater than 1, then this map has cycles that are not of minimal type. The following theorem is true:

Theorem 3.6. *A map $f \in C^0(I, I)$ has cycles of minimal type if and only if the period of any cycle is a power of two.*

Assume that a map f has a cycle of period $2^l k$, where k is odd and greater than one. It follows from Lemma 3.3 and 3.5 that the map $g = f^{2^l}$ has a cycle of type

$$\pi_8 = \begin{pmatrix} 1 & 2 & 3 & 4 & 5 & 6 & 7 & 8 \\ 5 & 8 & 7 & 6 & 4 & 3 & 2 & 1 \end{pmatrix}$$

(to within the inverse permutation). The cycle of type π_8 is not minimal. Therefore, the map f has a cycle of period 2^{l+3}, which is not of minimal type.

Consider a cycle of any map with cycles whose periods are equal only to powers of two. Let π_{2^l}, $l > 1$, be the type of this cycle. By virtue to Lemma 3.5, we have $\pi_{2^l} \notin \mathfrak{A}$. Therefore, $\max\{i \in |1, 2^l| : \pi_{2^l}(i) > i\} = 2^{l-1}$ and each set $|1, 2^{l-1}|$, $|2^l + 1, 2^l|$ is invariant under $\pi^2_{2^l}$. Hence, the cycle under consideration is of minimal type if $l = 2$. For $l > 2$, the argument presented above must be repeated for $\pi^2_{2^l}$, and so on.

Parallel with the problem of coexistence of periods of cycles, it is natural to consider the problem of coexistence of their types. To do this, we equip the set of cyclic permutations with relation of ordering ($\prec$) as follows: we say that $\pi \prec \pi'$ if, for any $f \in C^0(I, I)$, the condition that the map f has a cycle of type π implies that it has a cycle of type π'. The relation $\prec$ is not linear. For example, the map f_1 in Fig. 26 has a cycle of type

$$\pi_4^{(1)} = \begin{pmatrix} 1 & 2 & 3 & 4 \\ 3 & 1 & 4 & 2 \end{pmatrix}$$

but has no cycles of type

$$\pi_4^{(2)} = \begin{pmatrix} 1 & 2 & 3 & 4 \\ 2 & 3 & 4 & 1 \end{pmatrix}.$$

On the contrary, the map f_2 has a cycle of type $\pi_4^{(2)}$ but has no cycles of type $\pi_4^{(1)}$. One can formulate general theorems on the coexistence of cycles of various types but, unfortunately, these theorems are cumbersome (Fedorenko [1]) and we do not present them here. We restrict ourselves to the investigation of cycles of some special types.

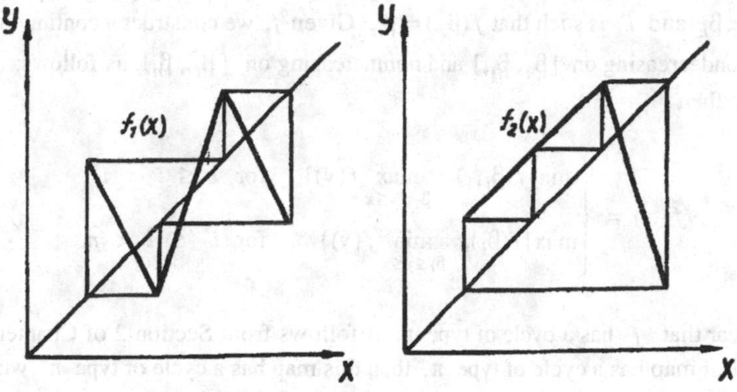

Fig. 26

Let

$$\pi = \begin{pmatrix} 1 & \ldots & i & \ldots & n \\ s_1 & \ldots & s_i & \ldots & s_n \end{pmatrix}$$

be a cyclic permutation such that $s_1 = n$ and let i^* be an element of the set $\{2, \ldots, n\}$ such that $\pi(i^*) = n$. The permutation π is called unimodal if $s_i > s_{i+1}$ for $1 \le i < i^*$ and $s_i < s_{i+1}$ for $i^* < i \le n$. The set of permutations of this sort is denoted by Σ.

It turns out that the relation of ordering $\prec$ induces the relation of linear ordering in the set Σ. Let us prove this assertion. For any permutation $\pi \in \Sigma$, we define a sequence $\Theta(\pi) = (\theta_0(\pi), \theta_0(\pi), \ldots)$ by setting

$$\varepsilon_k(\pi) = \begin{cases} +1, & \text{if } \pi^k(i^*) \le i^*, \\ -1, & \text{if } \pi^k(i^*) > i^* \end{cases}$$

for $k = 0, 1, 2, \ldots$ and

$$\theta_r(\pi) = \prod_{k=0}^{r} \varepsilon_k(\pi).$$

For any $\pi \in \Sigma$, the sequence $\Theta(\pi)$ is admissible in the sense of the definition introduced in Section 2 of Chapter 2. Recall that the set $\{\Theta(\pi), \pi \in \Sigma\}$ is lexicographically ordered and this relation is denoted by $<$.

Theorem 3.7. *If a map $f \in C^0(I, I)$ has a cycle of type $\pi \in \Sigma$, then f has a cycle of type $\pi' \in \Sigma$ whenever $\Theta(\pi) < \Theta(\pi')$.*

Proof. Assume that f has a cycle of type $\pi \in \Sigma$ formed by points $\beta_1 < \beta_2 < \ldots < \beta_i < \ldots < \beta_n$ and i^* is such that $f(\beta_{i^*}) = \beta_n$. Given f, we construct a continuous function $\tilde{f}$ nondecreasing on $[\beta_1, \beta_{i^*}]$ and nonincreasing on $[\beta_{i^*}, \beta_n]$ as follows: If $x \in [\beta_i, \beta_{i+1}]$, then

$$\tilde{f}(x) = \begin{cases} \min\{f(\beta_{i+1}), \max\limits_{\beta_i \le y \le x} f(y)\} & \text{for } 1 \le i < i^*, \\ \max\{f(\beta_i), \min\limits_{\beta_i \le y \le x} f(y)\} & \text{for } i^* \le i < n. \end{cases}$$

It is clear that $\tilde{f}$ has a cycle of type π. It follows from Section 2 of Chapter 2 that if a unimodal map has a cycle of type π, then this map has a cycle of type π' whenever $\Theta(\pi) < \Theta(\pi')$. This result remains true for the map $\tilde{f}$. Therefore, to complete the proof it remains to note that every cycle of the map $\tilde{f}$ is, at the same time, a cycle of f.

In conclusion, it should be noted that the relation of ordering $\prec$ in the set of unimodal permutations is closely related to the location of periodic points of the map

$$x \to f(x) = \begin{cases} 2x, & 0 \le x < \dfrac{1}{2}, \\ 2 - 2x, & \dfrac{1}{2} \le x < 1. \end{cases}$$

Any $\pi \in \Sigma$ corresponds to a cycle of type π of this map, and vice versa. Denote the minimal point of a cycle of type π of the map f by $x_{\min}(\pi)$. As can be proved by direct computation, for any $\pi, \pi' \in \Sigma$, we have $x_{\min}(\pi) < x_{\min}(\pi')$ if and only if $\pi \prec \pi'$.

4. SIMPLE DYNAMICAL SYSTEMS

As shown in previous chapters, maps of the interval onto itself exhibit fairly diverse dynamical behavior. Therefore, in studying dynamical systems of this sort, it is natural to decompose the entire set of maps into classes exhibiting "similar" dynamical behavior.

At present, dynamical systems are usually regarded as simple or complex, depending on whether their topological entropy is equal to zero or not. The dynamics of complex maps is characterized by the following property: There exists a subset of the interval on which some iteration of a map is semiconjugate to the Bernoulli shift on the set of all unilateral sequences with two-symbol alphabet. It was shown in Chapter 3 that complex maps form an open dense set in $C^0(I, I)$ and, in this sense, they are generic. Moreover these maps with infinite topological entropy are generic in $C^0(I, I)$.

Simple maps, in turn, form a closed nowhere dense subset in $C^0(I, I)$. However, they are not exceptional in the space $C^r(I, I)$, $r > 0$, and form a closed set that contains an open subset.

The dynamics of simple maps is described by using the notion of splitting (see Section 2). Then they are classified according to the criterion of coincidence of different types of return and various criteria for a map to belong to a certain class are applied. Note that, by establishing criteria of simplicity for dynamical systems, we, in fact, also established criteria of their complexity (i.e., generic properties of maps from $C^0(I, I)$), which can be obtained by "converting" the simplicity criteria.

1. Maps without Periodic Points

Let $\mathcal{F}_m$ denote the set $\{f \in C^0(I, I) \mid \text{Per}(f) = \text{Fix}(f^m)\}$. The set $\mathcal{F}_m$ consists of maps such that the period of any their cycle does not exceed m. In this case, $m = 2^k$, $k < \infty$, and, consequently, any set $\mathcal{F}_m$ $(m = 1, 2, 2^2, \ldots)$ consists of maps generating simple systems.

First, we consider the case where a map has only fixed points but no periodic points, i.e., $f \in \mathcal{F}_1$. The fact that $\text{Per}(f) = \text{Fix}(f)$ can easily be verified. Indeed, it suffices to show that $\mathrm{Fix}(f^2) = \mathrm{Fix}(f)$. Note that, for maps from $\mathcal{F}_1$, any closed set can play a role of their set of fixed points. Indeed, for any closed set $F \subset \mathbb{R}$, we have $\mathrm{Per}(f) = \mathrm{Fix}(f) = F$ for the map $x \mapsto x + \rho(x, F)$, where $\rho(x, F)$ denotes the distance from x to F, because $f(x) \geq x$ for $x \in \mathbb{R}$ and $f(x) = x$ only for $x \in F$.

What kinds of behavior can be demonstrated by the trajectories $x_1, x_2, \ldots, x_m, \ldots, x_{m+1} = f(x_m)$ of maps from $\mathcal{F}_1$? How to describe the relative arrangement of points of a single trajectory in I?

If a function f is monotonically increasing, i.e., if $f(x') \geq f(x'')$ for any $x' \geq x''$, then $f^m(x') \geq f^m(x'')$ for any $m > 0$. Any trajectory is monotone: If $x_1 \geq x_2 = f(x_1)$, then $x_1 \geq x_2 \geq \ldots \geq x_m \geq \ldots$; if $x_1 \leq x_2$, then $x_1 \leq x_2 \leq \ldots \leq x_m \leq \ldots$. Hence, any trajectory is attracted by a fixed point of f, i.e., the ω-limit set of any trajectory is a fixed point.

The following theorems describe the behavior of trajectories of arbitrary maps in $\mathcal{F}_1$:

Theorem 4.1 (Sharkovsky [4]). $f \in \mathcal{F}_1 \Leftrightarrow$ *for any* $x \in I$ *and* $m > 0$, *there are no points* x_i *with* $i < m$ *between* x_m *and* x_{m+1} *provided that* x_m *and* x_{m+1} *are distinct or, equivalently, for* $i > m$, *all points* x_i *lie on the same side of the point* x_m *as* x_{m+1}.

Theorem 4.2 (Sharkovsky [4]). $f \in \mathcal{F}_1 \Leftrightarrow$ *the* ω*-limit set of any trajectory is a fixed point.*

Let us prove these theorems.

In Section 1 of Chapter 3, we proved the following assertion:

A map f *has a cycle of period 2* $\Leftrightarrow$ *there exists a point* $x \in I$ *such that* $f(x) \neq x$ *and* $x \in f(\langle x, f(x)\rangle)$, *where* $\langle x, f(x)\rangle$ *is the closed interval with ends at* x *and* $f(x)$.

Below, we present another formulation of this lemma.

Lemma 4.1. $f \in \mathcal{F}_1 \Leftrightarrow$ *for any* $x \in I$, *either* $f(x) = x$ *or* $x \notin f(\langle x, f(x)\rangle)$.

Since the interval $f(\langle x, f(x)\rangle)$ contains points $f(x)$ and $f^2(x)$, one can replace this interval in Lemma 4.1 by the interval $\langle f(x), f^2(x)\rangle$ contained in it.

Lemma 4.2. *Let* $f \in C^0(I, I)$. *If, for* $x \in I$, *the relations* $f^{m'}(x) \in \langle f^m(x), f^{m+1}(x)\rangle$ *and* $f^m(x) \neq f^{m+1}(x)$ *are true for some* m *and* m', $m' < m$, *then the collection of points* $f^i(x)$, $i = m', m'+1, \ldots, m$, *contains a point* x' *such that* $x' \in f(\langle x', f(x')\rangle)$.

Proof. Let m'' be such that $f^i(x) \geq f^m(x)$ for $i = m', m'+1, \ldots, m''$ and $f^{m''+1}(x) <$

$f^m(x)$. Denote by F the set $\{f^i(x), i = m', \ldots, m''\}$. If $f(y) < y$ for every $y \in F$, then $x' = f^m(x)$. If $f(y') > y'$ for some $y' \in F$, then $x' = \max\{y \in F \mid f(y) > y\}$.

Theorem 4.1 follows from Lemmas 4.1 and 4.2.

For any $x \in I$, we define

$$T_m(x) = \left[\inf_{i>m} f^i(x), \sup_{i>m} f^i(x)\right] = \overline{\bigcup_{i=m}^{\infty} \langle f^i(x), f^{i+1}(x)\rangle}, \quad m = 0, 1, 2, \ldots.$$

It follows from this definition that $T_0(x) \supseteq T_1(x) \supseteq T_2(x) \supseteq \ldots \supset \omega(x)$. Let $T_\infty(x) = \bigcap_{m \geq 0} T_m(x)$. Clearly, $T_\infty(x)$ is either a closed interval or a point and $T_\infty(x) \supseteq \omega(x)$.

Theorem 4.1 immediately implies the following assertion:

Lemma 4.3. *$f \in \mathcal{F}_1 \Leftrightarrow f^i(x) \notin T_{i+1}(x)$ for any $x \in I$ and $i \in \mathbb{N}$ provided that $f^{i+1}(x) \neq f^i(x)$.*

We now prove Theorem 4.2. Assume that $f \in \mathcal{F}_1$ and x is an arbitrary point of I. Let us show that $\omega(x)$ is a fixed point. If $T_\infty(x)$ is a point, then $\omega(x)$ is also a point. Consider another possibility: $T_\infty(x)$ is an interval. It follows from Lemma 4.3 that $\operatorname{int} T_\infty(x)$ does not contain points of the trajectory of x. Hence, $\omega(x) = \partial T_\infty(x)$, i.e., $\omega(x)$ consists of two points. We know that $\omega(x)$ is an invariant set which cannot be formed by two fixed points. Therefore, $\omega(x)$ must be a cycle of period 2 but this is impossible.

Assume that a trajectory approaches a fixed point β. We are interested in the behavior of this trajectory near the point β. If f is a monotonically increasing function, then all trajectories are monotonically convergent (increasing or decreasing) as mentioned above. If the function f is differentiable in β, then the trajectories convergent to β approach this point monotonically β provided that $f'(\beta) > 0$. At the same time, if $f'(\beta) < 0$, then the trajectories approach β spasmodically so that the points with odd numbers are located on the same side of the point β, while the points with even numbers are located on the other side. In the case where $f'(\beta) = 0$ or the derivative does not exist at the point β, the trajectory may exhibit extremely irregular behavior approaching this point.

Consider an example: Let

$$f(x) = \begin{cases} e^{-1/x^2} \sin 1/x, & x \neq 0, \\ 0, & x = 0. \end{cases}$$

In this case, $|f(x)| < |x|$ as $x \neq 0$ and, consequently, the map has a single fixed point $x = 0$ and all trajectories approach this point. At the same time, for any partition

$\{N_1, N_2\}$ of the set of natural numbers, one can indicate a point x such that $f^n(x) > 0$ for $n \in N_1$ and $f^n(x) < 0$ for $n \in N_2$.

This property of f follows from the fact that, in any neighborhood U of $x = 0$, there are intervals U_1, U_2, $\overline{U}_1 \subset \{x < 0\} \cap U$, $\overline{U}_2 \subset \{x > 0\} \cap U$ such that $f(U_1)$ and $f(U_2)$ are neighborhoods of the point $x = 0$ (in our example, we can choose the intervals $2k\pi < |1/x| < 2(k+1)\pi$, $0 < k < \infty$, i.e., $(-\gamma_k, -\gamma_{k+1})$ and (γ_k, γ_{k+1}), where $\gamma_k = 1/2k\pi$). Due to the possibility of passing from one half neighborhood of the point $x = 0$ into the other under the action of f, we can apply a version of symbolic dynamics (with symbols +, −) that admits sequences generated by trajectories $\{x_i\}$ with arbitrary collections of $\operatorname{sign} x_i$ (the matrix of admissible transitions is $\begin{pmatrix} 1 & 1 \\ 1 & 1 \end{pmatrix}$).

Let us now dwell upon the rate of convergence of various trajectories. It is well known that typical trajectories of smooth maps converge as geometric progressions, namely, if $x_i \to \beta$ and $|f'(\beta)| = b < 1$, then $|x_{i+1} - \beta| / |x_i - \beta| \approx b$. However, if $|f'(\beta)| = 1$, then the rate of convergence is substantially smaller. Thus, for $f(x) = x - a(x-\beta)^{2m+1}(1 + o(1))$ as $x \to \beta$ and $a > 0$, we have $|x_{i_2} - \beta| / |x_{i_1} - \beta| \approx b$ only if $i_2 - i_1 \approx (b|x_{i_1} - \beta|)^{-2m} / 2am$.

In general, for $f \in \mathcal{F}_1$, the rate of convergence may vary from arbitrarily high to arbitrarily low. In particular, for the map presented above with $f'(0) = 0$, there are trajectories that converge arbitrarily rapidly (among trajectories that do not hit the point $x = 0$ after finitely many steps). Let us now formulate the corresponding general assertions.

Let β be a fixed point of f with an invariant half neighborhood, i.e., there exists a neighborhood U of β such that $fU^- \subset U^-$, where $U^- = U \cap \{x \le \beta\}$. For the sake of simplicity, the required assertions are formulated only for this case.

We say that f has trajectories that approach β *arbitrarily rapidly* if, for any sequence $\alpha_1 < \alpha_2 < \ldots \to \beta$, one can indicate a trajectory $x_1 < x_2 < \ldots \to \beta$, $x_i \ne \beta$, such that $x_i > \alpha_i$ for $i = 1, 2, \ldots$. If we replace the last inequality by the inverse one: $x_i < \alpha_i$ beginning with some $i \ge 1$, then we say that f has trajectories that converge *arbitrarily slowly*.

The following assertions are true:

A. *A map f has trajectories that converge to a fixed point β arbitrarily rapidly if and only if there exists a sequence of points $y_1 < y_1' \le y_2 < y_2' \le y_3 < y_3' \le \ldots \to \beta$ such that $f(y_i) = f(y_i') = \beta$ and $f(x) < \beta$ for $x \in (y_i, y_i')$, $i = 1, 2, \ldots$.*

B. *A map f has trajectories that converge to a fixed point β arbitrarily slowly if and only if there exists a sequence of fixed points $y_1 < y_1' \le y_2 < y_2' \le y_3 < y_3' \le \ldots \to \beta$ such that $y_{i+1} \in \operatorname{int} f([y_i, y_i'])$, $i = 1, 2, \ldots$.*

Thus, for the existence of trajectories whose convergence is arbitrarily slow, it is necessary that any neighborhood of the point β contain trajectories that do not converge to this point (e.g., fixed points). For the existence of trajectories whose convergence to β is arbitrarily rapid, it is necessary that any neighborhood of the point β contain trajectories whose points immediately hit the point β (e.g., the trajectories starting at the points $y_i, y_i', i = 1, 2, \ldots$).

The proof of assertions A and B is quite simple and we do not present it here.

As an example that illustrates assertion B, we consider a C^0-map $f: [-1, 0] \to [-1, 0]$ generated by $f(x) = x + a\rho(x, F)$, where $F = \{-1, -1/2, -1/3, \ldots, 0\}$ is the set of fixed points, ρ is the distance from a point x to the set F, and $1 < a < 3$.

The properties of maps from $\mathcal{F}_1$ established in this section can be summarized as follows: For any map $f \in C^0(I, I)$, the following statements are equivalent:

(i) $\mathrm{Per}(f) = \mathrm{Fix}(f)$ (i.e., $f \in \mathcal{F}_1$);

(ii) $\mathrm{Fix}(f^2) = \mathrm{Fix}(f)$;

(iii) $\forall x \in I$, $\omega(x)$ is a fixed point;

(iv) $\forall x \in I$ and $m > 1$, the points $f^i(x)$ with $i > m$ are located on the same side of the point $f^m(x)$,

(v) $\forall x \in I$, either $x \notin f(\langle x, f(x)\rangle)$ or $x = f(x)$.

The following two statements established in Section 6 of this chapter are equivalent to properties (i)–(v):

(vi) $\mathrm{NW}(f) = \mathrm{Fix}(f)$;

(vii) $\mathrm{CR}(f) = \mathrm{Fix}(f)$,

where $\mathrm{NW}(f)$ and $\mathrm{CR}(f)$ denote the sets of nonwandering and chain recurrent points, respectively.

If $f \in \mathcal{F}_{2^k}$, $k > 0$, then $g = f^{2^k} \in \mathcal{F}_1$. Therefore, all statements formulated above are true for g and one can reformulate them for f. In particular, the following statements are equivalent:

(i) $\mathrm{Per}(f) = \mathrm{Fix}(f^{2^k})$ (i.e., $f \in \mathcal{F}_{2^k}$);

(ii) $\mathrm{Fix}(f^{2^{k+1}}) = \mathrm{Fix}(f^{2^k})$;

(iii) $\forall x \in I$, $\omega(x)$ is a cycle of period 2^i, $0 \le i \le k$;

(iv) $\forall x \in I$ and $\forall s \in \{0, 1, \dots, m-1\}$, the points $f^{i2^k+s}(x)$ with $i > m$ are located on the same side of the point $f^{m2^k+s}(x)$;

(v) $\forall x \in I$, either $x \notin f^{2^k}((x, f^{2^k}(x)))$ or $x = f^{2^k}(x)$;

(vi) $\mathrm{NW}(f) = \mathrm{Fix}(f^{2^k})$;

(vii) $\mathrm{CR}(f) = \mathrm{Fix}(f^{2^k})$.

Each of the sets $\mathcal{F}_{2^k}$ is closed in $C^0(I, I)$ (if $f_i \in \mathcal{F}_1$ and $f_i \to f_*$, then $f_* \in \mathcal{F}_1$ because it follows from $x_i \in \mathrm{Fix}(f_i)$ and $x_i \to x_*$ that $x_* \in \mathrm{Fix}(f_*)$).

The set of all simple systems is not exhausted by the sets $\mathcal{F}_{2^k}$, $k = 0, 1, 2, \dots$. The maps from the set $\mathcal{F}_{2^\infty} \setminus \bigcup_{k=0}^{\infty} \mathcal{F}_{2^k}$, where

$$\mathcal{F}_{2^\infty} = \overline{\bigcup_{k=0}^{\infty} \mathcal{F}_{2^k}} = \left\{ f \in C^0(I, I) \mid \mathrm{Per}(f) = \bigcup_{k=0}^{\infty} \mathrm{Fix}(f^{2^k}) \right\},$$

i.e., simple maps with cycles of arbitrarily large periods have not been studied yet. These maps are investigated in the remaining part of this chapter.

2. Simple Invariant Sets

Let us now study the structure of ω-limit sets of dynamical systems with cycles whose periods are necessarily equal to 2^i, $i \ge 0$. As we have already explained, for maps from $\mathcal{F}_{2^k}$, every ω-limit set is a cycle. At the same time, in the case where simple maps have cycles of arbitrary large periods, one may encounter much more complicated situations. Thus, there are maps such that each their ω-limit set is also a cycle. However, the maps whose ω-limit sets are Cantor sets are more typical. In addition, there are maps whose ω-limit sets are composed of a Cantor set and a countable set of points. It can be shown that infinite ω-limit sets of maps from the set $\mathcal{F}_{2^\infty}$ are characterized by the properties similar to the properties of cycles.

The maps from $\mathcal{F}_{2^\infty}$ are usually called simple maps. It is also convenient to say that the cycles of simple maps are simple cycles. (Thus, the topological entropy of a map is equal to zero if and only if all cycles of this map are simple.) As stated in Theorem 3.5, any simple cycle is a cycle of the minimal type, which means that simple maps may have only cycles of the minimal type with periods 2^i. Therefore, simple cycles possess the following property: A cycle

$$B = \{\beta_i,\ i = 1, \dots, 2^k;\ \beta_i < \beta_{i+1}\}$$

of a map f is simple if and only if either B is a fixed point or the sets $\{\beta_1, \dots, \beta_{2^{k-1}}\}$ and $\{\beta_{2^{k-1}+1}, \dots, \beta_{2^k}\}$ are invariant under the action of f^2 and the restriction of f^2 to each of these sets is a simple cycle.

Hence, each simple cycle B which is not a fixed point can be decomposed into two subsets B_1 and B_2 such that $f(B_1) = B_2$ and $f(B_2) = B_1$. By generalizing this property to the case of arbitrary invariant sets, we arrive at the following notion of splitting:

We say that a closed invariant set $M \subset I$ admits *splitting* under the map f^2 if it can be decomposed into sets M_1 and M_2 such that

(i) M_1 and M_2 belong to two different closed disjoint intervals;

(ii) $f(M_1) = M_2$ and $f(M_2) = M_1$.

We say that a closed invariant set M admits *n-fold splitting* $(n > 1)$ under the map f^2 if it can be split under f^2 and each of the sets M_1 and M_2 admits $(n-1)$-fold splitting under the map g^2, where $g = f^2$ (the terms *1-fold splitting* and simply *splitting* are synonyms).

In the case where M admits splitting, the sets M_1 and M_2 are determined uniquely; therefore, in the case where M admits n-fold splitting, the set M is decomposed, in a unique manner, into subsets $M_i^{(n)}$, $i = 1, 2, \dots, 2^n$, such that

(a) there exist 2^n mutually disjoint intervals $J_1, \dots, J_{2^n}$ ordered on $\mathbb{R}$ by increasing of their subscripts and such that $M_i^{(n)} = J_i \cap M$;

(b) $M = \bigcup_{i=1}^{2^n} M_i^{(n)}$;

(c) $f^{2^{n-1}}(M_{2j-1}^{(n)}) = (M_{2j}^{(n)})$ and $f^{2^{n-1}}(M_{2j}^{(n)}) = (M_{2j-1}^{(n)})$, $j = 1, 2, \dots, 2^{n-1}$.

Note that the permutation

$$\pi = \begin{pmatrix} 1 & 2 & \dots & 2^n \\ s_1 & s_2 & \dots & s_{2^n} \end{pmatrix}$$

defined by condition (c) (where $s_i = j$ if $f(M_i) = M_j$) determines a simple cycle, i.e., it is a minimal permutation of length 2^n.

Any decomposition of M into subsets $\{ M_i^{(n)},\ i = 1, \dots, 2^n \}$ is called a *simple decomposition of rank* n provided that this decomposition satisfies conditions (a)–(c).

Propositions 4.1 and 4.2 presented below are obvious consequences of condition (c).

Proposition 4.1. *If* $\{ M_i^{(n)},\ i = 1, 2, \dots, 2^n \}$ *and* $\{ M_i^{(n+1)},\ i = 1, 2, \dots, 2^{n+1} \}$ *are simple decompositions of the set* M *of ranks* n *and* $n + 1$, *respectively, then* $M_i^{(n)} = M_{2i-1}^{(n+1)} \cup M_{2i}^{(n+1)},\ i = 1, \dots, 2^n$.

Proposition 4.2. *An invariant set* M *admits n-fold splitting if and only if, for any* $n' \le n$, *there exists a simple decomposition of the set* M *of rank* n'.

A cycle of period 2^n is simple if and only if it admits n-fold splitting. It can be shown that infinite ω-limit sets of maps from $\mathfrak{F}_{2^\infty}$ also possess the property of splitting and, moreover, that these sets admit n-fold splitting for any n.

A closed invariant set M is called *simple* either if it is a fixed point or if it admits n-fold splitting for any $n \le \log_2 \operatorname{card} M$ (Fedorenko [3]). Simple sets have the following properties (some of these properties are obvious, and we give them without proofs):

(i) Any simple set cannot be decomposed (i.e., cannot be represented as a union of two closed disjoint invariant subsets).

(ii) Any finite simple set is a simple cycle.

(iii) Any infinite simple set has the cardinality of continuum.

(iv) Any infinite simple set contains no periodic points.

Proof. Let $\{ M_i^{(n)},\ i = 1, \dots, 2^n \}$ be a simple decomposition of rank n of an infinite simple set M of a map f. Then $f^k(M_i^{(n)}) \cap M_i^{(n)} = \varnothing$ for $k = 1, \dots, 2^n - 1$. For any $j > 0$, there exists a simple decomposition of rank n such that $2^n > j$. Hence, $f^j(x) \ne x$ for all $x \in M$ and $j > 0$. Therefore, $x \notin \operatorname{Per}(f)$.

(v) Any simple set that contains a periodic point is a simple cycle.

(vi) If a simple set contains an open interval, then each point of this interval is a wandering point.

Proof. Any simple set M that contains an open interval U is infinite. Moreover, the interval U contained in M necessarily belongs to a single element of the decomposition $\{ M_i^{(n)},\ i = 1, \ldots, 2^n \}$. Therefore, by using the same argument as in the proof of (iv), we conclude that $f^j(U) \cap U = \varnothing$ for $j > 0$, i.e., each point of U is wandering.

(vii) Any simple invariant set contains an almost periodic point.

Proof. If M is a finite simple set, i.e., if it is a periodic trajectory, then property (vii) is evident. Let M be an infinite simple set of the map f and let $\{ M_i^{(n)},\ i = 1, \ldots, 2^n \}$ be a simple decomposition of the set M of rank n. It follows from Proposition 4.1 that $M_i^{(n)} = M_{2i-1}^{(n+1)} \cup M_{2i}^{(n+1)}$, $i = 1, 2, \ldots, 2^n$. Let $\tilde{M}^{(n)}$ denote the element of the decomposition $\{ M_i^{(n)},\ i = 1, \ldots, 2^n \}$ with the smallest diameter. Since $M^{(1)}$ and $M^{(2)}$ belong to two disjoint closed intervals, we have $\operatorname{diam} \tilde{M}^{(1)} < (1/2) \operatorname{diam} M$. Then, for any $n > 1$ (beginning with $n = 2$), we successively choose a single element $\tilde{M}^{(n)}$ in a simple decomposition of rank n as follows: Assume that the element $\tilde{M}^{(n)}$ is chosen. Then $\tilde{M}^{(n)}$ contains exactly two elements of the simple decomposition of rank $n + 1$. In this pair of elements, the element with smaller diameter is denoted by $\tilde{M}^{(n+1)}$. Obviously, $\tilde{M}^{(n)} \supset \tilde{M}^{(n+1)}$ for any $n > 1$ and $\operatorname{diam} \tilde{M}^{(n)} < (1/2^n) \operatorname{diam} M$. Therefore, $\bigcap_{n>1} \tilde{M}^{(n)}$ must be a point, and we denote it by x.

Let $\varepsilon > 0$. We choose n such that $\operatorname{diam} \tilde{M}^{(n)} < \varepsilon$. Since $f^{2^n}(\tilde{M}^{(n)}) = \tilde{M}^{(n)}$, we have $| f^{2^n i}(x) - x | < \varepsilon$ for all $i > 0$. Therefore, $x \in \mathrm{AP}(f)$.

(viii) Each point of a simple invariant set is a chain recurrent point.

Proof. Let M be a simple invariant set of the map f. If M is a finite set, then M is a simple periodic trajectory. Therefore, $M \in \mathrm{CR}(f)$.

Now let M be an infinite simple invariant set. By virtue of Proposition 4.2, for any $n \geq 0$, there exists a simple decomposition of the set M of rank n, namely, $\{ M_i^{(n)},\ i = 1, \ldots, 2^n \}$. Since the elements of a simple decomposition of rank n belong to 2^n mutually disjoint intervals, we have $\min_i \operatorname{diam} M_i^{(n)} \to 0$ as $n \to \infty$. Let $\tilde{M}^{(n)}$ denote an element of the decomposition $\{ M_i^{(n)},\ i = 1, \ldots, 2^n \}$ such that

$$\operatorname{diam} \tilde{M}^{(n)} = \min_i \operatorname{diam} M_i^{(n)}.$$

We fix $\varepsilon > 0$ and choose n_0 such that $\operatorname{diam} \tilde{M}^{(n_0)} < \varepsilon$. Let x be an arbitrary point of the set M. It follows from the invariance of the set M, that there exist $k_1 \geq 0$, $k_2 > 0$, and a point $y \in \tilde{M}^{(n_0)}$ such that $k_1 + k_2 = 2^{n_0}$, $f^{k_1}(x) \in \tilde{M}^{(n_0)}$, and $f^{k_2}(x) = x$. Since $\operatorname{diam} \tilde{M}^{(n_0)} < \varepsilon$, the points $x_0 = x$, $f^r(x_0)$, and $f^s(y)$, $r = 1, \ldots, k_1$, $s = 1, \ldots, k_2$, are an ε-trajectory of x. Hence, $x \in \mathrm{CR}(f)$.

(ix) Any recurrent point of a simple invariant set is regularly recurrent.

Proof. Let M be a simple invariant set of the map f. If M is finite, then property (ix) is obvious.

Let M be infinite and let $x \in M \cap R(f)$. Also let $M^{(n)}(x)$ be the element of the simple decomposition of the set M of rank n which contains the point x. Denote $M(x) = \bigcap_{n \geq 1} M^{(n)}(x)$. Since $M^{(n)}(x)$, $n = 1, 2, \ldots$, are closed sets, $M(x)$ is a nonempty closed set. Denote by α and β the maximal and minimal points of $M(x)$, respectively. If $\alpha = \beta = x$, then $x \in AP(f)$. Indeed, let $\varepsilon > 0$. Since $\bigcap_{n \geq 1} M^{(n)}(x)$ is a point, one can find n_1 such that $\operatorname{diam} M^{(n_1)}(x) < \varepsilon$. This implies that $x \in AP(f)$ because $f^{2^{n_1}}(M^{(n_1)}(x)) = M^{(n_1)}(x)$.

Now assume that $\alpha \neq \beta$. It follows from property (vi) that $(\alpha, \beta) \cap NW(f) = \varnothing$. Therefore, x coincides either with α or with β.

For definiteness, we assume that $x = \alpha$. Let ε be an arbitrary number satisfying the inequality $0 < \varepsilon < \beta - \alpha$. Since $M(x) = \bigcap_{n \geq 1} M^{(n)}(x)$, there exists n_2 such that the last point of the set $M^{(n_2)}(x)$ lies to the right of the point $x - \varepsilon$, i.e., $\inf M^{(n_2)}(x) > x - \varepsilon$. Since $x \in R(f)$, $f^{2^{n_2}}(M^{(n_2)}(x)) = M^{(n_2)}(x)$, and $(x, x + \varepsilon] \cap R(f) = \varnothing$, the interval $[x - \varepsilon, x]$ contains infinitely many points of the trajectory f_x. Therefore, there exists n_3 for which some element $M_{i_0}^{(n_3)}$ of the simple decomposition of M of rank n_3 lies in the interval $[x - \varepsilon, x]$. Since $f^{2^{n_3}}(M_{i_0}^{(n_3)}) = M_{i_0}^{(n_3)}$, one can indicate $k \leq 2^{n_3}$ such that $f^{k + i2^{n_3}}(x) \in (x - \varepsilon, x)$ for all $i \geq 0$. Thus, any segment of the trajectory of the point x of length 2^{n_3} contains a point from the ε-neighborhood of x, i.e., $x \in RR(f)$.

There are many other dynamical properties of simple invariant sets. We restrict ourselves to properties (i)–(ix) necessary for what follows and proceed to the principal theorem of this chapter.

3. Separation of All Maps into Simple and Complicated

The following theorem clarifies common features in the dynamical behavior of all simple maps and common features in the dynamical behavior of all complicated maps:

Theorem 4.3. *Any map $f \in C^0(I, I)$ possesses exactly one of the following properties:*

(i) *the set of all chain recurrent points of the map f coincides with the union of all simple sets of this map;*

(ii) *there exist* $n \geq 0$ *and closed intervals* I_1 *and* I_2 *with* $\text{int}\, I_1 \cap \text{int}\, I_2 = \varnothing$ *such that* $f^{2^n}(I_1) \supseteq I_1 \cup I_2$ *and* $f^{2^n}(I_2) \supseteq I_1 \cup I_2$.

Property (i) is a general property of all simple maps, and property (ii) is a general property of all complicated maps.

Here, we present the proof of Theorem 4.3 suggested by Fedorenko [5]; this proof is based on the classification of trajectories of maps of an interval and on the properties of the set of chain recurrent points. To describe this classification, we consider the simplest class of dynamical systems, namely, cyclic permutations, as an example.

Recall that the set $\{ i \in \mathbb{N} \mid i_1 \leq i \leq i_2 \}$ is denoted by $| i_1, i_2 |$.

Lemma 4.4. *Each cyclic permutation* π *of length* $i > 1$ *possesses exactly one of the following properties:*

(i) *there are elements* $i_1, i_2 \in |1, n|$ *such that either* $i_1 < \pi(i_1) \leq i_2 < \pi(i_2)$ *or* $i_1 > \pi(i_1) \geq i_2 > \pi(i_2)$,

(ii) n *is even,* $\pi |1, n/2| = |n/2 + 1, n|$, *and* $\pi |n/2 + 1, n| = |1, n/2|$.

Proof. Since π is a cyclic permutation, we have $\pi(1) > 1$ and $\pi(n) < n$. Hence, the sets $N^+ = \{ i \in |1, n| \mid \pi(i) > i \}$ and $N^- = \{ i \in |1, n| \mid \pi(i) < i \}$ are nonempty. We denote $i_0 = \max N^+$ and $i_0^- = \min N^-$.

There are two possible cases, namely, $i_0^- < i_0$ and $i_0^- > i_0$. We consider each of these cases separately.

Suppose that $i_0^- < i_0$. Let $\tilde{i}$ be the preimage of i_0^-. We have either $\tilde{i} < i_0^-$ or $\tilde{i} > i_0^-$. In the first case, we have $\tilde{i} < \pi(\tilde{i}) < i_0 < \pi(i_0)$. In the second case, the elements i_0^- and $\tilde{i}$ satisfy the inequality $\tilde{i} > \pi(\tilde{i}) \geq i_0^- > \pi(i_0^-)$.

Suppose that $i_0^- > i_0$. Due to the choice of the elements i_0^- and i_0, we have $i < \pi(i)$ if $i \in |1, i_0|$ and $i > \pi(i)$ whenever $i \in |i_0^-, n|$, where $i_0^- = i_0 + 1$. Consider two possible cases:

(i) there exists an element $\tilde{\tilde{i}}$ such that $\tilde{\tilde{i}}$ and $\pi(\tilde{\tilde{i}})$ simultaneously belong either to $|1, i_0|$ or to $|i_0^-, n|$;

(ii) $\pi(i) \in |i_0^-, n|$ if $i \in |1, i_0|$ and $\pi(i) \in |1, i_0|$ if $i \in |i_0^-, n|$.

In the first case, either $\tilde{\tilde{i}} < \pi(\tilde{\tilde{i}}) < \pi^2(\tilde{\tilde{i}})$ or $\tilde{\tilde{i}} > \pi(\tilde{\tilde{i}}) > \pi^2(\tilde{\tilde{i}})$ and, therefore, we arrive at property (i) of Lemma 4.4. In the second case, in view of the fact that π is a one-to-one map, we conclude that n is even, $i_0 = n/2$, and, hence, the permutation π possesses property (ii).

It follows from Lemma 4.4 that any cyclic permutation π is either a minimal permutation of length 2^n or possesses the following property: There exists $k = 0$ such that one of the orbits of the permutation π^{2^k} has property (i) of Lemma 4.4.

Similar classification is also applicable to trajectories of maps of an interval.

Lemma 4.5. *Let $f \in C^0(I, I)$, $a \in I$. The trajectory* $\text{orb}(a)$ *has exactly one of the following properties:*

(i) $f(a) = a$, i.e., a is a fixed point;

(ii) one can indicate points a', $a'' \in \text{orb}(a)$ such that either $a' < f(a') \le a'' \le f(a'')$ or $a' > f(a') \ge a'' \ge f(a'')$;

(iii) the trajectory $\text{orb}(a)$ *can be decomposed into sets* $\text{orb}'(a)$ *and* $\text{orb}''(a)$ *such that*

(a) $\text{orb}'(a)$ *and* $\text{orb}''(a)$ *belong to closed intervals I' and I'' such that* $\text{int}\, I' \cap \text{int}\, I'' = \varnothing$, *$f'_a \subset I'$, and $f''_a \subset I''$;*

(b) $f(\text{orb}'(a)) \subseteq \text{orb}''(a)$ *and* $f(\text{orb}''(a)) \subseteq \text{orb}'(a)$.

Proof. Suppose that $f(a) \ne a$. Let us show that, in this case, we have either (ii) or (iii).

If the inequalities $f^{i'-1}(a) \ne f^{i'}(a)$ and $f^{i'}(a) = f^{i'+1}(a)$ hold for some $i' > 0$, then f_a possesses property (ii). In this case, $a' = f^{i'-1}(a)$ and $a'' = f^{i'}(a)$.

Now suppose that $f^i(a) \ne f^{i+1}(a)$ for all $i > 0$. Denote $\text{orb}^-(a) = \{x \in \text{orb}(a) \mid f(x) < x\}$ and $\text{orb}^+(a) = \{x \in \text{orb}(a) \mid f(x) > x\}$. If one of these sets is empty, then the trajectory $\text{orb}(a)$ possesses property (ii).

It remains to consider the case where $\text{orb}^-(a) \ne \varnothing$ and $\text{orb}^+(a) \ne \varnothing$.

If there are points $x_1 \in \text{orb}^-(a)$ and $x_2 \in \text{orb}^+(a)$ such that $x_1 < x_2$, then the trajectory $\text{orb}(a)$ has property (ii). Indeed, since $x_1, x_2 \in \text{orb}(a)$, the preimage of at least one of these points (x_1 or x_2) belongs to $\text{orb}(a)$. For definiteness, we assume that this is the preimage of the point x_1. Denote it by x_3. Thus, $x_3 \in \text{orb}(a)$, $f(x_3) = x_1$, and $f(x_1) < x_1 < x_2 < f(x_2)$. If $x_3 < x_1$, one should take $a' = x_3$ and $a'' = x_2$. If $x_1 < x_3$, then $a' = x_3$ and $a'' = x_1$.

Now assume that the inequality $x_1 > x_2$ holds for any x_1 and x_2 such that $x_1 \in \text{orb}^-(a)$ and $x_2 \in \text{orb}^+(a)$. If there exists a point $x \in \text{orb}^-(a)$ such that $f(x) \in$

$\text{orb}^-(a)$, then the trajectory $\text{orb}(a)$ has property (ii). The trajectory $\text{orb}(a)$ also possesses the same property in the case where both x and $f(x)$ belong to the set $\text{orb}^+(a)$.

Consider the last possibility: $f(\text{orb}^+(a)) \subseteq \text{orb}^-(a)$ and $f(\text{orb}^-(a)) \subseteq \text{orb}^+(a)$. In this case, the trajectory $\text{orb}(a)$ has property (iii). Indeed, one can choose $\text{orb}'(a) = \text{orb}^+(a)$ and $\text{orb}''(a) = \text{orb}^-(a)$ and, hence, $I' = [\inf \text{orb}(a), \ \sup \text{orb}^+(a)]$ and $I'' = [\inf \text{orb}^-(a), \sup \text{orb}(a)]$.

To complete the proof, it remains to note that the trajectory cannot have properties (ii) and (iii) simultaneously.

Lemmas 4.6 and 4.7 describe the properties of chain recurrent points. Lemma 4.6 can be regarded as a consequence of the incompressibility of the set $\text{CR}(f)$.

Lemma 4.6. *Let $a \in \text{CR}(f) \setminus \text{Fix}(f)$. Then one can indicate points $b, c \in I$ such that $f(b) = a$, $f(c) = b$, and $a \le c < b$ if $a < f(a)$, or $b < c \le a$ if $a > f(a)$.*

Proof. Assume that $a < f(a)$ (the proof for the case $a > f(a)$ is similar). Denote $p = \min_{x \ge a} f(x)$.

Assume that $p > a$. Let $\varepsilon = (p-a)/3$. Then $f(x) \in [p, 1]$ for all $x \in [a, 1]$. Hence, any sequence of points $\{x_i \in I, \ i = 0, 1, \ldots\}$ such that $x_0 = a$ and $|f(x_i) - x_{i+1}| < \varepsilon$ belongs to the interval $[p - \varepsilon, 1]$ and, consequently, $|x_i - a| > \varepsilon$ for all $i > 0$, i.e., $a \notin \text{CR}(f)$.

Thus, $p \le a$ whenever $a \in \text{CR}(f) \setminus \text{Fix}(f)$. Therefore, the set $\{x \ge a \,|\, f(x) = a\}$ is nonempty. Let b be the least point in this set. If $f(a) > b$, then $f([a, b]) \supset [a, b]$. Hence, there exists a point $c \in [a, b]$ such that $f(c) = b$. If $f(a) = b$, then $c = a$.

Consider the case where $f(a) < b$. Let $q = \max_{a \le x \le b} f(x)$. We now prove that the inequality $q < b$ implies that $a \notin \text{CR}(f)$. Let $p_1 = \min_{a \le x \le q} f(x)$. By the choice of the point b, we have $f(a) > a$ for all $x \in [a, b)$. Consequently, $p_1 > 0$. Let $\delta = (p_1 - a)/4$. Since $f[q, b] \supset [a, p]$, the set $\{x \in [q, b] \,|\, f(x) = p_1 - \delta\}$ is nonempty. Let d be the least point in this set. Denote $\varepsilon_1 = \min\{\delta, d - q\}$. Then

$$f([p_1 - 2\varepsilon_1, q + \varepsilon_1]) \subset [p - \varepsilon_1, q].$$

Hence, any sequence $\{x_i\}_{i=0}^{\infty}$ with $x_0 = a$ such that $|f(x_i) - x_{i+1}| < \varepsilon_1$ satisfies the inequality $|x_i - a| > \varepsilon_1$ for all $i > 0$.

Thus, $q \ge b$ and, consequently, there exists a point $c \in (a, b)$ such that $f(c) = b$.

Lemma 4.7. *Let $a \in \text{CR}(f)$. If the trajectory $\text{orb}(a)$ of the point a contains points a' and a'' such that either $a' < f(a') \le a'' \le f(a'')$ or $a' > f(a') \ge a'' \ge f(a'')$, then one can indicate closed intervals I_1 and I_2 such that* $\text{int}\, I_1 \cap$

$\operatorname{int} I_2 = \varnothing$, $f^{2^n}(I_1) \supseteq I_1 \cup I_2$, and $f^{2^n}(I_2) \supseteq I_1 \cup I_2$, where n is equal to either 0 or 1.

Proof. We consider only the case where $a' < f(a') \le a'' \le f(a'')$ (for the second case, the proof is similar). There are four different possibilities:

(I) $a' < f(a') = a'' \le f(a'')$;

(II) $a' < f(a') < a'' < f(a'')$;

(III) $a' < f(a') = a'' = f(a'')$;

(IV) $a' < f(a') < a'' = f(a'')$.

We prove Lemma 4.7 in each of these cases.

I. $a' < f(a') = a'' \le f(a'')$. By virtue of Lemma 4.6, one can find a point b such that $a' < b$ and $f(b) = a'$. If $b < a''$, then $I_1 = [a', b]$ and $I_2 = [b, a'']$ are just the required intervals and $n = 0$.

Now let $a'' < b$. By Lemma 4.6, there exists a point c such that $a' < c < b$ and $f(c) = b$. Since $f(b) < c < b = f(c)$, the interval $[c, b]$ contains at least one fixed point. Let α be the greatest fixed point of this sort. It is clear that $f(a') < \alpha$ because $f(a'') > a''$. Denote the intervals $[a', c]$, $[c, \alpha]$, and $[\alpha, b]$ by I_1, I_2, and I_3, respectively. By the choice of the points b, c, and α, we get $f(I_1) \supset I_3$, $f(I_2) \supseteq I_3$, and $f(I_3) \supseteq I_1 \cup I_2$. This means that $f^2(I_1) \supseteq I_1 \cup I_2$ and $f^2(I_2) \supseteq I_1 \cup I_2$.

II. $a' < f(a') < a'' < f(a'')$. If $f(a') < f^2(a')$, we proceed as in case I. Assume that $f(a') > f^2(a')$. In this case, the set of fixed points from the interval $[f(a'), a'']$ is nonempty. We denote by β_1 and β_2 the least and the greatest points of this sort, respectively, i.e., $\beta_1 = \min [f(a'), a''] \cap \operatorname{Fix}(f)$ and $\beta_2 = \max [f(a'), a''] \cap \operatorname{Fix}(f)$. Since the points a' and a'' belong to the same trajectory, one can indicate either i_1 such that $f^{i_1}(a'') = a'$ or i_2 such that $f^{i_2}(a') = a''$. Suppose that $f^{i_1}(a'') = a'$ (the case where $f^{i_2}(a') = a''$ can be investigated in exactly the same way). Since $f^{i_1}(a'') = a'$, there are points $\beta' = \min \{x \in I \mid a'' < x \text{ and } f(x) = \beta_2\}$ and β'' such that $\beta_2 < \beta'' < \beta'$ and $f(\beta'') = \beta'$. By the choice of the points β_2, β', and β'', the intervals $I_1 = [\beta_2, \beta'']$ and $I_2 = [\beta'', \beta']$ satisfy Lemma 4.7 with $n = 0$.

III. $a' < f(a') = a'' = f(a'')$. It follows from Lemma 4.6 that $f(b) = a'$ for some $b > a'$. If $b < a''$, then $f([b, a'']) \supseteq [a', a'']$ and $f([a', b]) \supseteq [a', a'']$. If $a'' > b$, then Lemma 4.6 implies the existence of a point c such that $c \in (a', b)$ and $f(c) = b$. If $c > a''$, then the intervals $[a'', c]$ and $[c, b]$ satisfy Lemma 4.7 with $n = 0$. If $c > a''$, then $f^2([a', c]) \supseteq [a', a'']$ and $f^2([c, a'']) \supseteq [a', a'']$.

IV. $a' < f(a') < a'' = f(a'')$. Since the points a' and a'' belong to the same trajectory $f_{a'}$, one can indicate a point $a''' \in f_a$ such that $a''' \neq a''$ and $f(a''') = a''$. By repeating the reasoning used in case III for the points a'' and a''', we complete the proof of Lemma 4.7.

Remark 1. It is clear that Lemma 4.7 is not true without the condition $a \in \mathrm{CR}(f)$. Actually, let $f_0 \in C^0(I, I)$ be such that $f_0(x) < x$ for $x \in \mathrm{int}\, I$, $f_0(0) = 0$, and $f_0(1) = 1$. Then, for any point x from $\mathrm{int}\, I$, we have $x > f_0(x) > f_0^2(x)$. At the same time, $\mathrm{CR}(f_0) = \mathrm{Fix}(f_0) = \{0, 1\}$. Let I_1 and I_2 be some intervals such that $\mathrm{int}\, I_1 \cap \mathrm{int}\, I_2 = \varnothing$. For definiteness, we assume that I_2 is located to the right of I_1. Then the intervals $f^n(I_2)$ are located to the right of I_1 for all $n > 0$, i.e., $f^n(I_2) \cap I_1 = \varnothing$.

Theorem 4.4. *For* $f \in C^0(I, I)$, *the following assertions are equivalent:*

I. $\mathrm{CR}(f)$ *does not coincide with the union of all simple sets of the map* f.

II. There exists a chain recurrent point which does not belong to a simple invariant set.

III. There exists a chain recurrent point a *whose trajectory contains points* a' *and* a'' *such that either* $a' < f^{2^k}(a') \leq a'' \leq f^{2^k}(a'')$ *or* $a' > f^{2^k}(a') \geq a'' \geq f^{2^k}(a'')$ *for some* $k \geq 0$.

IV. There exists $l \geq 0$ *and closed intervals* I_1 *and* I_2 *in* I *such that* $\mathrm{int}\, I_1 \cap \mathrm{int}\, I_2 = \varnothing$, $f^{2^l}(I_1) \supseteq I_1 \cup I_2$, *and* $f^{2^l}(I_2) \supseteq I_1 \cup I_2$ *for some* $l \geq 0$.

V. There exists a cycle whose period is not a power of 2, *i.e.,* $f \notin \mathfrak{F}_{2^\infty}$.

VI. There exists a cycle which is not simple.

VII. There exists an ω*-limit set which is not a simple invariant set.*

Proof. I $\Leftrightarrow$ II. This is a consequence of property (viii) of simple invariant sets which can be formulated as follows: $\mathrm{CR}(f) \supseteq \bigcup_\alpha M_\alpha$, where $\bigcup_\alpha M_\alpha$ is the union of all simple sets of the map f.

II $\Rightarrow$ III. Let $a \in \mathrm{CR}(f) \setminus \bigcup_\alpha M_\alpha$. It is clear that $a \notin \mathrm{Fix}(f)$ because $\mathrm{Fix}(f) \subset \bigcup_\alpha M_\alpha$. Since $\mathrm{CR}(f)$ is an invariant set, for any $i > 0$, one can find a point $a_{-i} \in \mathrm{CR}(f)$ such that $f^i(a_{-i}) = a$. Let $M = \bigcup_{i=0}^{\infty} \{f^i(a), a_{-i}\}$, where $a_0 = a$. By definition, M is an invariant set and $M \subset \mathrm{CR}(f)$. If, for some $i > 0$, the trajectory $\mathrm{orb}(a_{-i})$ has property (ii) of Lemma 4.5, then a_{-i} is a required chain recurrent point for statement III with $k = 0$. If property (ii) of Lemma 4.5 does not hold for any $i > 0$, then, by virtue of Lemma 4.5, for every $i > 0$, the trajectory $\mathrm{orb}(a_{-i})$ admits a decomposition $\{\mathrm{orb}'(a_{-i}), \mathrm{orb}''(a_{-i})\}$ such that

(1) $\mathrm{orb}'(a_{-i})$ and $\mathrm{orb}''(a_{-i})$ belong to closed intervals I_i' and I_i'' such that $\mathrm{int}\, I_i' \cap \mathrm{int}\, I_i'' = \varnothing$.

(2) $f(\mathrm{orb}'(a_{-i})) \subseteq \mathrm{orb}''(a_{-i})$ and $f(\mathrm{orb}''(a_{-i})) \subseteq \mathrm{orb}'(a_{-i})$.

Therefore, the set M admits a decomposition $\{M_1, M_2\}$ such that

(1) M_1 and M_2 belong to closed intervals I' and I'' such that $\mathrm{int}\, I' \cap \mathrm{int}\, I'' = \varnothing$, $M_1 \subset I'$, and $M_2 \subset I''$;

(2) $f(M_1) = M_2$ and $f(M_2) = M_1$.

Moreover, $\mathrm{orb}'(a_{-i}) \subset M_1$, $\mathrm{orb}''(a_{-i}) \subset M_2$, $I_i' \subset I'$, and $I_i'' \subset I''$ for all $i > 0$.

Consider f^2. The map f^2 decomposes the set M into two trajectories M_1 and M_2. Since $\mathrm{Fix}(f^2) \subset \bigcup_\alpha M_\alpha$, every trajectory possesses either property (ii) or property (iii) in Lemma 4.5. If at least one of these trajectories has property (ii), then assertion III holds with $k = 1$. Assume that both M_1 and M_2 possess property (iii) in Lemma 4.5. In this case, $\overline{M}$ admits a decomposition into the sets $\overline{M}_1$ and $\overline{M}_2$. For the sets M_1 and M_2, we repeat the same argument as for the set M. Then the entire procedure is repeated once again, and so on. After finitely many (k) steps, we arrive at a simple set. This yields III.

III $\Rightarrow$ IV. It suffices to apply Lemma 4.7 to the map f^{2^k}.

IV $\Rightarrow$ V. The map f^{2^k} possesses a cycle of period 3 (see, e.g., Lemma 3.3 in Section 3.1). Therefore, f possesses a cycle of period $2^l 3$, $l \le k$.

V $\Rightarrow$ VI. Note that any cycle whose period is not a power of two is not simple by definition.

VI $\Rightarrow$ VII. Note that any cycle is an ω-limit set.

VII $\Rightarrow$ I. It follows from VII that $\mathrm{CR}(f)$ contains a closed invariant indecomposable set which is not simple. Hence, $\mathrm{CR}(f) \neq \bigcup_\alpha M_\alpha$.

Note that Theorem 4.3 is equivalent to the statement "I ⇔ IV". Moreover, in order to prove "I ⇔ IV", it suffices to show that V ⇔ I (by analogy with the proof of the equivalence VII ⇔ I). However, we have added assertions VI and VII because these are quite useful and their proofs are very simple. It should also be noted that the equivalence "V ⇔ IV" was proved by Sharkovsky [3]; the equivalence "V ⇔ VI" was established by Block [2], and the assertions similar to "VII ⇔ V" can be found in Barkovsky and Levin [1], Blokh [1], Fedorenko [3], Li, Misiurewicz, and Yorke [1], Misiurewicz [2], and Smital [1].

At the end of this section, we present a lemma used in what follows. Recall that $f \in \mathfrak{F}_{2^\infty} \Leftrightarrow \mathrm{CR}(f) = \bigcup_\alpha M_\alpha$.

Lemma 4.8. *Let M be a simple invariant set of a map $f \in \mathfrak{F}_{2^\infty}$ and let* card $M > 2^n$, $n > 0$. *Then there exist closed mutually disjoint intervals I_i, $i = 1, \ldots, 2^n$, each of which contains an element of a simple decomposition of the set M of rank n, $f^k(I_i) \cap I_i = \varnothing$ for $k = 1, \ldots, 2^n - 1$, and $f^{2^n}(I_i) \subset I_i$, $i = 1, \ldots, 2^n$.*

Proof. For $n > 1$, the proof of the lemma is a simple consequence of its assertion with $n = 1$. Thus, it suffices to prove Lemma 4.8 for $n = 1$.

Since card $M > 2$, the set M admits a simple decomposition $\{M_i^{(2)},\ i = 1, \ldots, 2^2\}$ of rank 2. Denote $\alpha_i' = \min\{x \mid x \in M_i^{(2)}\}$ and $\alpha_i'' = \max\{x \mid x \in M_i^{(2)}\}$, $i = 1, \ldots, 2^2$. Consider the map $g = f^2$. It is clear that the map g has fixed points in the interval $[\alpha_3'', \alpha_4']$. Let β be one of these points. Moreover, since $\beta \in g([\alpha_2'', \alpha_3'])$, there exists a point $\gamma = \max\{x \in [\alpha_2'', \alpha_3'] \mid g(x) = \beta\}$.

Let us now indicate a closed interval I_2 that contains the element $M_2^{(1)} = M_3^{(2)} \cup M_4^{(2)}$ of the simple decomposition of rank 1. If the point γ has no preimages to the right of the point γ, then $I_2 = \{x \in I \mid x \geq \gamma\}$. Now suppose that there exists $\gamma_1 = \min\{x \in I \mid x \geq \gamma \text{ and } g(x) = \gamma\}$. If $\gamma_1 \in [\alpha, \beta]$, then $g([\gamma, \gamma_1]) \supseteq [\gamma, \beta]$ and $g([\gamma_1, \beta]) \supseteq [\gamma_1, \beta]$. Hence, by virtue of Theorem 4.3, we have $f \notin \mathfrak{F}_{2^\infty}$. If $\gamma_1 \in [\beta, \alpha_4'']$, then there exists a point $x_1 \in [\gamma, \alpha_3'']$ such that $g^2(x_1) = \gamma_1$ (the existence of x_1 is a consequence of the inclusion $g([\gamma, \beta]) \supset [\beta, \alpha_4''])$. Therefore, $g^2([\gamma, x_1]) \supseteq [\gamma, \beta]$, $g^2(x_1, \beta) \supseteq [\gamma, \beta]$, and $f \notin \mathfrak{F}_{2^\infty}$.

Consequently, $\gamma_1 > \alpha_4''$. If the point γ_1 has a preimage, then, by the same argument, one can also prove that $f \notin \mathfrak{F}_{2^\infty}$. Hence, $g(x) < \gamma_1$ for $x_1 \in [\gamma, \gamma_1]$ and $g(I_2) \subset I_2$ for the interval $I_2 = [\gamma, \gamma_1]$.

Similarly, one can prove the existence of a closed interval $I_1 \subset [\inf I, \gamma]$ such that $I_1 \supset M_1^{(2)} \cup M_2^{(2)}$, $g(I_1) \subset I_1$, and $I_1 \cap I_2 = \varnothing$.

4. Return for Simple Maps

As shown in the previous section, a map is simple if and only if its set of almost returning points is the union of all simple sets of this map. However, the set of almost returning points of any map in $\mathfrak{F}_{2^\infty}$ always contains points with the stronger property of return than chain recurrence.

Recall (see Theorem 1.5) that the following chain of inclusions is valid for $f \in C^0(I, I)$:

$$\mathrm{Per}(f) \subseteq \mathrm{APB}(f) \subseteq \mathrm{AP}(f) \subseteq \mathrm{RR}(f) \subseteq \mathrm{R}(f) \subseteq \mathrm{C}(f) \subseteq \Omega(f) \subseteq \mathrm{NW}(f) \subseteq \mathrm{CR}(f).$$

What types of return appearing in this chain may simple maps have? This problem is completely solved by the following two theorems:

Theorem 4.5. *The set of all chain recurrent points of the map* f *coincides with the union of all simple sets of this map (i.e.,* f *is a simple map) if and only if* $\mathrm{RR}(f) = \mathrm{R}(f)$.

Proof. Let $\mathrm{CR}(f) = \bigcup_\alpha M_\alpha$. Since any recurrent point of a simple invariant set is regularly recurrent (see property (ix) in Section 2) and $\mathrm{CR}(f) \supseteq \mathrm{R}(f)$, we have $\mathrm{RR}(f) = \mathrm{R}(f)$.

Now suppose that $\mathrm{CR}(f) \neq \bigcup_\alpha M_\alpha$. By virtue of Theorem 4.3, in this case, there exist $n \geq 0$ and closed intervals I_1 and I_2 such that $\mathrm{int}\, I_1 \cap \mathrm{int}\, I_2 = \varnothing$, $f^{2^n}(I_1) \supseteq I_1 \cup I_2$, and $f^{2^n}(I_2) \supseteq I_1 \cup I_2$.

By using standard methods of symbolic dynamics (see the proofs of Propositions 1.1 and 1.2), one can show that f possesses an infinite closed invariant set F that contains an everywhere dense trajectory and an everywhere dense subset of periodic points. This dense trajectory consists of recurrent points. However, F is not a minimal set. Hence, by virtue of the Birkhoff theorem (Birkhoff [1]), the points of this trajectory are not regularly recurrent, i.e., $\mathrm{RR}(f) \neq \mathrm{R}(f)$.

Theorem 4.6. (Sharkovsky and Fedorenko [1]). *There exists a map* $f_0 \in C^0(I, I)$ *such that*

$$\mathrm{Per}(f_0) \neq \mathrm{APB}(f_0) \neq \mathrm{AP}(f_0) \neq \mathrm{RR}(f_0)$$

$$= \mathrm{R}(f_0) \neq \mathrm{C}(f_0) \neq \Omega(f_0) \neq \mathrm{NW}(f_0) \neq \mathrm{CR}(f_0). \qquad (4.1)$$

Proof. To prove the theorem, we use the following considerations: Let I_i, $i = 1, \ldots, k$, be closed mutually disjoint intervals on I. Assume that maps $f_i \in C^0(I_i, I_i)$, $i =$

$1, \ldots, k$ possess properties A_i invariant under topological conjugation. In addition, we suppose that the left end of the interval I_1 coincides with the left end of the interval I, i.e., $\inf I_1 = \inf I$, and that the right end of I_k coincides with the right end of I, i.e., $\sup I_k = \sup I$. Furthermore, let the map $f \in C^0(I, I)$ be such that

(i) $f|_{I_i}$ is topologically conjugate to f_i, $i = 1, \ldots, k$;

(ii) f is a linear function in each component of $I \setminus \bigcup_{i=1}^{k} I_i$.

In this case, the map f possesses all the properties A_i, $i = 1, \ldots, k$. Hence, in order to prove the theorem, it suffices to construct a series of maps with the following common property: For each of these maps, the set of recurrent points coincides with the set of almost periodic points (i.e., each of these maps belongs to $\mathfrak{F}_{2^\infty}$). Note that, for any two maps in this series, some pair of neighboring sets in (4.1) does not coincide (these pairs are different for different maps).

Any map in the series constructed below is a modification of two fixed maps. The first of these maps is

$$f(x) = \begin{cases} 2x, & x \in [0, 1/2], \\ -2x + 2, & x \in [1/2, 1], \end{cases} \tag{4.2}$$

and the second one was introduced by Sharkovsky in [7].

I. Consider the map f given by (4.2).

Here, we use the binary representation of points in $I = [0, 1]$ instead of their decimal representation. Let $0.a_1 \ldots a_i \ldots$, where a_i is either 0 or 1, be the coordinate of a point $x \in I$. Then

$$f(0.a_1 a_2 \ldots a_i \ldots) = \begin{cases} 0.a_2 a_3 \ldots a_i \ldots, & \text{if} \quad a_1 = 0, \\ 0.\bar{a}_2 \bar{a}_3 \ldots \bar{a}_i \ldots, & \text{if} \quad a_1 = 1, \end{cases} \tag{4.3}$$

where $\bar{a}_i = 1 - a_i$.

The following property is a generalization of this representation for the nth power of the map f.

Property 1.

$$f^n(0.a_1 a_2 \ldots a_n a_{n+1} \ldots a_i \ldots) = \begin{cases} 0.a_{n+1} \ldots a_i \ldots, & \text{if} \quad a_n = 0, \\ 0.\bar{a}_{n+1} \ldots \bar{a}_i \ldots, & \text{if} \quad a_n = 1. \end{cases} \tag{4.4}$$

Proof. Since $\overline{\overline{a}}_i = a_i$, it follows from (4.3) that $f^{n-1}(0.a_1\, a_2 \ldots a_i \ldots)$ is equal either to $0.a_n a_{n+1} \ldots a_i \ldots$ or to $0.\overline{a}_n\, \overline{a}_{n+1} \ldots \overline{a}_i \ldots$. In the first case, by substituting $0.a_n a_{n+1} \ldots a_i \ldots$ in (4.3), we obtain (4.4). In the second case, we also arrive at (4.4). Indeed, if $a_n = 0$, then $\overline{a}_n = 1$ and (4.3) implies that

$$f^n(0.a_1\, a_2 \ldots a_i \ldots) = 0.a_{n+1} \ldots a_i \ldots .$$

Further, if $a_n = 1$, then $\overline{a}_n = 0$ and, thus,

$$f^n(0.a_1\, a_2 \ldots a_i \ldots) = 0.\,\overline{a}_{n+1} \ldots \overline{a}_i \ldots .$$

We now introduce several definitions and notation necessary for what follows. Any finite ordered sequence that consists of 0 and 1 is called a block. Let B be a block that consists of elements $a_1 \ldots a_n$. Then $\overline{B}$ denotes the block formed by the elements $\overline{a}_1 \ldots \overline{a}_n$. The infinite sequence of blocks B is denoted by (B), i.e., $(B) = BBB\ldots$.

Any positive integer i can be represented in the form $i = \sum_{j\geq 0} s_j 2^j$, where $s_j \in \{0, 1\}$. We set

$$p(i) = \sum_{j\geq 0} s_j, \qquad q(i) = \min\{j \mid s_j \neq 0\},$$

and $R(i) = p(i) + q(i)$.

The block formed by a single 1 is denoted by B_{2^0}. Beginning with B_{2^0}, we construct blocks for any $k > 0$ according to the formula

$$B_{2^{k+1}} = B_{2^k}\overline{B}_{2^k}. \tag{4.5}$$

Let $\alpha = 0.a_1 \ldots a_i \ldots$ be a point of the interval I such that $a_1 \ldots a_{2^k} = B_{2^k}$ for any $k \geq 0$.

Property 2. For any fixed $k \geq 0$, $\alpha = 0.\,C_1^k \ldots C_i^k \ldots$, where

$$C_i^k = \begin{cases} B_{2^k}, & \text{if } R(i) \text{ is odd,} \\ \overline{B}_{2^k}, & \text{if } R(i) \text{ is even.} \end{cases} \tag{4.6}$$

Proof. It follows from (4.5) that, for any $k > 0$ and $k' < k$, B_{2^k} is representable as an ordered sequence of blocks $B_{2^{k'}}$ and $\overline{B}_{2^{k'}}$. Therefore, $C_1^k \ldots C_i^k \ldots$, where C_i^k is either B_{2^k} or $\overline{B}_{2^k}$ for any fixed k. Let us determine the block occupying the ith position. Let $i = \sum_{j\geq 0} s_j 2^j$, where $s_j \in \{0, 1\}$. Denote $k_1 = \max\{j \mid s_j \neq 0\} + 1$.

Then it follows from the definition of α and (4.5) that

$$\alpha = 0.\, B_{2^{k+k_1}} \dots = 0.\, C_1^k \dots C_{2^{k_1}-1}^k \overline{C}_1^k \dots \overline{C}_{2^{k_1}-1}^k \dots$$

and, consequently, $C_i^k = \overline{C}_{i-2^{k_1}}^k$.

We set

$$k_2 = \max_{j \in k_1 - 1} \{ j \mid s_j \neq 0 \} + 1.$$

By repeating the same argument for the block occupying the $(i - 2^{k_1})$th position, we obtain $C_i^k = \overline{\overline{C}}_{i-2^{k_1}-2^{k_2}}^k$. Iterating this procedure $R(i)$ times, we arrive at the equality

$$C_i^k = \overset{\left.\begin{matrix}=\\ \vdots \\ =\end{matrix}\right\} l \text{ times}}{C_1^k}, \qquad \text{where } l = R(i). \tag{4.7}$$

Since $\overline{\overline{C}}_1^k = C_1^k = B_{2^k}$, relation (4.7) yields (4.6). By Property 2 with $k = 0$, we obtain, in particular, $\alpha = 0.a_1 \dots a_i \dots$, where

$$a_j = \begin{cases} 0, & \text{if } R(i) \text{ is odd,} \\ 1, & \text{if } R(i) \text{ is even.} \end{cases}$$

Property 3. $\alpha \in \mathrm{RR}(f)$.

Proof. Let $U(\alpha)$ be an arbitrary neighborhood of the point α. It follows from the definition of α that there exists an odd number k such that each point of the form $0.B_{2^k} a_{2^k+1} \dots a_i \dots$ belongs to the neighborhood $U(\alpha)$. We fix an arbitrary number i and prove that there exists $i_0 \in \{ i, i+1, \dots, i + 2^{k+2} - 1 \}$ such that $f^{i_0}(\alpha) = 0.B_{2^k} \dots$. This, in fact, means that $\alpha \in \mathrm{RR}(f)$.

Indeed, let $\alpha = 0.\, C_1^{k+2} \dots C_i^{k+2} \dots$. Then $f^{2^{k+2}j}(\alpha)$, $j = 1, 2, \dots$, is either the point $0.B_{2^{k+2}} \dots$ or the point $0.\overline{B}_{2^{k+2}} \dots$. Formula (4.5) implies that

$$0.B_{2^{k+2}} \dots = 0.B_{2^k} \overline{B}_{2^k} \overline{B}_{2^k} B_{2^k} \dots$$

and

$$0.\overline{B}_{2^{k+2}} \dots = 0.\overline{B}_{2^k} B_{2^k} B_{2^k} \overline{B}_{2^k} \dots$$

Since k is odd, $R(2^k)$ is even and, therefore, the last element of the block B_{2^k} is 0.

Hence, $f^{2^{k+2}j+2^{k+1}}(\alpha) = 0.B_{2^k}\ldots$ for any j. Consider the representation $i = 2^{k+2}j_1 + j_2$, where $0 \le j_2 < 2^{k+2}$. If $j_2 \le 2^{k+1}$, then we take $i_0 = 2^{k+2}j_1 + 2^{k+1}$. If $j_2 > 2^{k+1}$, then $i_0 = 2^{k+2}(j_1+1) + 2^{k+1}$. Here, the inequality $i \ge i_0 \ge i + 2^{k+2} - 1$ holds in both cases.

Property 4. $\alpha \notin \mathrm{AP}(f)$.

Proof. Let $\alpha = 0.C_1^0 \ldots C_i^0 \ldots$, and let N be an arbitrary fixed positive integer. Since $\alpha = 0.1\ldots$, it suffices to show that there exists $j > 0$ such that $f^{N_j}(\alpha) = 0.0\ldots$ We set $j = 2^{q(N)+1}$. By the definition of the function $R(j)$, we can write $R(N_j) = p(N) + 1 + 2q(N)$ and $R(N_j + 1) = p(N) + 1$. This means that the numbers $R(N_j)$ and $R(N_j+1)$ are either both even or both odd. Thus, in the first case, $C_{N_j}^0 = 0$ and $C_{N_{j+1}}^0 = 0$, while in the second case, $C_{N_j}^0 = 1$ and $C_{N_{j+1}}^0 = 1$. However, in any case $f^{N_j}(\alpha) = 0.0\ldots$.

Property 5. If k is odd, then $x = 0.(B_{2^k})$ is a periodic point with period 2^{k-1}.

Proof. For odd k, the last element of the block $B_{2^{k-1}}$ is 1 and the last element of the block $B_{2^{k-2}}$ is 0. Consequently,

$$f^{2^{k-1}}(0.(B_{2^k})) = f^{2^{k-1}}(0.(B_{2^{k-1}}\overline{B}_{2^{k-1}})) = 0.(B_{2^k})$$

and

$$f^{2^{k-2}}(0.(B_{2^k})) = f^{2^{k-2}}(0.(B_{2^{k-2}}\overline{B}_{2^{k-2}}\overline{B}_{2^{k-2}}B_{2^{k-2}})) = 0.\overline{B}_{2^{k-2}}\ldots.$$

The first equality implies that the point $0.(B_{2^k})$ is periodic and the second one means that its period is equal to 2^{k-1} because the number 2^{k-2} and, hence, any number 2^{k_1} with $k_1 < k-1$ cannot be a period of this point.

Property 6. $\alpha < 0.(B_{2^{k+2}}) < 0.(B_{2^k})$ for any odd k.

Proof. The proof follows from the equality $B_{2^{k+2}} = B_{2^k}\overline{B}_{2^k}\overline{B}_{2^k}B_{2^k}$.

Property 7. $f^{2^k+1}(\alpha) > f(0.(B_{2^k}))$ for any odd k.

Proof. Let $\alpha = 0.C_1^k\ldots C_i^k\ldots$ and let $a_1\ldots a_k$ be a part of the block B_{2^k}. Then $f^{2^k+1}(\alpha) = 0.\bar{a}_2\ldots\bar{a}_{2^k}\overline{B}_{2^k}B_{2^k}\ldots$ and $f(0.(B_{2^k})) = 0.\bar{a}_2\ldots\bar{a}_{2^k}(\overline{B}_{2^k})$. This means that the first $2^{k+1}-1$ elements of the points under consideration are equal, while the (2^{k-1})th element of the points $f^{2^k+1}(\alpha)$ and $f(0.(B_{2^k}))$ is equal to 1 or 0, respectively. This, in fact, proves Property 7.

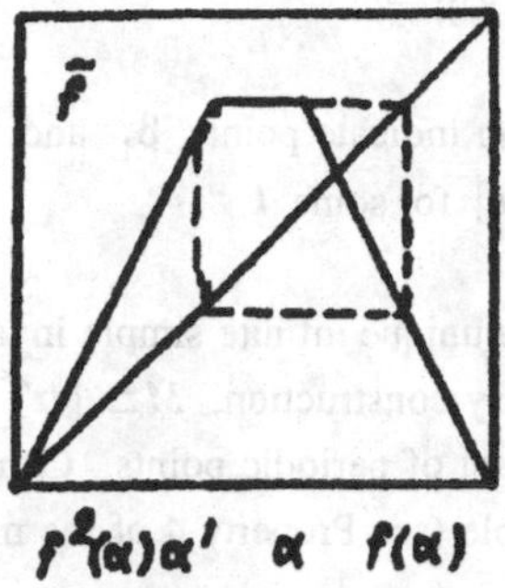

Fig. 27

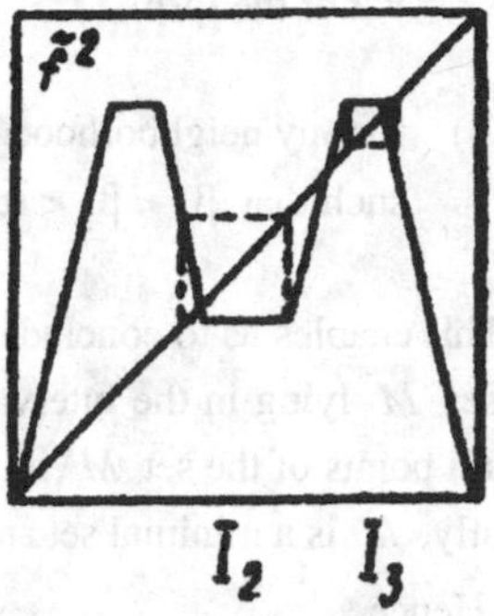

Fig. 28

Let α' be the point of I such that $\alpha' = 1 - \alpha$. Consider a map $\tilde{f} \in C^0(I, I)$ (Fig. 27) such that

$$\tilde{f}(x) = \begin{cases} f(\alpha), & \text{if } x \in [\alpha', \alpha], \\ f(x), & \text{if } x \in I \setminus [\alpha', \alpha]. \end{cases}$$

The map $\tilde{f}$ possesses the invariant interval $I_1 = [f^2(\alpha), f(\alpha)]$. Consider $\tilde{f}^2|_{I_1}$. It follows from Property 7 with $k=1$ that $f^3(\alpha) > 0.(B_{21})$. This condition implies the following property of the map $\tilde{f}^2$: The interval I_1 contains two intervals $I_2 = [f^2(\alpha), f^4(\alpha)]$ and $I_3 = [f^3(\alpha), f(\alpha)]$ invariant under $\tilde{f}^2$ and such that $\tilde{f}(I_2) = I_3$ and $\tilde{f}(I_3) = I_2$. This enables us to conclude that

(i) the map $\tilde{f}$ has no periodic trajectories with odd periods;

(ii) the interval I_3 contains points β_1 and β_2 such that $\tilde{f}^2[\beta_1, \beta_2] = [\alpha', \alpha]$ (Fig. 28).

In view of the fact that Property 7 holds for any odd k, one can repeat this reasoning for $k = 3$ and the interval I_3, etc. As a result, we obtain the following properties of the map $\tilde{f}$:

(1) $\tilde{f}$ possesses a periodic trajectory of period 2^k, $k = 0, 1, \ldots$;

(2) $\tilde{f}$ has no periodic trajectories whose periods are not powers of 2;

(3) $\tilde{f}(\alpha) \in \mathrm{AP}(f)\setminus\mathrm{Per}(f)$ (this property is a consequence of the facts that $0.(B_{2^k}) \to \alpha$ as $k \to \infty$ and that, for any neighborhood $U(\tilde{f}(\alpha))$, there exists an interval of the form $[f^{2^k+1}(\alpha), \alpha]$ invariant under $\tilde{f}^{2^k}$);

(4) for any neighborhood of the point $\tilde{f}(\alpha)$, one can indicate points β_1 and β_2 such that $\beta_1 < \beta_2 < \alpha$ and $\tilde{f}^k[\beta_1, \beta_2] = [\alpha', \alpha]$ for some k.

This enables us to conclude that the map $\tilde{f}$ possesses a unique infinite simple invariant set M lying in the interval $[0, \alpha'] \cup [\alpha, \tilde{f}(\alpha)]$. By construction, $M \supset [\alpha', \alpha]$ and all points of the set $M \setminus \mathrm{int}\, M$ are limit points of the set of periodic points. Consequently, M is a minimal set and $f|_M$ is Lyapunov unstable (see Property 4 of the map $\tilde{f}$). Hence,

$$\mathrm{RR}(\tilde{f}) \supset \mathrm{AP}(\tilde{f}) \supset \mathrm{APB}(\tilde{f}) = \mathrm{Per}(\tilde{f}).$$

Let f_1 be a continuous function defined on the interval $[\alpha', \alpha]$ and such that

(i) $f_1(\alpha') = f_1(\alpha) = \tilde{f}(\alpha)$;

(ii) $f_1(x) > \tilde{f}(\alpha)$ for any $x \in (\alpha', \alpha)$ and $\max\limits_{x \in [\alpha', \alpha]} f_1(x) = \gamma$.

Consider the map

$$\hat{f}(x) = \begin{cases} f_1(x), & \text{if } x \in [\alpha', \alpha], \\ \tilde{f}(x), & \text{if } x \in [0, \tilde{f}(\alpha)] \setminus [\alpha', \alpha], \\ \tilde{f}^2(\alpha), & \text{if } x \in [\tilde{f}(\alpha), \gamma], \\ f_2(x), & \text{if } x > \gamma, \end{cases}$$

where $f_2(x)$ is a continuous function such that

(i) $f_2(\gamma) = \tilde{f}^2(\alpha)$;

(ii) $f_2(x) > \tilde{f}^2(\alpha)$ for $x > \gamma$ (Fig. 29).

Since the maps $\hat{f}$ and $\tilde{f}$ coincide on the set $[0, \alpha'] \cup [\alpha, \tilde{f}(\alpha)]$, the set M is invariant both under $\hat{f}$ and $\tilde{f}$. Moreover, Property 4 of the map $\tilde{f}$ implies that each point of the interval $(\tilde{f}(\alpha), \gamma)$ belongs to the set $\mathrm{CR}(f)\setminus \mathrm{NW}(f)$ and $\gamma \in \mathrm{NW}(f)$. Finally, since any sufficiently small neighborhood of the point γ contains at most two points of each trajectory of the map $\hat{f}$, we have $\gamma \notin \Omega(f)$. Hence,

$$\mathrm{APB}(\hat{f}) \neq \mathrm{AP}(\hat{f}) \neq \mathrm{RR}(\hat{f}) = \mathrm{R}(\hat{f}) \neq \mathrm{NW}(\hat{f}) \neq \mathrm{CR}(\hat{f}).$$

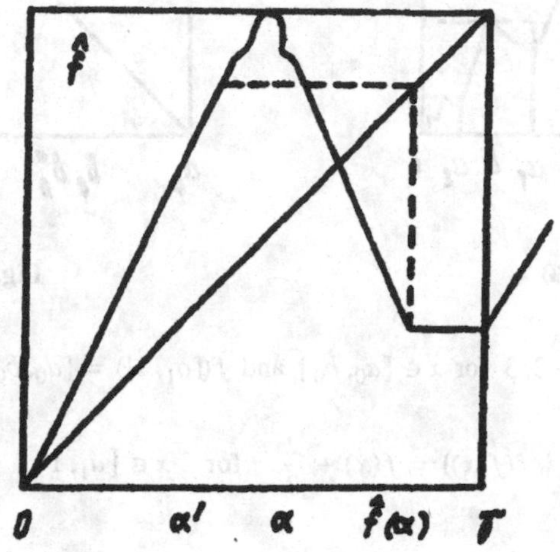

Fig. 29

II. The second series of maps is formed by modifications of the map introduced by Sharkovsky in [7]. First, we describe this map.

Let $h(x) = (x+2)/3$. In the segment I, we choose two sequences $\{a_n\}$ and $\{b_n\}$, $n = 0, 1, 2, \ldots$, such that $a_{n+1} = h(a_n)$, $b_{n+1} = h(b_n)$, $a_0 = 0$, and $b_0 = 1/3$. It is clear that $a_n < b_n < a_{n+1}$ for $n = 0, 1, 2, \ldots$, and $a_n \to 1$, $b_n \to 1$ as $n \to \infty$.

Let f be a continuous function piecewise linear in I, linear in each of the segments $[a_n, b_n]$ and $[b_n, a_{n+1}]$, and satisfying the relations

$$f(x)|_{[a_{n+1}, b_{n+1}]} = \frac{1}{3} f(x)|_{[a_n, b_n]}, \tag{4.8}$$

$f(x)|_{[a_0, b_0]} = x + 2/3$, and $f(1) = 0$ (Fig. 30).

It follows from the definition of the function f that the equality

$$f(h(x)) = \frac{1}{3} f(x) \tag{4.9}$$

holds for all $x \in I$.

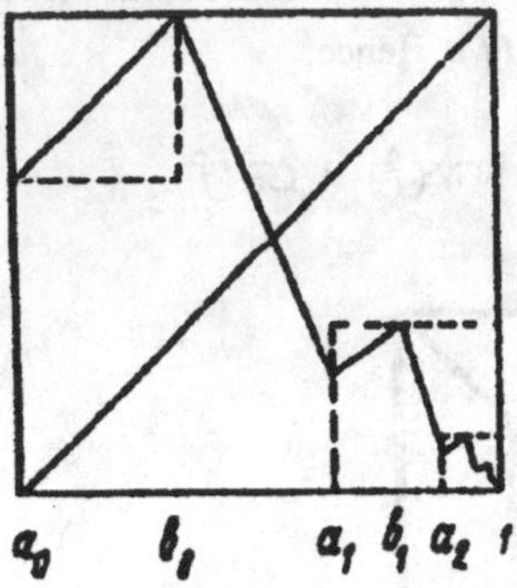

Fig. 30

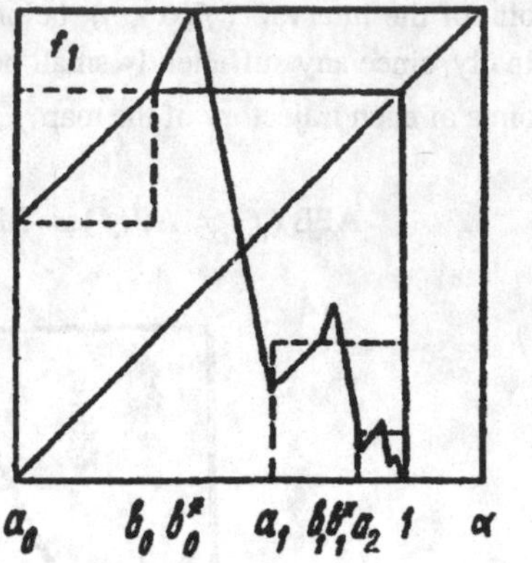

Fig. 31

Since $f(x) = x + 2/3$ for $x \in [a_0, b_0]$ and $f([a_1, 1]) = [a_0, b_0]$, we have

$$f(f(x)) = f(x) + \frac{2}{3} \quad \text{for} \quad x \in [a_1, 1],$$

$$f(f(x)) = f\left(x + \frac{2}{3}\right) \quad \text{for} \quad x \in [a_0, b_0]. \tag{4.10}$$

It follows from (4.9) and (4.10) that the maps $f|_{[0,1]}$ and $f^2|_{[a_0, b_0]}$ ($f|_{[0,1]}$ and $f^2|_{[a_1,1]}$) are topologically conjugate; furthermore,

$$h(f(x)) = f^2(h(x)),$$

$$g(f(x)) = f^2(g(x)), \tag{4.11}$$

where $g(x) = x/3$.

It follows from (4.11) that the following relations hold for any $n = 1, 2, \ldots$:

$$f^{2^n}(x) = h^n(f(h^{-n}(x))) \quad \text{for} \quad x \in [a_n, 1],$$

$$f^{2^n}(x) = g^n(f(g^{-n}(x))) \quad \text{for} \quad x \in [0, n/3]. \tag{4.12}$$

Let us establish some properties of the map f.

(1) $f \in \mathfrak{F}_{2^\infty}$. Indeed, we have $f([a_0, b_0]) = [a_1, 1]$, $f([a_1, 1]) = [a_0, b_0]$, and there are no periodic points in the interval $[b_0, a_1]$ except a fixed point.

Hence, f possesses a periodic trajectory of period 2 and has no periodic trajectories with odd periods. Therefore, the fact that f and f^2 are topologically conjugate implies that $f \in \mathfrak{F}_{2^\infty}$.

(2) $CR(f) = AP(f)$. The fact that f^{2^n} and f are topologically conjugate implies that the intervals $[a_n, 1]$ and $[0, n/3]$ are periodic with period 2^n. Furthermore,

$$CR(f) \subset \bigcup_{i=0}^{2^n-1} f^i[a_n, 1] \cup \text{Fix}(f^{2^{n-1}}), \quad n = 1, 2, \ldots .$$

Hence,

$$CR(f) \subset \bigcap_{n\geq 1} \bigcup_{i=0}^{2^n-1} f^i[a_n, 1] \cup \text{Per}(f)$$

The set $\bigcap_{n\geq 1} \bigcup_{i=0}^{2^n-1} f^i[a_n, 1]$ is the standard Cantor set. This means that each its point is almost periodic in the sense of Bohr.

Relations presented below are necessary for what follows.

It follows from (4.12) that

$$f^{2^n}(x) = x + \frac{2}{3^{n+1}} \quad \text{for} \quad x \in \left[0, \frac{1}{3^{n+1}}\right]. \tag{4.13}$$

Therefore, $f^{2^n}(x) \in [0, 1/3^n]$ if $x \in [0, 1/3^{n+1}]$. This yields

$$f^{2^n-1}(x) = x + 1 - \frac{1}{3^n} \quad \text{for} \quad x \in \left[0, \frac{1}{3^n}\right]. \tag{4.14}$$

We also note that if $x \in [0, 1/3^{n+1}]$, then $f^i(x) \in \bigcup_{n=0}^{\infty} [a_n, b_n]$ for $i = 1, 2, \ldots, 2^{n+1} - 1$ and

$$f(x)|_{[a_n, b_n]} = x + 1 - b_n - a_n, \quad f^{2^n}(1) = a_n,$$

$$f^{2^n}(a_n) = a_{n+1}, \quad f^{2^n}(b_n) = 1.$$

By using the map constructed above, we can now present an example of a map f such that $CR(f) \neq NW(f) \neq \Omega(f) = APB(f) \neq \text{Per}(f)$.

1. Let $b_n^* = b_n + 1/3^{n+3}$. Clearly, $b_n^* < a_{n+1}$. Consider a piecewise linear map f_1 defined on $[0, 1]$ and such that

$$f_1(x) = \begin{cases} f(x) & \text{if} \quad x \in [a_n, b_n], \\ 3(x + b_n) + f(b_n) & \text{if} \quad x \in [b_n, b_n^*], \\ f(a_{n+1}) - \dfrac{f(a_{n+1}) - f(b_n) - 3(b_n^* - b_n)}{a_{n+1} - b_n^*}(a_{n+1} - x) & \text{if} \quad x \in [b_n^*, a_{n+1}], \\ 0 & \text{if} \quad x \geq 1, \end{cases}$$

for any $n = 0, 1, 2, \ldots$ (see Fig. 31).

By using (4.14), one can easily show that

$$f_1^{2^n}(b_n^*) = \alpha, \qquad n = 1, 2, \ldots,$$

where $\alpha = 3(b_0^* - b_0) + f(b_0)$. Thus, for any $n = 0, 1, 2, \ldots$, the interval $[a_n, \alpha]$ is periodic with period 2^n. Moreover, each interval $f_1^i([a_n, \alpha])$, $i = 0, 1, \ldots, 2^n - 1$, can be split into three intervals so that f_1^{2n} is a homeomorphism defined on the central interval with a single fixed point and the other two intervals form a cycle of intervals with period 2. Therefore, f_1 is a continuous map from $\mathfrak{F}_{2^\infty}$ and

$$.\mathrm{NW}(f_1) \subset \bigcap_{n\geq 1} \bigcup_{i=0}^{2^n-1} f^i[a_n, \alpha] \cup \mathrm{Per}(f_1).$$

We also note that $\bigcap_{n\geq 1} \bigcup_{i=0}^{2^n-1} f^i[a_n, \alpha]$ is the standard Cantor set with a system of closed intervals attached to each unilateral point of the Cantor set which is a left limit point for points of this set (these intervals are preimages of the interval $[1, \alpha]$). Since each point of the interval $[1, \alpha]$ is wandering, we have $\mathrm{NW}(f_1) = \mathrm{APB}(f_1)$.

Consider a map

$$f_2(x) = \begin{cases} f_1(x), & \text{if} \quad x \in [0, \alpha]; \\ x - \alpha, & \text{if} \quad x > \alpha. \end{cases}$$

The point α is nonwandering for f_2. This follows from the fact that there exists a sequence of points $\{\alpha_n\}_{n=1}^{\infty}$ such that

(i) $\alpha_n \geq \alpha$;

(ii) $\alpha_n \to \alpha$ as $n \to \infty$;

(iii) $f_2(\alpha_n) = f_2(b_n^*)$, i.e., α is the limit point for its own preimages.

Hence, $\mathrm{NW}(f_2) = \mathrm{AP}(f_2) \cup \{\alpha\}$.

2. By using f_2, we can construct an example of a simple map f_3 such that $C(f_3) \neq R(f_3)$.

Let $f_3(x)|_{\bigcup_{n\geq 0}[a_n, b_n^*]} = f_1(x)$. We fix a sequence of points $a_n^* = a_n - 1/3^{n+2}$ and set $f^3(x) = f_1(x) + (1/3)(x - a_n)$ for $x \in [a_n^*, a_n]$, $n \geq 0$. We also consider a sequence of points

$$a_n' = a_n - \frac{5}{3^{n+3}}, \quad b_n' = b_n + \frac{1}{3^{n+2}}, \quad c_n = b_n + \frac{2}{3^{n+2}}, \quad n \geq 0.$$

Obviously,

$$a_n' < a_n^* < a_n < b_n < b_n^* < b_n' < c_{n+1} < a_{n+1}.$$

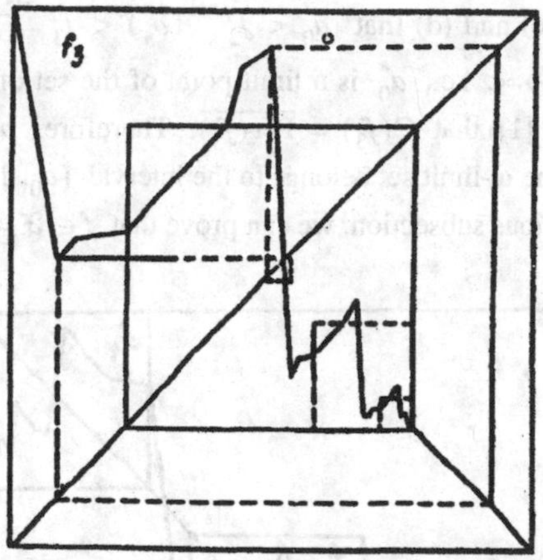

Fig. 32

The function f_3 can be extended by continuity to $\mathbb{R}^+$ as follows (see Fig. 32):

$$f_3(x) = \begin{cases} x - a_n^* + f_3(a_n), & \text{for} \quad x \in [a_n', a_n^*] \cup [b_n^*, b_n'], \\ 1 - x, & \text{for} \quad x \in \{c_n\} \cup \{x \mid x \geq 1\}, \\ \text{linear}, & \text{for} \quad x \in [b_n', c_n] \cup [c_n, a_n'], \ n = 0, 1, 2, \ldots. \end{cases}$$

Let us now describe the properties of the map f_3. By virtue of the definition of f_3 and (4.13), we can write

(a) $f_3^{2^n}(c_n) = b_n'$;

(b) $f_3^{2^n}(a_n') = c_n$;

(c) $f_3^{2^n-2}(x) = x + b_0^* - b_{n+1}^*$ for $x \in [b_{n+1}^*, b_{n+1}']$;

(d) $f_3^{2^n}(x) = a_n^* - x$ for $x \in [a_0^* - b_{n+1}' + b_{n+1}^*, a_0^*]$;

(e) $f_3^2(x) = a_0^* - x + b_0^*$ for $x \in [b_0^*, b_0']$, $n \geq 0$.

It follows from (c)–(e) that $f_3^{2^{n+1}}(b_n') = a_n'$. Then (a) and (b) imply that the points a_n', b_n', and c_n' belong to the same periodic trajectory with period 2^{n+2}, $n \geq 0$. Moreover, it follows from (c) and (d) that $a_0' < f_3^{2^{n-1}}(b_n') < f_3^{2^{n+1}-1}(b_{n+1}')$, $n \geq 0$, and $f_3^{2^{n-1}}(b_n') \to a_0^*$ as $n \to \infty$, i.e., a_0^* is a limit point of the set of periodic points. It is known (see Sharkovsky [1]) that $C(f_3) = \overline{\mathrm{Per}(f_3)}$. Therefore, $a_0^* \in C(f_3)$. However, $a_0^* \notin R(f_3)$, because the ω-limit set belongs to the interval $[a_0, 1]$. By using the same reasoning as in the previous subsection, we can prove that $f \in \mathfrak{F}_{2^\infty}$.

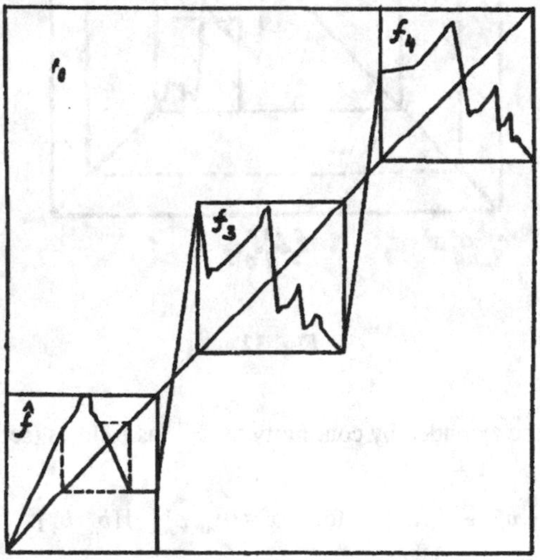

Fig. 33

3. Let us now construct a map f_4 with $\Omega(f_4) \neq C(f_4)$.
Assume that $f_4(x)$ is a continuous piecewise linear function such that

$$f_4(x) = \begin{cases} f_3(x), & \text{for} \quad x \geq a_0^*, \\ f_3^{2^{n+1}}(b'_{n+1}), & \text{for} \quad x = f_3^{2^n - 1}(b'_n), \\ \text{linear}, & \text{for} \quad x \in [f_3^{2^n-1}(b'_n), f_3^{2^{n+1}-1}(b'_{n+1})]. \end{cases}$$

The map f_4 belongs to $\mathfrak{F}_{2^\infty}$ and the point a_0^* belongs to the ω-limit set of the point a'_0. However, the point a_0^* does not belong to the center of the map f_4 because the interval $[a'_0, a_0^*]$ contains no periodic points of the map f_4.

The graph of the function appearing in the theorem is displayed in Fig. 33.

5. Classification of Simple Maps According to the Types of Return

According to Theorem 4.6, simple maps may have a large variety of types of "returning" points. This is why it is reasonable to construct a classification of one-dimensional dynamical systems based on the coincidence of different types of "returning" points from the following chain of inclusions:

$$\mathrm{Per}(f) \subseteq \mathrm{APB}(f) \subseteq \mathrm{AP}(f) \subseteq \mathrm{RR}(f) \subseteq \mathrm{R}(f)$$
$$\subseteq \mathrm{C}(f) \subseteq \Omega(f) \subseteq \mathrm{NW}(f) \subseteq \mathrm{CR}(f) \qquad (4.15)$$

A class of maps $\{f \in C(I, I) \mid A_1(f) = A_2(f)\}$, where $A_1(f)$ and $A_2(f)$ are two arbitrary sets from (4.15), is denoted by $\underline{A_1(f) = A_2(f)}$. As follows from (4.15), there are 36 classes of maps of the type $\underline{A_1(f) = A_2(f)}$. All these classes are depicted in Diagram 1, where "$\rightarrow$" denotes (replaces) the sign of inclusion "$\subset$". Recall that a map is simple if and only if it belongs to the class $\underline{\mathrm{RR}(f) = \mathrm{R}(f)}$ (see Theorem 4.5). This and Diagram 1 imply that the problem of classification of all simple maps is reduced to the problem of selection of all classes of maps in Diagram 1 that belong to the class $\underline{\mathrm{RR}(f) = \mathrm{R}(f)}$ followed by the identification of coinciding classes in the group thus selected. This problem is solved in Theorems 4.7–4.10 and corollaries to these theorems.

Theorem 4.7. *The class of maps* $\underline{\mathrm{RR}(f) = \mathrm{R}(f)}$ *contains the following classes of maps:*

(i) $\underline{\mathrm{R}(f) = \mathrm{C}(f)}$;

(ii) $\underline{\mathrm{AP}(f) = \mathrm{RR}(f)}$;

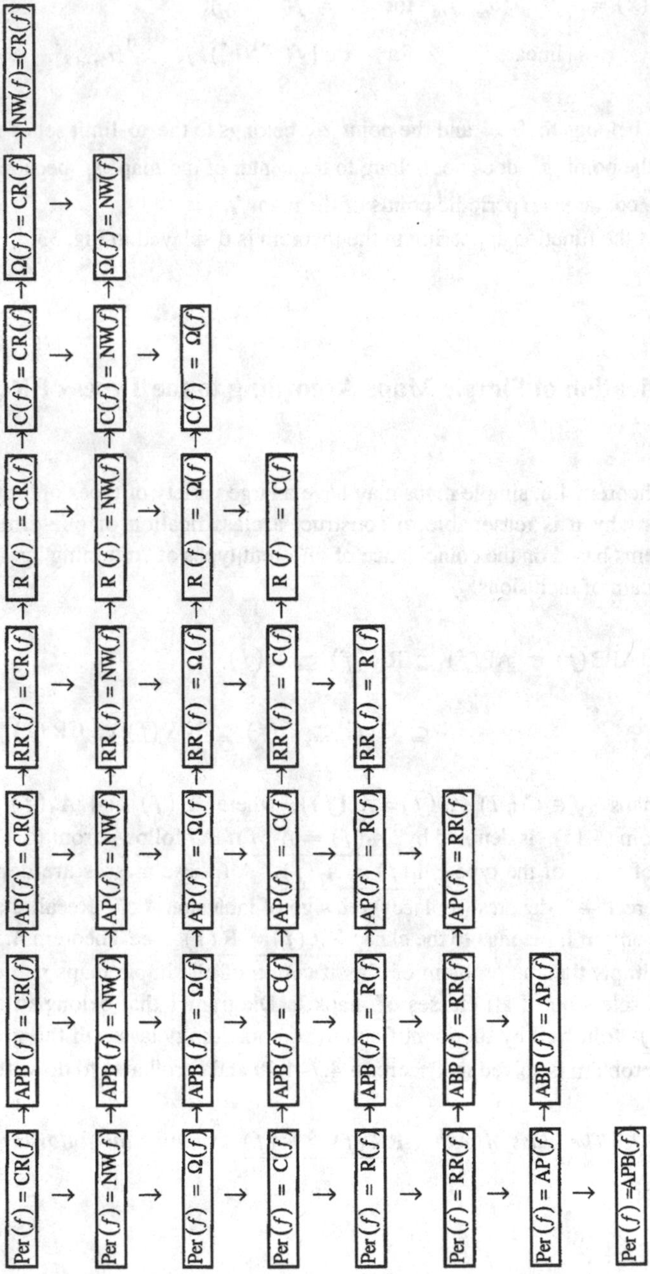
Per(f) = CR(f)
APB(f) = CR(f)
AP(f) = CR(f)
RR(f) = CR(f)
R(f) = CR(f)
C(f) = CR(f)
Ω(f) = CR(f)
NW(f) = CR(f)
Per(f) = NW(f)
APB(f) = NW(f)
AP(f) = NW(f)
RR(f) = NW(f)
R(f) = NW(f)
C(f) = NW(f)
Ω(f) = NW(f)
Per(f) = Ω(f)
APB(f) = Ω(f)
AP(f) = Ω(f)
RR(f) = Ω(f)
R(f) = Ω(f)
C(f) = Ω(f)
Per(f) = C(f)
APB(f) = C(f)
AP(f) = C(f)
RR(f) = C(f)
R(f) = C(f)
Per(f) = R(f)
APB(f) = R(f)
AP(f) = R(f)
RR(f) = R(f)
Per(f) = RR(f)
ABP(f) = RR(f)
AP(f) = RR(f)
Per(f) = AP(f)
ABP(f) = AP(f)
Per(f) =APB(f)

Diag. 1

(iii) APB(f) = AP(f).

Proof. Instead of proving the theorem itself, we shall prove the equivalent assertion, i.e., the fact that $C(f) \supset R(f)$, $RR(f) \supset AP(f)$, and $AP(f) \supset APB(f)$ whenever $RR(f) \neq R(f)$.

Since $RR(f) \neq R(f)$, Theorems 4.5 and 4.4 imply the existence of $l \geq 0$ and closed intervals I_1 and I_2 such that $\operatorname{int} I_1 \cap \operatorname{int} I_2 = \varnothing$, $f^{2^l}(I_1) \supseteq I_1 \cup I_2$, and $f^{2^l}(I_2) \supseteq I_1 \cup I_2$. By using standard methods of symbolic dynamics, we can now establish the existence of a periodic point whose preimage is not a periodic point but belongs to the closure of the set of periodic points. Therefore, there exists a point $x \in I$ such that $x \notin \omega(x)$ and $x \in \overline{\operatorname{Per}(f)}$. By virtue of the fact that $C(f) = \overline{\operatorname{Per}(f)}$ (Theorem 1.4), we conclude that $x \in C(f) \backslash R(f)$.

Actually, one can complete the proof of Theorem 4.7 by the methods of symbolic dynamics but, for diversity, we present another version of the proof.

According to Theorems 4.5 and 4.4, the fact that $RR(f) \neq R(f)$ is equivalent to the existence of a periodic trajectory which is not a simple cyclic permutation. Denote this trajectory, its type, and period by B, π, and n, respectively. Since π is not a simple cyclic permutation, there exists k, $1 \leq k < \log_2 n$, such that one of the orbits of the permutation π^k possesses Property (i) of Lemma 4.4.

Assume that $k = 1$ (for $k > 1$, the proof is similar). Let $\mathcal{A}$ denote the set of cyclic permutations with Property (i) of Lemma 4.4.

If $\pi \in \mathcal{A}$, then the map f possesses a periodic trajectory of the type

$$\pi_1 = \begin{pmatrix} 1 & 2 & 3 & 4 & 5 & 6 \\ 4 & 6 & 5 & 3 & 2 & 1 \end{pmatrix}.$$

The points of this trajectory are denoted by α_i, $i = 1, \ldots, 6$, according to the natural ordering on $\mathbb{R}$. Denote $\alpha_1' = \max\{x \in [\alpha_1, \alpha_2] \mid f(x) = \alpha_6\}$. It follows from the form of the cyclic permutation π_1 that $f([\alpha_1, \alpha_3]) \supset [\alpha_4, \alpha_6]$ and $f([\alpha_4, \alpha_6]) \supset [\alpha_1, \alpha_3]$, where $[\alpha_1', \alpha_2] \supset [\alpha_1, \alpha_3]$. This means that the interval $[\alpha_1, \alpha_6]$ contains a closed invariant set which admits splitting (e.g., a periodic trajectory of period 2).

Consider the map $g = f^2$. A periodic trajectory of the type π_1 admits splitting into two periodic trajectories of period 3. The type of 3-periodic trajectories belongs to the set $\mathcal{A}$. Therefore, in each of the intervals $[\alpha_1, \alpha_3]$ and $[\alpha_4, \alpha_6]$, the map g possesses a periodic trajectory of the type π_1. As above, this means that the segment $[\alpha_1, \alpha_3]$ contains two intervals I_1 and I_2 whose ends are points of a periodic trajectory of type π_1 of the map g. Moreover, these intervals are such that $g(I_1) \supset I_2$, $g(I_2) \supset I_1$, and $I_1 \supset [\alpha_1', \alpha_2]$. Similarly, the segment $[\alpha_4, \alpha_6]$ contains two closed disjoint intervals I_3 and I_4 such that $g(I_3) \supset I_4$ and $g(I_4) \supset I_3$. Consequently, the interval $[\alpha_1, \alpha_6]$ contains a closed invariant set which can be split 2 times.

Consider the map $h = f^{2^2}$. Each periodic trajectory of type π_1 of the map g can be split into two periodic trajectories (of period 3) of the map h, etc. This reasoning can be repeated infinitely many times. Hence, the interval $[\alpha_1, \alpha_6]$ contains an infinite simple invariant set, which is denoted by M.

Let $\{ M_i^{(n)},\ i = 1, \dots, 2^n\}$ be a simple decomposition of the set M. It follows from the construction that this decomposition includes an element $\tilde{M}^{(n)}$ such that the interval whose ends coincide with the minimal and maximal points of the set $\tilde{M}^{(n)}$ contains the interval $[\alpha_1', \alpha_2]$. Hence, the set $\bigcap_{n \geq 0} \tilde{M}^{(n)}$ consists of two points. This enables us to conclude that

(i) $f|_M$ is Lyapunov unstable (because for any element $M_i^{(n)}$, one can find $k < 2^n$ such that $f^k(M_i^{(n)}) = \tilde{M}^{(n)}$);

(ii) there exists $k > 0$ such that $f^k(\bigcap_{n \geq 0} \tilde{M}^{(n)})$ is a point (because M is a minimal set).

It follows from (i) (see Sibirsky [1]) that M contains no almost periodic in the sense of Bohr points. Moreover, (ii) implies that $f|_M$ is not a homeomorphism. Hence, f possesses an almost periodic point which is not almost periodic in the sense of Bohr. Furthermore, since $f|_{AP(f)}$ is a homeomorphism and M is a minimal set, the map f possesses a regularly recurrent point which is not almost periodic.

Corollary 4.1. *Let* $A(f)$ *be an arbitrary set from (4.15) other than* RR(f) *or* R(f). *Then the following classes of maps coincide:*

(i) $\underline{\mathrm{RR}(f) = A(f)}$;

(ii) $\underline{\mathrm{R}(f) = A(f)}$.

Proof. We split the proof into two parts. First, we consider the case where $A(f)$ is equal to Per(f), APB(f), or AP(f) and then the case where $A(f)$ is equal to C(f), $\Omega(f)$, NW(f), or CR(f).

Let $A(f)$ be either Per(f), or APB(f), or AP(f). As follows from (4.15), the class of maps $\underline{\mathrm{R}(f) = A(f)}$ is contained in the class $\underline{\mathrm{RR}(f) = A(f)}$ (see Diagram 1). Furthermore, it follows from (4.15) that the class of maps $\underline{\mathrm{RR}(f) = A(f)}$ is contained in the class $\underline{\mathrm{RR}(f) = \mathrm{AP}(f)}$. This and Theorem 4.7 together imply that the class of maps $\underline{\mathrm{RR}(f) = A(f)}$ is contained in the class $\underline{\mathrm{RR}(f) = \mathrm{R}(f)}$. Hence, the class of maps $\underline{\mathrm{RR}(f) = A(f)}$ belongs to the class $\underline{\mathrm{R}(f) = A(f)}$. This means that the first part of the proof of Corollary 4.1 is completed.

Now suppose that $A(f)$ is either C(f), or $\Omega(f)$, or NW(f), or CR(f). According to Diagram 1, the class $\underline{\mathrm{R}(f) = A(f)}$ contains the class $\underline{\mathrm{RR}(f) = A(f)}$. Moreover, the class $\underline{R(f) = A(f)}$ is contained in the class $\underline{R(f) = C(f)}$. Therefore, by virtue of Theorem 4.7, the class of maps $\underline{R(f) = A(f)}$ is contained in the class $\underline{RR(f) = R(f)}$ and, hence, in the class $\underline{RR(f) = A(f)}$. The proof of Corollary 4.1 is completed.

Theorem 4.8. *The following classes of maps coincide:*

(i) $\underline{\mathrm{Per}(f) = \mathrm{CR}(f)}$;

(ii) $\underline{\mathrm{Per}(f) = \mathrm{NW}(f)}$;

(iii) $\underline{\mathrm{Per}(f) = \Omega(f)}$;

(iv) $\underline{\mathrm{Per}(f) = \mathrm{C}(f)}$;

(v) $\underline{\mathrm{Per}(f) = \mathrm{R}(f)}$;

(vi) $\underline{\mathrm{Per}(f) = \mathrm{RR}(f)}$;

(vii) $\underline{\mathrm{Per}(f) = \mathrm{AP}(f)}$.

Proof. According to Diagram 1, to prove Theorem 4.8, it suffices to show that the class of maps $\underline{\mathrm{Per}(f) = \mathrm{AP}(f)}$ is contained in the class $\underline{\mathrm{Per}(f) = \mathrm{CR}(f)}$. To do this, we now establish the fact that the map f such that $\mathrm{Per}(f) = \mathrm{AP}(f)$ possesses the property $\mathrm{Per}(f) = \mathrm{CR}(f)$.

Suppose that the map f is such that $\mathrm{Per}(f) = \mathrm{AP}(f)$. Then it follows from Diagram 1 and Theorem 4.7 that $f \in \mathfrak{F}_{2^\infty}$. By virtue of Theorem 4.4, each chain recurrent point of the map f belongs to a simple invariant set. It follows from Properties 4 and 7 of simple invariant sets that each infinite simple invariant set of the map f contains an almost periodic point which is not periodic.

Since $\mathrm{Per}(f) = \mathrm{AP}(f)$, we conclude that the map f has no infinite simple invariant sets. Therefore, each simple invariant set of the map f is a periodic trajectory. This yields $\mathrm{Per}(f) = \mathrm{CR}(f)$.

Theorem 4.9. *The following classes of maps coincide:*

(i) $\underline{\mathrm{APB}(f) = \mathrm{AP}(f)}$;

(ii) $\underline{\mathrm{AP}(f) = \mathrm{RR}(f)}$.

Proof. Theorems 4.7, 4.5, and 4.4 imply that any map f from one of the classes in the formulation of Theorem 4.9 belongs to $\mathfrak{F}_{2^\infty}$. Let M be a minimal infinite set of the map f (if M is a finite set, then the proof is evident). Since $f \in \mathfrak{F}_{2^\infty}$, the set M is a simple infinite invariant set of the map f.

Let $\{M^{(n)}, i = 1, \ldots, 2^n\}$ be a simple decomposition of the set M of rank n. Assume that f is such that $\mathrm{APB}(f) = \mathrm{AP}(f)$. Since M contains an almost periodic point (Property 7), it also contains a almost periodic in the sense of Bohr point. Taking into account the fact that the closure of the trajectory of any almost periodic in the sense of Bohr point is Lyapunov stable (see Sibirsky [1]), we find that the map $f|_M$ is Lyapunov stable. Hence,

$$\max_{i \in |1, 2^n|} \operatorname{diam} M_i^{(n)} \to 0 \quad \text{as} \quad n \to \infty.$$

Therefore, each point of any minimal set is almost periodic, i.e., $\mathrm{AP}(f) = \mathrm{RR}(f)$. Thus, the class of maps $\underline{\mathrm{APB}(f) = \mathrm{AP}(f)}$ contains the class $\underline{\mathrm{AP}(f) = \mathrm{RR}(f)}$.

Assume that f is such that $\mathrm{AP}(f) = \mathrm{RR}(f)$. Let x be an arbitrary point from M and let $M^{(n)}(x)$ be an element of the simple decomposition of the set M of rank n which contains the point x. Suppose that the set $\bigcap_{n>0} M^{(n)}(x)$ is not a point. Denote by x_1 and x_2 the minimal and the maximal points of this set, respectively. Since M is a minimal set, and $\mathrm{AP}(f) = \mathrm{RR}(f)$, the points x_1 and x_2 are almost periodic. We choose $\varepsilon < |x_2 - x_1|/3$. Since $x_1, x_2 \in \mathrm{AP}(f)$, one can indicate N_1 and N_2 such that $|f^{N_k i}(x_k) - x_k| < \varepsilon$ for any $i > 0$ and $k = 1, 2$. Consequently, $f^{N_1 N_2}(x_1) < x_1 < x_2 < f^{N_1 N_2}(x_2)$. Hence, the interval $[x_1, x_2]$ contains a periodic point but this is impossible by Lemma 4.8 and Property 4 of simple sets. Therefore, for any point $x \in M$, the set $\bigcap_{n>0} M^{(n)}(x)$ consists of a single point and this means that each almost periodic point is almost periodic in the sense of Bohr, i.e., $\mathrm{AP}(f) = \mathrm{APB}(f)$.

Corollary 4.2. *Let* $A(f)$ *be an arbitrary set from (4.15) but not* Per(f), APB(f), *or* AP(f). *Then the following classes of maps coincide:*

(i) $\underline{\mathrm{APB}(f) = A(f)}$;

(ii) $\underline{\mathrm{AP}(f) = A(f)}$.

Proof. According to Diagram 1, the class of maps $\underline{\mathrm{APB}(f) = A(f)}$ is contained in the class $\underline{\mathrm{AP}(f) = A(f)}$. Moreover, the class of maps $\underline{\mathrm{AP}(f) = A(f)}$ is contained in the class $\underline{\mathrm{APB}(f) = \mathrm{RR}(f)}$ which, in turn, coincides with the class $\underline{\mathrm{APB}(f) = \mathrm{AP}(f)}$ (by virtue of Theorem 4.9). Hence, the class $\underline{\mathrm{AP}(f) = A(f)}$ is contained in the class $\underline{\mathrm{APB}(f) = A(f)}$. Therefore, the classes $\underline{\mathrm{APB}(f) = A(f)}$ and $\underline{\mathrm{AP}(f) = A(f)}$ coincide.

Theorem 4.10. *The following classes of maps coincide:*

(i) $\underline{\mathrm{AP}(f) = \mathrm{CR}(f)}$;

(ii) $\mathrm{RR}(f) = \mathrm{CR}(f)$.

Proof. To prove the theorem, it suffices to show that the class of maps $\mathrm{RR}(f) = \mathrm{CR}(f)$ is contained in the class $\mathrm{AP}(f) = \mathrm{CR}(f)$.

Actually, let f be such that $\mathrm{RR}(f) = \mathrm{CR}(f)$. Then $f \in \mathfrak{F}_{2^\infty}$. Let $x \in \mathrm{CR}(f)$. For $x \in \mathrm{Per}(f)$, the proof is obvious. Assume that $x \in \mathrm{CR}(f) \setminus \mathrm{Per}(f)$.

Let M be a simple invariant set (maximal by inclusion) that contains the point x, let $\{M_i^{(n)},\ i = 1, \dots, 2^n\}$ be a simple decomposition of this set of rank n, and let $M^{(n)}(x)$ be the element of this decomposition that contains the point x. Consider the set $\bigcap_{n>0} M^{(n)}(x)$. For $f \in \mathfrak{F}_{2^\infty}$, this set may be a segment. Every interior point of the set $\bigcap_{n>0} M^{(n)}(x)$ is chain recurrent; it is also a wandering point. Therefore, for any point $x \in M$, the set $\bigcap_{n>0} M^{(n)}(x)$ is a point provided that $\mathrm{RR}(f) = \mathrm{CR}(f)$. This means that each point of a simple set of the map f is almost periodic.

The results of this section are displayed in Diagram 2. This diagram includes all classes presented in Diagram 1 that belong to the class of maps $\mathrm{RR}(f) = \mathrm{R}(f)$. All classes united in a single block coincide.

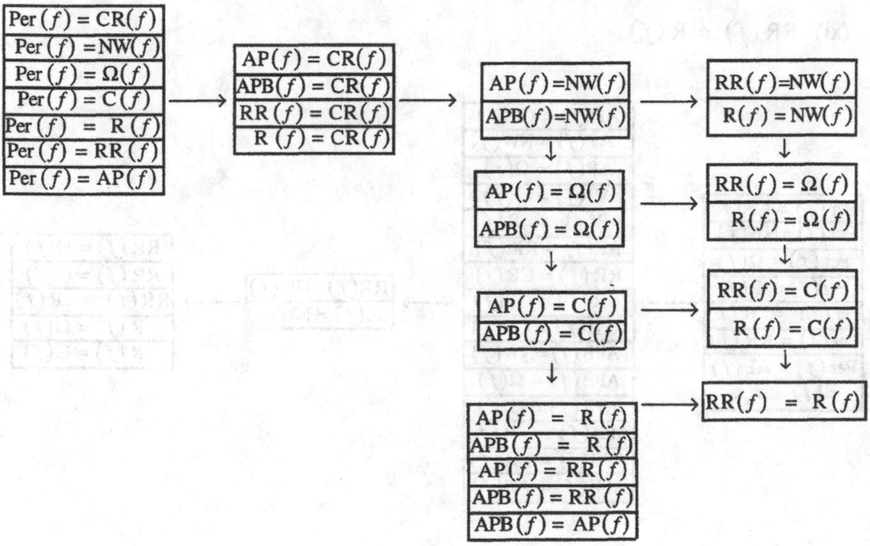

Diagram 2

The examples constructed to prove Theorem 4.6 demonstrate that the classes of maps that are not united in a single block in Diagram 2 do not coincide. Therefore, the space $f \in C^0(I, I)$ admits a decomposition into the following classes of maps:

(i) $\mathrm{Per}(f) = \mathrm{AP}(f)$;

(ii) $\mathrm{Per}(f) \neq \mathrm{AP}(f) = \mathrm{CR}(f)$;

(iii) $\mathrm{AP}(f) = \mathrm{NW}(f) \neq \mathrm{CR}(f)$;

(iv) $\mathrm{AP}(f) = \Omega(f) \neq \mathrm{NW}(f)$;

(v) $\mathrm{AP}(f) = \mathrm{C}(f) \neq \Omega(f)$;

(vi) $\mathrm{AP}(f) = \mathrm{R}(f) \neq \mathrm{C}(f)$;

(vii) $\mathrm{AP}(f) \neq \mathrm{RR}(f) \neq \mathrm{NW}(f)$;

(viii) $\mathrm{AP}(f) \neq \mathrm{RR}(f) \neq \Omega(f) \neq \mathrm{NW}(f)$;

(ix) $\mathrm{AP}(f) \neq \mathrm{RR}(f) \neq \mathrm{C}(f) \neq \Omega(f)$;

(x) $\mathrm{AP}(f) \neq \mathrm{RR}(f) = \mathrm{R}(f) \neq \mathrm{C}(f)$;

(xi) $\mathrm{RR}(f) \neq \mathrm{R}(f)$.

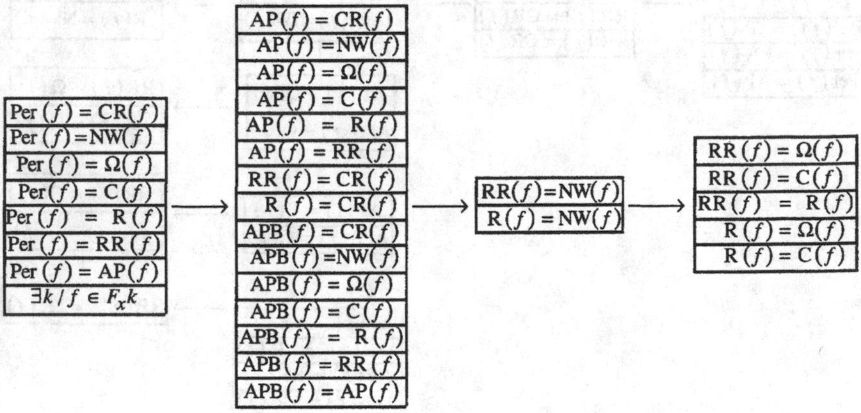

Diagram 3

Note that for smooth or piecewise monotone maps, the classification displayed in Diagram 2 takes the form presented in Diagram 3.

6. Properties of Individual Classes

The classification described above is more or less complete from the following point of view: Topological dynamics deals not only with the property of return but also with many other important concepts, namely, with topological entropy, Lyapunov stability, homoclinic trajectories, etc. Many of these concepts can be regarded as criteria that enable one to attribute a given map to a certain class in Diagram 2. Therefore, there is no need in more detailed classifications of simple maps. Theorems 4.11–4.19 presented below clarify this observation. For simplicity, in Diagram 4, we depict a single representative of each class of equivalence from Diagram 2 and present the numbers of the corresponding theorems.

Th.11 $\boxed{\mathrm{Per}\,(f) = \mathrm{CR}(f)}$ $\longrightarrow$ Th.12 $\boxed{\mathrm{AP}(f) = \mathrm{CR}(f)}$

$\qquad\qquad\qquad\qquad\qquad\qquad\qquad\qquad\qquad\downarrow$

Th.13 $\boxed{\mathrm{AP}(f) = \mathrm{NW}(f)}$ $\longrightarrow$ $\boxed{\mathrm{RR}(f) = \mathrm{NW}(f)}$

$\qquad\qquad\qquad\quad\downarrow\qquad\qquad\qquad\qquad\downarrow$

Th.14 $\boxed{\mathrm{AP}(f) = \Omega(f)}$ $\longrightarrow$ $\boxed{\mathrm{RR}\,(f) = \Omega(f)}$ Th.17

$\qquad\qquad\qquad\quad\downarrow\qquad\qquad\qquad\qquad\downarrow$

Th.15 $\boxed{\mathrm{AP}(f) = \mathrm{C}(f)}$ $\longrightarrow$ $\boxed{\mathrm{RR}\,(f) = \mathrm{C}(f)}$ Th.18

$\qquad\qquad\qquad\quad\downarrow\qquad\qquad\qquad\qquad\downarrow$

Th.16 $\boxed{\mathrm{AP}(f) = \mathrm{R}\,(f)}$ $\longrightarrow$ $\boxed{\mathrm{RR}\,(f) = \mathrm{R}\,(f)}$ Th.19

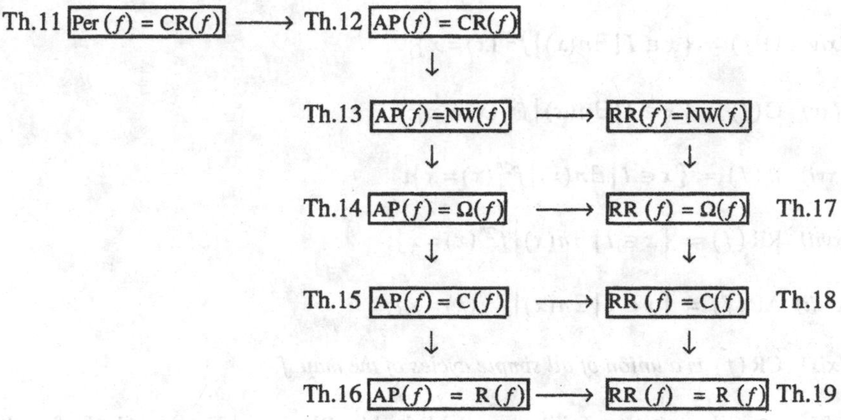

Diagram 4

Theorem 4.11. *Let $f \in C^0(I, I)$. Then the following statements are equivalent:*

(i) $\mathrm{Per}\,(f) = \mathrm{CR}\,(f)$;

(ii) $\mathrm{Per}\,(f) = \mathrm{NW}(f)$;

(iii) $\mathrm{Per}\,(f) = \Omega(f)$;

(iv) $\mathrm{Per}\,(f) = \mathrm{C}(f)$;

(v) $\mathrm{Per}\,(f) = \mathrm{R}\,(f)$;

(vi) $\mathrm{Per}\,(f) = \mathrm{RR}\,(f)$;

(vii) $\mathrm{Per}(f) = \mathrm{AP}(f)$;

(viii) $\mathrm{Per}(f) = \overline{\mathrm{Per}(f)}$;

(ix) *for any* $x \in I$, $\omega_f(x)$ *is a cycle;*

(x) *any invariant ergodic measure is concentrated on a cycle;*

(xi) *for any* $x \in I$, $\omega_f(x)$ *is a simple cycle;*

(xii) $\mathrm{CR}(f) = \{x \in I \mid \exists n(x) \mid f^{2^n}(x) = x\}$;

(xiii) $\mathrm{NW}(f) = \{x \in I \mid \exists n(x) \mid f^{2^n}(x) = x\}$;

(xiv) $\Omega(f) = \{x \in I \mid \exists n(x) \mid f^{2^n}(x) = x\}$;

(xv) $\mathrm{C}(f) = \{x \in I \mid \exists n(x) \mid f^{2^n}(x) = x\}$;

(xvi) $\mathrm{R}(f) = \{x \in I \mid \exists n(x) \mid f^{2^n}(x) = x\}$;

(xvii) $\mathrm{RR}(f) = \{x \in I \mid \exists n(x) \mid f^{2^n}(x) = x\}$;

(xviii) $\mathrm{AP}(f) = \{x \in I \mid \exists n(x) \mid f^{2^n}(x) = x\}$;

(xix) $\mathrm{CR}(f)$ *is a union of all simple cycles of the map* f.

The equivalence (i) ⇔ (viii) was established by Block and Franke [1], the fact that (ii) ⇔ (viii) was proved by Blokh [1] and Fedorenko and Sharkovsky [2], and the facts that (viii) ⇔ (ix) and (viii) ⇔ (iv) were established by Sharkovsky [3] and Blokh [1], respectively.

Let $A \subset I$ be such that $f(A) \subseteq A$. We say that $f|_A$ is *Lyapunov stable* if, for any $\varepsilon > 0$, there exists δ such that the inequality $|f^n(x) - f^n(y)| < \varepsilon$ holds for any $|x - y| < \delta$, $x, y \in A$, and all $n > 0$.

Theorem 4.12. *Let* $f \in C^0(I, I)$. *Then the following statements are equivalent:*

(i) $\mathrm{APB}(f) = \mathrm{CR}(f)$;

(ii) $\mathrm{AP}(f) = \mathrm{CR}(f)$;

(iii) $\mathrm{RR}(f) = \mathrm{CR}(f)$;

(iv) $\mathrm{R}(f) = \mathrm{CR}(f)$;

(v) $f|_{\mathrm{CR}(f)}$ *is Lyapunov stable;*

(vi) $\mathrm{CR}(f) = \{x \in I \mid \lim_{n\to\infty} f^{2^n}(x) = x\}$;

Theorem 4.13. *Let* $f \in C^0(I, I)$. *Then the following assertions are equivalent:*

(i) $\mathrm{APB}(f) = \mathrm{NW}(f)$;

(ii) $\mathrm{AP}(f) = \mathrm{NW}(f)$;

(iii) $f|_{\mathrm{CR}(f)}$ *is Lyapunov stable and* $f(\mathrm{NW}(f)) = \mathrm{NW}(f)$;

(iv) $\mathrm{NW}(f) = \{x \in I \mid \lim_{n\to\infty} f^{2^n}(x) = x\}$;

We say that $f|_A$ is chaotic if

$$\limsup_{n\to\infty} |f^n(x) - f^n(y)| > 0$$

and

$$\liminf_{n\to\infty} |f^n(x) - f^n(y)| = 0$$

for some $x, y \in A$.

For maps of an interval, the definition of chaotic maps is equivalent (see Jankova and Smital [1], Kuchta and Smital [1], and Smital [1]) to the definition of "Li–Yorke chaotic" maps (see Li and Yorke [1]).

We say that the trajectory of a point $x \in I$ is *approximated by periodic trajectories* if, for any $\varepsilon > 0$, there exists $p \in \mathrm{Per}(f)$ and $n > 0$ such that $|f^i(x) - f^i(p)| < \varepsilon$ for all $i > n$.

An interval $J \subseteq I$ is called *periodic* if there exists $m > 0$ such that $f^m(J) \subseteq J$ and $\mathrm{int}(J) \cap \mathrm{int} f^i(J) = \varnothing$ for $i = 1, \ldots, m-1$.

We say that a closed invariant indecomposable set $A \subset I$ *admits cyclic decomposition* if it can be represented as a union of closed mutually disjoint sets A_i, $i = 1, \ldots, n$, such that $f^n(A_i) = A_i$.

We say that a decomposition $\mathcal{A}$ of the set A *improves* a decomposition $\mathcal{B}$ of the same set (and write $\mathcal{A} \succ \mathcal{B}$) if each element of the decomposition $\mathcal{A}$ is contained in a single element of the decomposition $\mathcal{B}$.

We say that a closed invariant indecomposable set A possesses an exhausting sequence of cyclic decompositions if there exists a sequence of cyclic decompositions $\{\mathcal{A}_n\}$ of the set A such that $\mathcal{A}_{n+1} \succ \mathcal{A}_n$ for all n and the maximum diameter of the elements of $\mathcal{A}_n$ tends to zero as $n \to \infty$.

Theorem 4.14. *Let* $f \in C^0(I, I)$. *Then the following assertions are equivalent:*

(i) $\mathrm{APB}(f) = \Omega(f)$;

(ii) $\mathrm{AP}(f) = \Omega(f)$;

(iii) f *is not chaotic;*

(iv) $f|_{\Omega(f)}$ *is Lyapunov stable;*

(v) $f|_{\mathrm{NW}(f)}$ *is Lyapunov stable;*

(vi) $\Omega(f) = \{x \in I \mid \lim_{n\to\infty} f^{2^n}(x) = x\}$;

(vii) *any trajectory can be approximated by periodic trajectories;*

(viii) *for any two distinct points of an infinite* ω*-limit set, one can find two disjoint periodic intervals each of which contains one of these points;*

(ix) *every* ω*-limit set which is not a cycle possesses an exhausting sequence of cyclic decompositions.*

The relations (iii) ⇔ (vii) ⇔ (viii) were established by Fedorenko and Sharkovsky [2] and Smital [1]. The facts that (iii) ⇔ (iv) and (iv) ⇔ (v) were proved by Fedorenko Sharkovsky, and Smital [1] and Fedorenko and Smital [1], respectively.

Theorem 4.15. *Let* $f \in C^0(I, I)$. *Then the following statements are equivalent:*

(i) $\mathrm{APB}(f) = \mathrm{C}(f)$;

(ii) $\mathrm{AP}(f) = \mathrm{C}(f)$;

(iii) $\mathrm{APB}(f) = \overline{\mathrm{APB}(f)}$

(iv) $\mathrm{AP}(f) = \overline{\mathrm{AP}(f)}$;

(v) $\mathrm{C}(f) = \{x \in I \mid \lim_{n\to\infty} f^{2^n}(x) = x\}$;

(vi) $f|_{\mathrm{C}(f)}$ *is Lyapunov stable;*

(vi) $f|_{\mathrm{R}(f)}$ *is Lyapunov stable;*

(viii) $f|_{\mathrm{RR}(f)}$ *is Lyapunov stable;*

(ix) $f|_{\mathrm{AP}(f)}$ *is Lyapunov stable;*

(x) $f|_{\mathrm{APB}(f)}$ *is Lyapunov stable;*

(xi) $f|_{\mathrm{Per}(f)}$ *is Lyapunov stable.*

A closed invariant set is called *minimal* if it does not contain any proper closed invariant subset.

Theorem 4.16. *Let* $f \in C^0(I, I)$. *Then the following assertions are equivalent:*

(i) $\mathrm{APB}(f) = \mathrm{R}(f)$;

(ii) $\mathrm{AP}(f) = \mathrm{R}(f)$;

(iii) $\mathrm{APB}(f) = \mathrm{RR}(f)$;

(iv) $\mathrm{AP}(f) = \mathrm{RR}(f)$;

(v) $\mathrm{APB}(f) = \mathrm{AP}(f)$;

(vi) $\mathrm{R}(f) = \{x \in I \mid \lim_{n\to\infty} f^{2^n}(x) = x\}$;

(vii) $\mathrm{RR}(f) = \{x \in I \mid \lim_{n\to\infty} f^{2^n}(x) = x\}$;

(viii) f *is Lyapunov stable on every minimal set;*

(ix) *each trajectory of an arbitrary minimal set can be approximated by periodic trajectories;*

(x) *for any two distinct point of an infinite minimal set, one can indicate two disjoint periodic intervals each of which contains one of these points;*

(xi) *any minimal set which is not a cycle possesses an exhausting sequence of cyclic decompositions.*

The equivalence (i) ⇔ (viii) is an analog of the Markov theorem on the relationship between the type of return on a minimal set and Lyapunov stability on this set (see Sibirsky [1]); the other equivalences are established in (Fedorenko [4]).

Theorem 4.17. *Let $f \in C^0(I, I)$. Then the following assertions are equivalent:*

(i) $\mathrm{RR}(f) = \Omega(f)$;

(ii) $\mathrm{R}(f) = \Omega(f)$;

(iii) for any $x \in I$, $\omega_f(x)$ *is a minimal set;*

(iv) $\omega_f(x) = \sigma_f(x)$ *for any* $x \in I$;

(v) the map $x \mapsto \omega_f(x)$ *regarded as a function* $I \to 2^I$ *(with the Hausdorff metric) is not of the first Baire class.*

The equivalence (i) ⇔ (iii) is an analog of the Birkhoff theorem on the type of return on minimal sets.

Theorem 4.18. *Let $f \in C^0(I, I)$. Then the following assertions are equivalent:*

(i) $\mathrm{RR}(f) = \mathrm{C}(f)$;

(ii) $\mathrm{R}(f) = \mathrm{C}(f)$;

(iii) $\mathrm{RR}(f) = \overline{\mathrm{RR}(f)}$;

(iv) $\mathrm{R}(f) = \overline{\mathrm{R}(f)}$.

Theorem 4.19. *Let $f \in C^0(I, I)$. Then the following assertions are equivalent:*

(i) the period of every cycle is a power of two;

(ii) each cycle is simple;

(iii) there are no homoclinic trajectories;

(iv) $h(f) = 0$;

(v) $h(f|_{\mathrm{CR}(f)}) = 0$;

(vi) $h(f|_{\mathrm{NW}(f)}) = 0$;

(vii) $h(f|_{\Omega(f)}) = 0$;

(viii) $h(f|_{C(f)}) = 0$;

(ix) $h(f|_{R(f)}) = 0$;

(x) $h(f|_{RR(f)}) = 0$;

(xi) $h(f|_{AP(f)}) = 0$;

(xii) $h(f|_{APB(f)}) = 0$;

(xiii) $h(f|_{Per(f)}) = 0$;

(xiv) $f|_{CR(f)}$ *is not chaotic;*

(xv) $f|_{NW(f)}$ *is not chaotic;*

(xvi) $f|_{\Omega(f)}$ *is not chaotic;*

(xvii) $f|_{C(f)}$ *is not chaotic;*

(xviii) $f|_{R(f)}$ *is not chaotic;*

(xix) $f|_{RR(f)}$ *is not chaotic;*

(xx) $RR(f) = R(f)$;

(xxi) $Per(f)$ *is a* G_δ*-set;*

(xxii) $R(f)$ *is a* F_σ*-set;*

(xxiii) $AP(f) = \{x \in I \mid \lim_{n\to\infty} f^{2^n}(x) = x\}$;

(xxiv) $APB(f) = \{x \in I \mid \lim_{n\to\infty} f^{2^n}(x) = x\}$;

(xxv) *every minimal set is simple;*

(xxvi) *there are no minimal sets with positive topological entropy;*

(xxvii) $CR(f)$ *is a union of all simple invariant sets of the map* f;

(xxviii) *every ω-limit set contains a simple minimal set;*

(xxviv) *every ω-limit set is simple;*

(xxx) *if* $\omega_f(x) = \omega_{f^2}(x)$, *then* $\omega_f(x)$ *is a fixed point;*

(xxxi) *every ω-limit set which is not a cycle does not contain any cycle;*

(xxxii) *there are no countable ω-limit sets;*

(xxxiii) *trajectories of any two points are correlated, namely, for the two-dimensional map* $g: \begin{cases} x \mapsto f(x) \\ y \mapsto f(x) \end{cases}$, *the inequality* $\omega_g((x, y)) \neq \omega_f(x) \times \omega_f(y)$ *holds for any two points* $x, y \in I$ *provided that* $\omega_f(x)$ *and* $\omega_f(y)$ *are not fixed points;*

(xxxiv) *for any* $x \in I$, $\delta_f(x)$ *is a minimal set;*

(xxxv) *any finite δ-limit set is a cycle;*

(xxxvi) *there are no countable δ-limit sets;*

(xxxvii) *every δ-limit set is a simple set;*

(xxxviii) *if* $\delta_f(x) = \delta_{f^2}(x)$, *then* $\delta_f(x)$ *is a fixed point;*

(xxxiv) *any δ-limit set which is not a cycle contains no cycles;*

(xl) *for any* $x, y \in I$, *the inequality* $\delta_g((x, y)) \neq \delta_f(x) \times \delta_f(y)$ *holds for the map* $g: \begin{cases} x \mapsto f(x) \\ y \mapsto f(x) \end{cases}$ *provided that* $\delta_f(x)$ *and* $\delta_f(y)$ *are not fixed points;*

(xli) *any trajectory can be approximated by trajectories of periodic intervals;*

(xlii) *for any closed intervals* I_1 *and* I_2 *such that* $\operatorname{int} I_1 \cap \operatorname{int} I_2 = \varnothing$ *and any* $m > 0$, *either* $f^m(I_1) \not\supset I_1 \cup I_2$ *or* $f^m(I_2) \not\supset I_1 \cup I_2$;

(xliii) *for any closed invariant set* F *and any* $m > 0$, *the map* $f^m|_F$ *cannot be topologically semiconjugate to a shift in the space of unilateral sequences of two symbols;*

(xliv) *there are no* $m \geq 0$ *with the following property: for any* $k > 0$, *one can*

find a set $F_k \subset \operatorname{Fix}(f^{mk})$ *such that* $f^m|_{F_k} \sim \pi_k$, *where*

$$\pi_k(i) = \begin{cases} 2i, & \text{if } i \text{ is even and } 1 \le i \le 2^{k-1}, \\ 2i-1, & \text{if } i \text{ is odd and } 1 \le i \le 2^{k-1}, \\ 2^{k+1} - 2i + 2, & \text{if } i \text{ is even and } 2^{k-1} \le i \le 2^k, \\ 2^{k+1} - 2i + 1, & \text{if } i \text{ is odd and } 2^{k-1} \le i \le 2^k; \end{cases}$$

(xlv) for any ω*-limit set* F, *the set* $\{x \in I \mid \omega_f(x) = F\}$ *is at most of the second class according to the Baire–de la Vallée Poussin classification.*

The equivalence (i) ⇔ (ii) follows from (Block [2]); (i) ⇔ (iii) follows from (Sharkovsky [13]); (i) ⇔ (iv) follows from (Misiurewicz [1]); (iv) ⇔ ... ⇔ (viii) is a general fact; (viii) ⇔ ... ⇔ (xix) follows from (Fedorenko, Sharkovsky, and Smital [1]); (i) ⇔ (xx) follows from (Xiong [1]); (i) ⇔ (xxi) ⇔ (xlv) ⇔ (xliii) follows from (Sharkovsky [3]); (i) ⇔ (xxxi) ⇔ (xxv) ⇔ (xxviii) follows from (Sharkovsky [7, 10–12]).

For the first time, different statements of this type were put together by Sharkovsky in [17]. Note that the major part of the equivalences given in Theorem 4.19 can be proved on the basis of Theorem 4.3. Different proofs of the equivalence of certain statements from Theorem 4.3 can also be found in (Alseda, Llibre, and Misiurewicz [1]) and (Block and Coppel [2]).

5. TOPOLOGICAL DYNAMICS OF UNIMODAL MAPS

1. Phase Diagrams of Unimodal Maps

Let $f: I \to I$ be a unimodal map (U-map). We say that a finite family $\mathcal{A} = \{J_0, J_1, \ldots, J_{n-1}\}$ of subintervals of the interval I forms a cycle of intervals of period n if the interiors of J_i are mutually disjoint and $f(J_i) \subset J_{(i+1) \bmod n}$ for all $i \in \{0, 1, \ldots, n-1\}$. Denote by $\mathcal{A}_n = \mathcal{A}_n(f)$ the set of cycles of intervals of period n of the map f which contain the critical point c. Suppose that, for some $n \geq 1$, the set $\mathcal{A}_n(f)$ is not empty (it is clear that $\mathcal{A}_1$ is not empty because $f(I) \subset I$). The set $\mathcal{A}_n$ contains an element maximal by inclusion. Indeed, let $A_n^{(\alpha)} = \{ J_0^{(\alpha)}, J_1^{(\alpha)}, \ldots, J_{n-1}^{(\alpha)} \}$ and $A_n^{(\beta)} = \{ J_0^{(\beta)}, J_1^{(\beta)}, \ldots, J_{n-1}^{(\beta)} \}$ be cycles of intervals from $\mathcal{A}_n$. We say that $A_n^{(\alpha)}$ is bounded from above by the cycle of intervals $A_n^{(\beta)}$ if $J_i^{(\alpha)} \subset J_i^{(\beta)}$ for all $i \in \{0, 1, \ldots, n-1\}$. If $\mathfrak{F} = \{ A_n^{(\alpha)}, \alpha \in \mathfrak{A} \}$ is a completely ordered (in the indicated sense) subfamily of the set $\mathcal{A}_n$, then the elements of $\mathfrak{F}$ are bounded from above by the cycle of intervals

$$A_n = \left\{ \overline{\bigcup_{\alpha \in \mathfrak{A}} J_0^{(\alpha)}}, \overline{\bigcup_{\alpha \in \mathfrak{A}} J_1^{(\alpha)}}, \ldots, \overline{\bigcup_{\alpha \in \mathfrak{A}} J_{n-1}^{(\alpha)}} \right\}.$$

Consequently, by the Zorn lemma, the partially ordered set $\mathcal{A}_n$ contains a maximal element $A_n^* = \{ J_{n,0}^*, J_{n,1}^*, \ldots, J_{n,n-1}^* \}$. We can assume that $c \in J_{n,0}^*$. Therefore, the cycle of intervals A_n^* is defined unambiguously. Clearly, $A_1^* = \{I\}$. For $n \geq 2$, it follows from the unimodality of the function f and maximality of the cycle of intervals A_n^* that

(a) for any $i \in \{0, 1, \ldots, n-1\}$, J_i^* is a closed interval;

(b) for any $i \in \{1, 2, \ldots, n-1\}$, the mapping of the interval J_i^* onto $J_{(i+1) \bmod n}^*$ is bijective;

(c) $f(\partial J_0^*) \subset \partial J_1^*$; therefore, if $J_0^* = [y, y']$, then $f(y) = f(y')$ and either $f^n(y) = y$ or $f^n(y') = y'$;

(d) if $m > n$ and $\mathcal{A}_m \neq \varnothing$, then $m = kn$ for some $k \geq 2$ and $A_m^* \subset A_n^*$ in the following sense:

$$\{x \in J \mid J \in A_m^*\} \subset \{x \in J \mid J \in A_n^*\}.$$

Let $\{p_m\}_{m=1}^{m^*}$ be an increasing sequence that consists of all positive integers such that $\mathcal{A}_{p_m}(f) \neq \varnothing$. In this case, $m^* \leq \infty$. Let $\Phi_m^* = \{x \in J \mid J \in A_{p_m}^*\}$. It is clear that $f(\Phi_m^*) \subset \Phi_m^*$ and the sequence of closed sets $\{\Phi_m^*\}_{m=1}^{m^*}$ forms a kind of filtration, which can be used to decompose the set of all trajectories of a given map into finitely or countably many natural classes and study some problems of the dynamics of one-dimensional maps in detail.

Dynamics of Maps $f|_{\Phi_m^*}$ for $m < m^*$. Consider a U-map $g = f^{p_m}|_{J_{p_m,0}^*}$. The point c is the critical point of g. For definiteness, we assume that c is its maximum point. Define the sets

$$R_m = \Phi_m^* \setminus \bigcup_{i \geq 0} f^{-i}\left(\bigcup_{0 \leq j < p_{m+1}} \operatorname{int} J_{p_{m+1},j}^* \right), \quad \mathcal{B}_m = \operatorname{int} R_m, \quad \text{and} \quad \mathcal{R}_m = R_m \setminus \mathcal{B}_m.$$

The sets R_m and $\mathcal{R}_m$ are closed and the set $\mathcal{B}_m$ is open (note that $\mathcal{B}_m$ can be empty but $R_m \neq \varnothing$ because fixed points of g must belong to R_m and, therefore, $\mathcal{R}_m \neq \varnothing$). Obviously, $f(R_m) \subset R_m$. It follows from the strict monotonicity of f on any interval which does not contain the point of extremum that $f(\mathcal{B}_m) \subset \mathcal{B}_m$ and the components of the set $\mathcal{B}_m$ are bijectively mapped onto each other by the map f. Hence, $f(\mathcal{R}_m) \subset \mathcal{R}_m$. It follows from the definition that $\mathcal{R}_m$ is a nowhere dense set.

Thus, the set Φ_m^* is decomposed into three subsets characterized by different types of dynamics, namely, trajectories that hit the interior of Φ_{m+1}^* after finitely many steps, trajectories that belong to the set $\mathcal{B}_m$, and trajectories that belong to the set $\mathcal{R}_m$.

We investigate the dynamics of the map f on $\mathcal{B}_m$. Components U and V of this set are called equivalent if there exist $i, j \geq 0$ such that $f^i(U) \cap f^j(V) \neq \varnothing$ (and, consequently, $f^i(U) = f^j(V)$). It is clear that the class of components equivalent to a component U is formed by the family of components of the set $\{\bigcup_{-\infty < i < +\infty} f^i(U)\}$ lying in Φ_m^*. The set of all classes of equivalent components is at most countable because $\mathcal{B}_m$ is open.

The components of $\mathcal{B}_m$ may exhibit the following two types of dynamics:

(i) the trajectory of a component eventually forms a cycle of intervals;

(ii) the trajectory of a component consists of infinitely many intervals, i.e., this component is a wandering interval: $f^i(U) \cap f^j(V) = \varnothing$ whenever $i \neq j$.

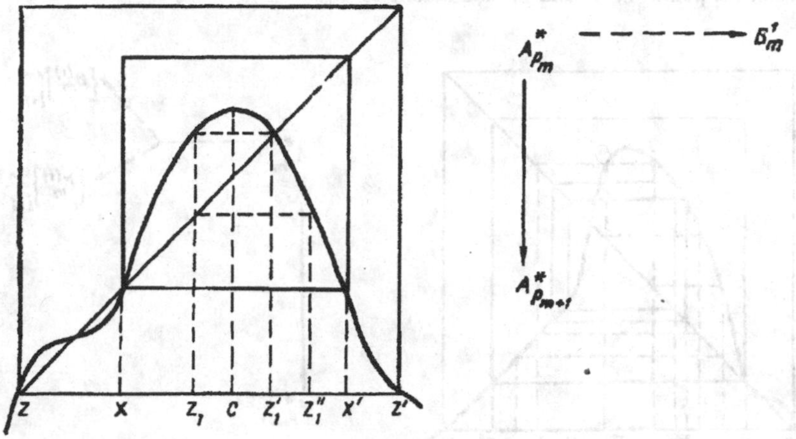

a b

Fig. 34

Let $\{B_m^{(i)}\}_{i=1}^{j_m}$ be the family of all different cycles of intervals formed by components of $\mathcal{B}_m$. The class of components equivalent to the components of a cycle of intervals $B_m^{(i)}$ is also denoted by $B_m^{(i)}$. Symbols $\Gamma_m^{(i)}$ denote classes of components formed by wandering intervals of $\mathcal{B}_m$; these classes form the set $\{\Gamma_m^{(i)}\}_{i=1}^{k_m}$. If there are no sets $B_m^{(i)}$ or $\Gamma_m^{(i)}$, then we assume that $j_m = 0$ or $k_m = 0$, respectively. In Fig. 35b, we present a formal illustration of the dynamics of the map on the set Φ_m^*. Arrows mean that there are trajectories of $A_{p_m}^*$ that hit intervals of the set $A_{p_{m+1}}^*$ and there are sets $B_m^{(i)}$, $1 \le i \le j_m$, and $\Gamma_m^{(i)}$, $1 \le i \le k_m$; dotted lines mean that the sets of types $B_m^{(i)}$ and $\Gamma_m^{(i)}$ may be absent for the map f.

Below, we consider two cases different from the dynamical point of view.

A. $p_{m+1}/p_m = 2$. Let $J_{p_m,0}^* = [z, z']$. We set $x = \sup\{y \in [z, c] : g(y) = y\}$ (Fig. 34). Since $m < m^*$, we have $g(c) > c$. Therefore, $z \le x < c$. For this point x, one can find a unique point $x' \in (c, z']$ such that $g(x') = x$. It follows from the inequality $m < m^*$ that $g([x, x']) \subset [x, x']$.

For $z < x$, we set $J_{m,0}^{(1)} = (z, x)$ and consider the cycle of intervals $B_m^{(1)} = \{ J_{m,0}^{(1)},$

$J^{(1)}_{m,1}, \ldots , J^{(1)}_{m,p_m-1}\}$, where $J^{(1)}_{m,i} = f^i(J^{(1)}_{m,0})$, $i = 1, 2, \ldots , p_m - 1$.

Since $p_{m+1}/p_m = 2$, we have $J^*_{p_{m+1},0} = [z_1, \; z_1']$, where $z_1 \in (x, x')$, $g(z_1') = z_1'$, $g(z_1) = z_1'$ (Fig. 34a), and $J^*_{p_{m+1}, p_m} = [z_1', \; z_1'']$, where $z_1'' \in (z_1', x')$ and $g(z_1'') = z_1$. In this case, the set R_m defined above consists of countably many points which are preimages of the points z, x, and z_1 and belong to R_m.

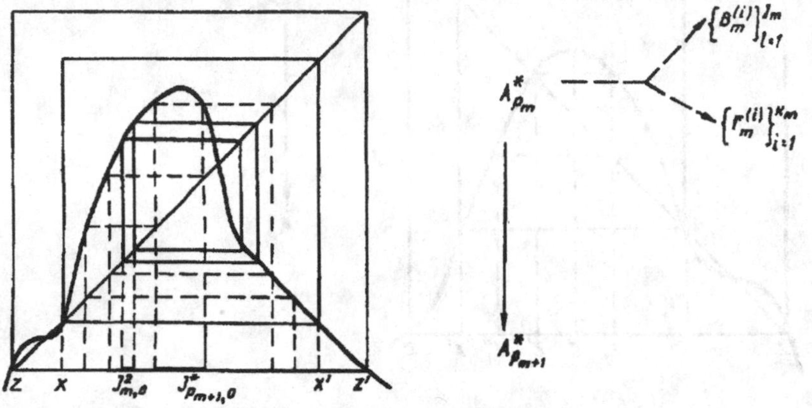

a b

Fig. 35

Thus, any trajectory of Φ^*_m hits either the interior of an interval from the set $A^*_{p_{m+1}}$ or an interval from $B^{(1)}_m$, or periodic points z, x, or z_1' after finitely many steps. Note that, for $p_{m+1}/p_m = 2$, the dynamics of the map $f|_{\Phi^*_m}$ is much simpler than in the case where $p_{m+1}/p_m > 2$, which is described below. In Fig. 34b, we display a formal diagram, which may be regarded as an illustration of these conclusions. It is much simpler than the diagram depicted in Fig. 35b.

B. $p_{m+1}/p_m > 2$. In Fig. 35, we present an example of the map g for $p_{m+1}/p_m = 3$. Unlike the first case, parallel with the cycles of intervals $A^*_{p_{m+1}}$ of period p_{m+1} and $B^{(1)}_m$ of period p_m, the map f possesses a cycle of intervals $B^{(2)}_m = \{ J^{(2)}_{m,0}, f(J^{(2)}_{m,0}), \ldots \}$ of period $2p_m$, which belongs to Φ^*_m but does not belong to Φ^*_{m+1}. For $p_{m+1}/p_m = 2$, this is impossible.

The dynamics of the map g in case B is schematically represented in Fig. 35b. As shown above, in this case, $j_m \le \infty$ and $k_m \le \infty$. In what follows, we construct examples of maps which illustrate some theoretical possibilities (according to the schematic diagram).

Dynamics of the Map $f|_{\Phi^*_{m^*}}$ for $m^* < \infty$. As in the case $m < m^*$, we consider the map $g = f^{p_{m^*}}|_{J^*_{p_{m^*},0}}$. We define an open set

$$\mathcal{B}_{m^*} = \{ x \in \Phi^*_{m^*} \colon \text{there exist a neighborhood } U \text{ of } x \\ \text{and an integer } N \geq 0 \text{ such that } c \notin f^i(U) \text{ for all } i > N \},$$

which may be empty, and the set $\mathcal{R}_{m^*} = \Phi^*_{m^*} \setminus \mathcal{B}_{m^*}$. As in the case $m < m^*$, the components of $\mathcal{B}_{m^*}$ can be decomposed into classes of equivalent components $\{B^{(i)}_{m^*}\}_{i=1}^{j_{m^*}}$ and $\{\Gamma^{(i)}_{m^*}\}_{i=1}^{k_{m^*}}$ (the first group of classes is formed by cycles of intervals, whereas the second group consists of wandering intervals). Moreover, $j_{m^*} \leq \infty$ and $k_{m^*} \leq \infty$. It should be noted that one of the components of the set $\mathcal{B}_{m^*}$ may contain the point c. Hence, for one of the classes of equivalent components, the image of a component is not necessarily a component (it may be a part of a component).

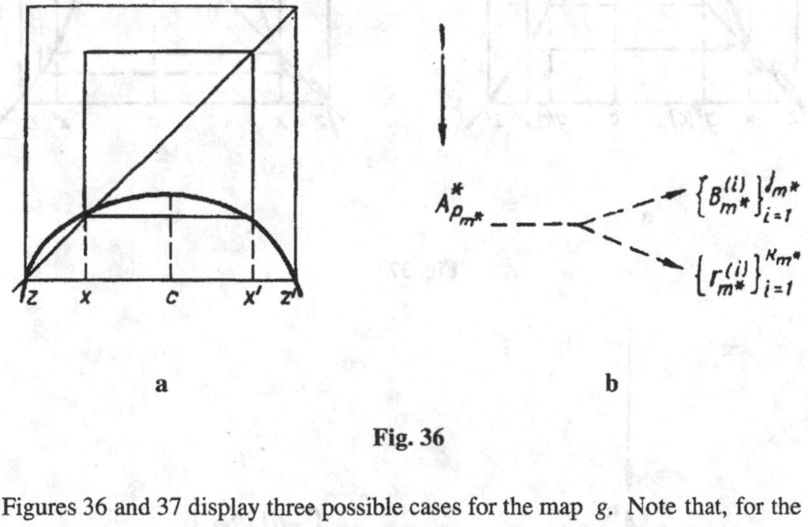

a b

Fig. 36

Figures 36 and 37 display three possible cases for the map g. Note that, for the map depicted in Fig 36a, $\mathcal{B}_{m^*}$ is the set of internal points of the intervals from $A_{p_{m^*}}$. Therefore, in this case, we have $j_{m^*} = 1$, $k_{m^*} = 0$, and $R_{m^*} = \bigcup_{0 \leq i \leq p_{m^*}} \partial J^*_{pm^*,i}$. In Fig. 36b, we present a schematic representation of the possible dynamics of the map f on the set $\Phi^*_{m^*}$ for $m^* < \infty$.

Case $m^* = \infty$. Consider the closed set $\Phi^*_\infty = \bigcap_{m \geq 1} \Phi^*_m$. This set is nonempty and $f(\Phi^*_\infty) \subset \Phi^*_\infty$ because $f(\Phi^*_m) \subset \Phi^*_m$ and $c \in \Phi^*_m$ for any $m < \infty$. Denote $\mathcal{B}_\infty = \operatorname{int} \Phi^*_\infty$. As in the case $m < \infty$, we define

$$\mathcal{B}_\infty = \{x \in \Phi_\infty^* \mid \text{there exist a neighborhood } U \text{ of } x$$
$$\text{and an integer } N \geq 0 \text{ such that } c \notin f^i(U) \text{ for } i > N\}.$$

The set $\mathcal{B}_\infty \cap \mathrm{Per}(f)$ is empty because the periods p_m of the cycles of intervals A_{p_m} approach infinity as m increases. Hence, the components of $\mathcal{B}_\infty$ are wandering intervals, i.e., the classes of equivalent components generate only the set $\{\Gamma_\infty^{(i)}\}_{i=1}^{k_\infty}$ and $j_\infty = 0$.

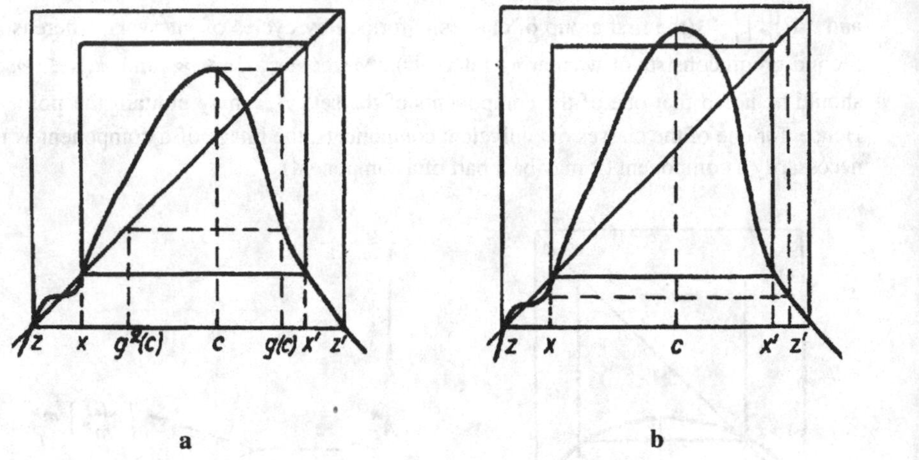

Fig. 37

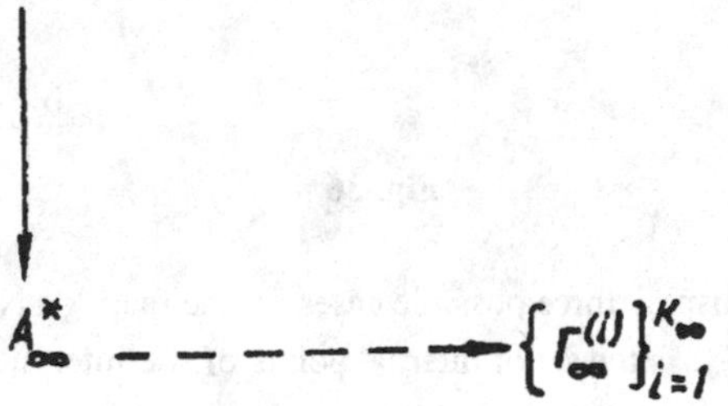

Fig. 38

By the definition of the set Φ_∞^*, its components are "almost periodic intervals", i.e., for any ε-neighborhood U_ε of a component S of the set Φ_∞^*, one can indicate an integer $m \geq 1$ such that $f^{p_m}(S) \subset U_\varepsilon$. This is why we write A_∞^* instead of Φ_∞^* in Fig. 38 illustrating the dynamics of the map f on Φ_∞^*.

Thus, the dynamics of an arbitrary unimodal map can be schematically represented in the form of a phase diagram, i.e., as an oriented graph whose vertices are cycles of intervals (maximal by inclusion) or classes of equivalence for wandering intervals. The general form of the phase diagram of a unimodal map is displayed in Fig. 39. In what follows, we demonstrate that the phase diagram contains essential (or almost all) information about the topological dynamics of a unimodal map.

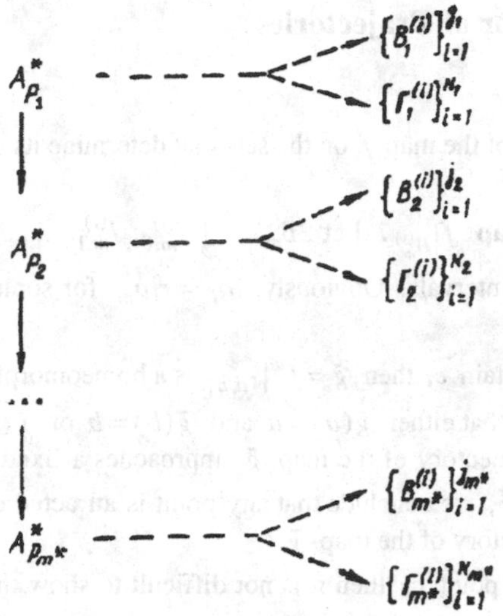

Fig. 39

Properties of Phase Diagrams of Unimodal Maps:

(i) The central vertices ($A^*_{p_{m^*}}$, $m \le m^*$) are linearly ordered.

(ii) If $m < m^*$ and $p_{m+1}/p_m = 2$, then $j_m \le 1$ and $k_m = 0$.

(iii) If $m < m^*$ and $p_{m+1}/p_m > 2$, then $j_m \le \infty$ and $k_m \le \infty$.

(iv) If $m^* < \infty$, then $j_{m^*} \le \infty$ and $k_{m^*} \le \infty$.

(v) If $m^* = \infty$, then $j_\infty = 0$ and $k_\infty \le \infty$.

It is thus natural to select the most "simple" and "complicated" (from the viewpoint of the shape of their phase diagrams) unimodal maps in the collection of maps of various smoothness. Thus, quadratic maps $x \to ax^2 + bx + c$, $a \ne 0$, prove to be simple in this sense; for these maps, $k_m = 0$ for any m and $j_m = 0$ for $m < m^*$; for $m = m^*$, either $j_m = 0$ or we arrive at the case displayed in Fig. 36a. Some examples of the most "complicated" C^∞-maps (e.g., such that $j_m = \infty$ and $k_m = \infty$ for some m) are constructed below, where we also discuss typical problems encountered in this case.

2. Limit Behavior of Trajectories

Consider the dynamics of the map f on the sets that determine its phase diagram.

Dynamics of the Map $f|_{B_m^{(i)}}$. Let $B_m^{(i)} = \{ J_{m,0}^{(i)}, J_{m,1}^{(i)}, \ldots, J_{m,q_i-1}^{(i)} \}$, where q_i is the period of a cycle of intervals. Obviously, $q_i = r_i p_m$ for some $r_i \geq 1$. Let $J_{m,0}^{(i)} = (a, b)$.

If $B_m^{(i)}$ does not contain c, then $\tilde{g} = f^{q_i}|_{[a,b]}$ is a homeomorphism. It follows from the maximality of $B_m^{(i)}$ that either $\tilde{g}(a) = a$ and $\tilde{g}(b) = b$ or $\tilde{g}(a) = b$ and $\tilde{g}(b) = a$. In the first case, any trajectory of the map $\tilde{g}$ approaches a fixed point. In the second case, by considering $\tilde{g}^2$, we conclude that any point is attracted either by a fixed point or by a 2-periodic trajectory of the map $\tilde{g}$.

If $B_m^{(i)}$ contains the point c, then it is not difficult to show that $m = m^* < \infty$ and $\mathcal{B}_{m^*} = \bigcup_{0 \leq i < p_{m^*}} \operatorname{int} J^*_{p_{m^*},i}$ (i.e., this is the case depicted in Fig. 36a). Moreover, all trajectories of the map $g = f^{p_{m^*}}|_{J^*_{p_{m^*},0}}$ approach its fixed points.

Thus, the dynamics of the map $f|_{B_m^{(i)}}$ is very simple. In particular, the set of nonwandering points of $f|_{B_m^{(i)}}$ is a subset of $\operatorname{Per}(f)$.

Dynamics of the Map f on $\Gamma_m^{(i)}$. Any interval U from the set $\Gamma_m^{(i)}$ is wandering, i.e., its different images are mutually disjoint. Consequently, the length of the interval $f^i(U)$ tends to zero as i increases. It follows from the definition of $\Gamma_m^{(i)}$ that the ends of the interval U belong to $\mathcal{R}_m$. Hence, the ω-limit set ω_y of any point $y \in U \in \Gamma_m^{(i)}$ does not depend on the choice of the point y and belongs to $\operatorname{NW}(f) \cap \mathcal{R}_m$.

Dynamics of the Map f on $\mathcal{R}_m$. We consider the case $m < m^*$.

A. $p_{m+1}/p_m = 2$. As follows from the analysis of the corresponding case in the previous section, the set $\mathcal{R}_m$ consists of finitely many preimages of periodic points z and x and infinitely many preimages of the periodic point z_1 (see Fig. 34a). Thus, for $p_{m+1}/p_m = 2$, every point $y \in \mathcal{R}_m$ hits one of the periodic points z, x, or z_1 within a finite period of time (i.e., after finitely many iterations). In this case, $\mathcal{R}_m \cap \mathrm{NW}(f) = \mathrm{NW}(f|_{\mathcal{R}_m}) \subset \mathrm{Per}(f)$. Denote the cycle that contains z by Ω_m^*, the cycle that contains x by Ω_m^{**}, and the cycle that contains z_1 by $\Omega_m^{(0)}$. It is clear that Ω_m^* and Ω_m^{**} may coincide and $\Omega_m^{(0)} = \Omega_{m+1}^*$. Hence,

$$\Omega_m \stackrel{\text{def}}{=} \mathcal{R}_m \cap \mathrm{NW}(f) = \Omega_m^* \cup \Omega_m^{**} \cup \Omega_m^{(0)},$$

where Ω_m^*, Ω_m^{**}, and $\Omega_m^{(0)}$ are cycles.

B. $p_{m+1}/p_m > 2$. We represent the set $\mathcal{R}_m$ as the union of sets $\mathcal{R}_m^*$, $\mathcal{R}_m^{**}$, and $\mathcal{R}_m^{(0)}$, where $\mathcal{R}_m^*$ and $\mathcal{R}_m^{**}$ are, respectively, the finite sets formed by the preimages of periodic points z and x in $\mathcal{R}_m$ (see Fig. 35b), and $\mathcal{R}_m^{(0)}$ denotes the set $\mathcal{R}_m \setminus (\mathcal{R}_m^* \cup \mathcal{R}_m^{**})$. Denote $\mathcal{R}_m \cap \mathrm{NW}(f)$ by Ω_m, $\mathcal{R}_m^* \cap \mathrm{NW}(f)$ by Ω^*_m, $\mathcal{R}_m^{**} \cap \mathrm{NW}(f)$ by Ω_m^{**}, and $\mathcal{R}_m^{(0)} \cap \mathrm{NW}(f)$ by $\Omega_m^{(0)}$. It is clear that the cycles Ω^*_m and Ω^{**}_m contain points z and x, respectively. The investigation of the structure and properties of the set $\Omega_m^{(0)}$ requires more detailed analysis.

Let $g = f^{p_m}|_{J^*_{p_m,0}}$, where $J^*_{p_m,0}$ is an interval from the cycle of intervals $A^*_{p_m}$ that contains the point c. Let $J^0_{p_m,0}$ be the closed interval with ends at $g(c)$ and $g^2(c)$ (see, e.g., Fig. 36a). It follows from the conditions $m < m^*$ and $p_{m+1}/p_m > 2$ that $c \in \mathrm{int}\, J^{(0)}_{p_m,0}$ and $g(J^{(0)}_{p_m,0}) = J^{(0)}_{p_m,0}$. Consider the cycle of intervals $A^{(0)}_{p_m} = \{J^{(0)}_{p_m,0}, J^{(0)}_{p_m,1}, \ldots, J^{(0)}_{p_m,p_m-1}\}$ and the invariant set $\Phi^{(0)}_m = \bigcup_{0 \le i < p_m} J^{(0)}_{p_m,i}$, where $J^{(0)}_{p_m,i} = f^i(J^{(0)}_{p_m,0})$. Let us show that, for any point $y \in \mathcal{R}_m^{(0)}$, its domain of influence $Q(y,f)$ coincides with the set $\Phi^{(0)}_m$.

Definition. *The asymptotic domain of influence of a point* $y \in I$ *under the map* $f\colon I \to I$ *is defined as the set*

$$Q(y,f) = \bigcap_{\substack{j \ge 0 \\ U \ni y}} \overline{\bigcup_{i \ge j} f^i(U)},$$

where U *denotes open neighborhoods of the point* y.

Indeed, every point $y \in \mathcal{R}_m^{(0)}$ hits the set $\Phi^{(0)}_m$ after finitely many steps and, consequently, $Q(y,f) \subset \Phi^{(0)}_m$. On the other hand, it follows from the definition of $\mathcal{R}_m$ that, for any neighborhood U of the point $y \in \mathcal{R}_m$, there exists $j = j(y)$ such that $f^j(U)$ contains the periodic point z_1, which is an end of the interval $J^*_{p_{m+1},0}$ from $A^*_{p_{m+1}}$

(for the proof, see Lemma 5.3). Hence, it suffices to show that $Q(z_1, f) = \Phi_m^{(0)}$. This equality is proved by the following two lemmas:

Lemma 5.1. $Q(z_1, f)$ *is a cycle of intervals for which* z_1 *is an internal point.*

Proof. The point z_1 is a periodic point whose period is at most p_{m+1}. Therefore, its domain of influence $Q(z_1, f)$ consists of at most p_{m+1} components permutable by the map f, i.e., it is a cycle of intervals. If z_1 belongs to the boundary of $Q(z_1, f)$, then, for any sufficiently small neighborhood U of z_1, either this neighborhood hits $\Phi_{m+1}^{(0)}$ after finitely many steps or it is a cycle of intervals that does not contain c. In the first case, z_1 must be a point of extremum of some iteration of the map f (recall that if c_0 is a point of extremum of the map f^j, then $f^k(c_0) = c$ for some $k < j$). Hence, this situation is impossible. The fact that the second case is also impossible follows from the proof of Lemma 5.2.

Lemma 5.2. $A_{p_m}^{(0)}$ *is the minimal cycle of intervals (with respect to the ordering of sets by inclusion) for which* z_1 *is an internal point.*

Proof. Let us prove that the intervals $J^*_{p_{m+1},i}$ of the cycle of intervals $A^*_{p_{m+1}}$ are mutually disjoint. Indeed, if this is not true, then, for some j, $0 < j < p_{m+1}$, we have $J^*_{p_{m+1},0} \cap J^*_{p_{m+1},j} \neq \varnothing$. Hence, $f^j(J^*_{p_{m+1},0}) \subset J^*_{p_{m+1},j}$ and $f^j(J^*_{p_{m+1},j}) \subset J^*_{p_{m+1},0}$. Therefore, $j = p_{m+1}/2$ and the intervals $J_i = J^*_{p_{m+1},i} \cup J^*_{p_{m+1},i+j}$, $i = 0, 1, \ldots, j-1$, form a cycle of intervals of period j that contains the point c. Moreover, $p_m < j = p_{m+1}/2 < p_{m+1}$, but this contradicts the assumptions made above.

Hence, z_1 is a periodic point with period p_{m+1}.

Let $F = \{F_0, F_1, \ldots, F_{p-1}\}$ be a cycle of intervals for which z_1 is an internal point and let $z_1 \in F_0$. Assume that c does not belong to any interval F_i, $0 \le i < p$. If F_0 has nonempty intersection only with the interval $J^*_{p_{m+1},0}$ from $A^*_{p_{m+1}}$, then $p = p_{m+1}$ (because the period of z_1 is p_{m+1}). Thus, the intervals $F_0 \cup J^*_{p_{m+1},0}$, $F_1 \cup J^*_{p_{m+1},1}, \ldots$, $F^*_{p_{m+1}-1} \cup J^*_{p_{m+1},p_{m+1}-1}$ form a cycle of intervals of period p_{m+1}. This contradicts the maximality of $A^*_{p_{m+1}}$.

Now assume that $F_0 \cap J^*_{p_{m+1},j} \neq \varnothing$ for some $j > 0$ $(j < p_{m+1})$. In this case, one can prove that $j = p_{m+1}/2$ and the images of the interval $J^*_{p_{m+1},j} \cup F_0 \cup J^*_{p_{m+1},0}$ form a cycle of intervals of period j. But this is impossible because $p_m < j < p_{m+1}$.

Thus, for the cycle of intervals F, we have $c \in F_0$. Therefore, $p = p_m$ because $A^*_{p_{m+1}}$ is maximal. Hence, $f^{ip_m}(c_0) \in F_0$ for all $i \ge 0$ and, consequently, $J^{(0)}_{p_m,0} \subset F_0$. This completes the proof of Lemma 5.2.

It follows from Lemmas 5.1 and 5.2 that $Q(z_1, f) = \Phi_m^{(0)}$ and $Q(y, f) = \Phi_m^{(0)}$ for all $y \in \mathcal{R}_m^{(0)}$.

Corollary 5.1. $\Omega_m^{(0)} = \mathcal{R}_m^{(0)} \cap \Phi_m^{(0)}$.

Proof. For $y \in \mathcal{R}_m^{(0)} \cap \Phi_m^{(0)}$, we have $y \in Q(y, f)$. Therefore, $y \in \mathrm{NW}(f)$. On the other hand, if $y' \in \mathcal{R}_m^{(0)}$ and V is a sufficiently small neighborhood of y', then $f^j(V) \subset \Phi_m^{(0)}$ for some $j \geq 0$. Hence, $\Omega_m^{(0)} = \mathrm{NW}(f) \cap \mathcal{R}_m^{(0)} \subset \Phi_m^{(0)}$. In view of the inclusion $\mathcal{R}_m^{(0)} \cap \Phi_m^{(0)} \subset \mathrm{NW}(f)$ established above, this yields the required equality.

Corollary 5.2. *For any point* $y \in \mathcal{R}_m$, *there exists* $j = j(y)$ *such that* $f^j(y) \in \mathrm{NW}(f)$.

Proof. This statement is a consequence of the inclusion

$$\mathcal{R}_m \subset \bigcup_{i \geq 0} f^{-i}(\Omega_m^* \cup \Omega_m^{**} \cup \Phi_m^{(0)}).$$

Lemma 5.3. *Assume that* $y \in \mathcal{R}_m^{(0)}$, V *is a sufficiently small neighborhood of* y, *and* $S = \bigcup_{0 \leq i < p_m} f^i(V)$. *Then there exists* $j = j(V)$ *such that* $f^j(S) = \Phi_m^{(0)}$.

Proof. Since

$$\mathcal{R}_m^{(0)} \subset \bigcup_{i \geq 0} f^{-i}(\Phi_m^{(0)})$$

and the boundary of the set $\Phi_m^{(0)}$ is a part of a trajectory, one can indicate N such that $f^i(y) \in \operatorname{int} \Phi_m^{(0)}$ for $i > N$. Hence, $f^i(S) \subset \Phi_m^{(0)}$ for $i > N$ and sufficiently small V.

As above, let $g = f^{p_m}|_{J^*_{p_m,0}}$ and let z_1 be a periodic point lying on the boundary of $J^*_{p_{m+1},0}$. It follows from the definition of the set $\mathcal{R}_m$ that $z_1 \in f^k(V)$ for some $k \geq 0$. If $f^k(V) \subset J^*_{p_{m+1},0}$, then $f^k(y) \in \partial J^*_{p_{m+1},0}$ because $y \in \mathcal{R}_m$. Hence, y is a point of extremum for f^k. In this case, $f^i(y) = c$ for some $i < k$ but this contradicts the condition $y \in \mathcal{R}_m$. Therefore, $f^k(V)$ contains either a neighborhood of the point z_1' (i.e., of the other end of the interval $J^*_{p_{m+1},0}$) or a half neighborhood V^- of the point z_1 which does not belong to $J^*_{p_{m+1},0}$. Note that, in the first case, V^- is contained in $f^{k+p_m}(V)$.

Thus, let V^- be a half neighborhood of the point z_1 lying in $f^k(V)$ and such that $V^- \cap J^*_{p_{m+1},0} = \{z_1\}$. It follows from the maximality of $J^*_{p_{m+1},0}$ that $V^- \subset g(V^-)$.

Let $W = \bigcup_{i\geq 0} g^i(V^-)$. Then W is an interval and $g(W) = W$.

Let us show that $W = J^{(0)}_{p_m,0}$, where $J^{(0)}_{p_m,0}$ is the closed interval with the ends $g(c)$ and $g^2(c)$. Indeed, the points $g(z_1)$ and $g^2(z_1)$ belong to W (because $z_1 \in W$) and cannot lie on the same monotone branch of the map g because this contradicts the condition $z_1 \in \operatorname{Per}(f)$ (one must take into account the fact that $g^2(z_1) \leq g^i(z_1) < g(z_1)$ for any $i \geq 0$; recall that c is regarded as the maximum point of the map g). Thus, $c \in W$. This implies the required equality $W = J^{(0)}_{p_m,0}$. Furthermore, $c \in \operatorname{int} J^{(0)}_{p_m,0}$ and, therefore, for some $i \geq 0$, we obtain $c \in g^i(V^-)$. Hence, $g^{i+2}(V^-) = J^{(0)}_{p_m,0}$, i.e., $f^j(V) = J^{(0)}_{p_m,0}$ and $f^j(\bigcup_{0\leq i<p_m} f^i(V)) = \Phi^{(0)}_m$ for $j = (i+2)p_m$. Lemma 5.3 is proved.

Corollary 5.3. $\Omega^{(0)}_m \cap \operatorname{int} \Phi^{(0)}_m \subset \overline{\operatorname{Per}(f)}$.

Proof. If a neighborhood V of the point $y \in \Omega^{(0)}_m$ lies inside the set $\Phi^{(0)}_m$, then, by Lemma 5.3, we have $V \subset f^j(V)$ for some $j \geq 0$. Hence, the map f^j possesses a fixed point in the interval V.

Denote the set $\overline{\Omega^{(0)}_m \cap \operatorname{Per}(f)}$ by $C^{(0)}_m$. The following statement is a consequence of Corollary 5.3 because the boundary of the set $\Phi^{(0)}_m$ is a part of the trajectory of the point c.

Corollary 5.4. $\Omega^{(0)}_m \setminus C^{(0)}_m \subset \left\{ \bigcup_{0<i\leq 2p_m} f^i(x) \right\}$.

Corollary 5.5. For any point $y \in \mathcal{R}_m$, there exists $j = j(y)$ such that $f^j(y) \subset \overline{\operatorname{Per}(f)}$.

The proof follows from Corollaries 5.2 and 5.4.

Lemma 5.4. $C^{(0)}_m$ *is the perfect part of the set* $\Omega^{(0)}_m$, *i.e.,* $C^{(0)}_m$ *is a perfect nowhere dense set (Cantor set).*

Proof. If a point $y \in \Omega^{(0)}_m$ is not isolated in $\Omega^{(0)}_m$, then it follows from Corollaries 5.1 and 5.3 that $y \in C^{(0)}_m$.

Now let y be an isolated point of the set $\Omega^{(0)}_m$ that belongs to $C^{(0)}_m$. Then $y \in \operatorname{Per}(f)$. If the period of y is p, then f^p is a homeomorphism in a certain neighborhood V of the point y. Therefore, $y \in \operatorname{int} f^{i\cdot p}(V)$ for all $i \geq 1$. Hence, y is not a boundary point of the set $\Phi^{(0)}_m$. By virtue of Lemma 5.3, the interval V contains infinitely many preimages of points of the set $\Omega^{(0)}_m$ and, by Corollary 5.1, the set $V \cap \Omega^{(0)}_m$ is infinite but this contradicts the assumption that the point y is isolated.

Corollary 5.6. $\mathrm{NW}(f|_{\Omega_m^{(0)}}) = C_m^{(0)}$.

Lemma 5.5. *If* $m+1<m^*$, *then* $\Omega_m^{(0)}\setminus C_m^{(0)} = \varnothing$; *if* $m+1=m^*$, *then the set* $\Omega_m^{(0)}\setminus C_m^{(0)}$ *is either empty or coincides with the set* $\{\bigcup_{0<i\le 2p_m} f^i(x)\}$, *i.e., with the boundary of the set* $\Phi_m^{(0)}$.

Proof. Assume that $f(c)\in \operatorname{int}\Phi_{m+1}^*$. Then the boundary of the set $\Phi_m^{(0)}$ does not belong to $\Omega_m^{(0)}$ because $\Omega_m^{(0)} = \mathcal{R}_m^{(0)} \cap \Phi_m^{(0)}$. Hence, in this case, $\Omega_m^{(0)} = C_m^{(0)}$. If $f(c)\in\partial\Phi_{m+1}^*$, then $\tilde g(J^*_{p_{m+1},0}) = J^*_{p_{m+1},0}$, where $\tilde g = f^{p_{m+1}}|_{J^*_{p_{m+1},0}}$. Therefore, f has no cycles of intervals that contain c with periods greater than p_{m+1}, i.e., $m^* = m+1$. This enables us to conclude that $\Omega_m^{(0)} = C_m^{(0)}$ whenever $m+1<m^*$.

Let $f(c)\in\partial\Phi_{m+1}^*$. For $i = 1, 2, \ldots, 2p_m$, we consider half neighborhoods V_i^- of the points $f^i(c)$ that do not belong to $\Phi_m^{(0)}$ and half neighborhoods V_i^+ of the points $f^i(c)$ lying in $\Phi_m^{(0)}$. We can assume that $V_i^+ = f^i(V)$, where V is a neighborhood of the point c. Then the neighborhood $V_i^- \cup f^i(c)\cup V_i^+$ of the point $f^i(c)$ contains no points of the set $\Omega_m^{(0)}$ other than $f^i(c)$ and the point $f^i(c)$ is isolated in $\Omega_m^{(0)}$. The statement of Lemma 5.5 now follows from Lemma 5.4.

By using Lemma 5.3 and the definition of the set $C_m^{(0)}$, one can prove the following assertion:

Lemma 5.6. *The set* $C_m^{(0)}$ *is invariant and the map possesses the mixing property on* $C_m^{(0)}$. *Moreover, for any subset* V *of the set* $C_m^{(0)}$ *open in* $C_m^{(0)}$, *one can find* $j=j(V)$ *such that* $f^j(\bigcup_{0\le i<p_m} f^i(V)) = C_m^{(0)}$.

Proof. Since $C_m^{(0)}$ is the closure of the set of periodic trajectories lying in $\Omega_m^{(0)}$, we have $f(C_m^{(0)}) = C_m^{(0)}$, i.e., $C_m^{(0)}$ is a closed invariant set. Further, by Corollary 5.3, the preimages of points of the set $C_m^{(0)}$ lying inside $\Phi_m^{(0)}$ also belong to $C_m^{(0)}$. Hence, by Lemma 5.3, the map f possesses the mixing property on $C_m^{(0)}$.

It follows from Lemma 5.6 that the set $C_m^{(0)}$ is, in a certain sense, indecomposable.

The results established above can be formulated as follows: For $m<m^*$ and $p_{m+1}/p_m > 2$, the set $\mathcal{R}_m\cap\mathrm{NW}(f)$ is nonempty and can be represented in the form

$$\mathcal{R}_m\cap\mathrm{NW}(f) = \Omega_m^*\cup\Omega_m^{**}\cup\Omega_m^{(0)}.$$

Moreover, for any point $y\in\mathcal{R}_m$, there exists j such that $f^j(y)\in\Omega_m^*\cup\Omega_m^{**}\cup C_m^{(0)}$, where $C_m^{(0)} = \overline{\Omega_m^{(0)}\cap\mathrm{Per}(f)}$. These sets have the following properties:

(i) Ω_m^* and Ω_m^{**} are cycles;

(ii) $C_m^{(0)}$ is a closed invariant set with the structure of the Cantor set and admits a decomposition into p_m closed subsets cyclically permutable by the map f;

(iii) f possesses the mixing property on $C_m^{(0)}$;

(iv) if $m+1 < m^*$, then $\Omega_m^{(0)} = C_m^{(0)}$; if $m+1 = m^*$, then either $\Omega_m^{(0)} = C_m^{(0)}$ or $\Omega_m^{(0)} \setminus C_m^{(0)} = \{f^i(c), i = 1, 2, \ldots, 2p_m\}$.

Dynamics of the Map f on the Set $\mathcal{R}_{m^*}$ for $m^* < \infty$. As above, we consider the map $g = f^{p_{m^*}}|_{J^*_{p_{m^*},0}}$ (for definiteness, we assume that c is its maximum point). If $g(c) \le c$, then $\mathcal{R}_{m^*} \cap \mathrm{NW}(f) = \Omega^*_{m^*} \cup \Omega^{**}_{m^*}$ (see Fig. 37a), where $\Omega^*_{m^*}$ and $\Omega^{**}_{m^*}$ are cycles that contain points z and x, respectively. Furthermore, $f^{p_{m^*}}(\mathcal{R}_{m^*}) = \Omega^*_{m^*} \cup \Omega^{**}_{m^*}$.

If $g(c) > c$, then $g^2(c) < c < g(c)$. (The case where $c < g^2(c)$ contradicts the condition $m = m^*$ because, in this case, g has a cycle of intervals of period two which contains c.) Note that the sets $\mathcal{R}_{m^*} \cap \mathrm{NW}(f)$ and $\mathcal{R}_{m^*} \cap \mathrm{NW}(f^{p_{m^*}}|_{\Phi^*_{m^*}})$ may be distinct: Indeed, if $z \ne x$ and $g(c) = z'$ (i.e., $g(c) \in \partial J^*_{p_{m^*},1}$), then the points $f^i(c)$, $i = 1, 2, \ldots, p_{m^*}$, belong to the set $\mathrm{NW}(f)$ but not to the set $\mathrm{NW}(f^{p_{m^*}}|_{\Phi^*_{m^*}})$. This follows from the argument used in the previous case. Moreover, in this case, the points $f^i(c)$, $i = 1, 2, \ldots, 2p_{m^*-1}$, belong to the set $\mathrm{NW}(f) \setminus \overline{\mathrm{Per}(f)}$.

Denote $\mathcal{R}_{m^*} \cap J^*_{p_{m^*},0}$ by $\mathcal{R}_{m^*,0}$ and consider the set $\mathcal{R}_{m^*,0} \cap \mathrm{NW}(g)$. This set can be represented in the form $\Omega^*_{m^*,0} \cup \Omega^{**}_{m^*,0} \cup \Omega^{(0)}_{m^*,0}$, where $\Omega^*_{m^*,0} = \{z\}$, $\Omega^{**}_{m^*,0} = \{x\}$, and $\Omega^{(0)}_{m^*,0} = \mathcal{R}_{m^*,0} \cap [g^2(c), g(c)]$. The proof of this fact is similar to the proof of the previous case. Consider the set $C^{(0)}_{m^*,0} = \overline{\Omega^{(0)}_{m^*,0} \cap \mathrm{Per}(g)}$. It is clear that $C^{(0)}_{m^*,0} = \overline{\Omega^{(0)}_{m^*,0} \cap \mathrm{Per}(f)}$.

Let $\Omega^*_{m^*}$ and $\Omega^{**}_{m^*}$ be the cycles of the map f that contain, respectively, the points z and x,

$$J^{(0)}_{p_{m^*},0} = [g^2(c), g(c)], \qquad \Phi^{(0)}_{m^*} = \bigcup_{0 \le i < p_{m^*}} f^i(J^{(0)}_{p_{m^*},0}),$$

$$\Omega^{(0)}_{m^*} = \mathcal{R}_{m^*} \cap \Phi^{(0)}_{m^*}, \qquad C^{(0)}_{m^*} = \bigcup_{0 \le i < p_{m^*}} f^i(C^{(0)}_{m^*,0}) = \overline{\Omega^{(0)}_{m^*} \cap \mathrm{Per}(f)}.$$

Then, for any point $y \in \mathcal{R}_{m^*}$, there exists $j = j(y)$ such that $f^j(y) \in \Omega^*_{m^*} \cup \Omega^{**}_{m^*} \cup C^{(0)}_{m^*}$.

An analog of the assertion of Lemma 5.3 holds for points of the set $\mathcal{R}_{m^*0} \cap (x, x')$ (see Fig. 37). Therefore, the sets defined above has the following properties:

(i) $C^{(0)}_{m*}$ is a closed invariant set of the map f, which can be decomposed into p_{m*} closed subsets cyclically permuted by the map f; this set has either the structure of cycle of intervals (i.e., coincides with $\Phi^{(0)}_{m*}$) or the structure of Cantor set;

(ii) the map f possesses the mixing property on $C^{(0)}_{m*}$;

(iii) the set $\Omega^{(0)}_{m*}$ either coincides with $C^{(0)}_{m*}$ or

$$\Omega^{(0)}_{m*} \setminus C^{(0)}_{m*} = \{f^i(c),\ i = 1, 2, \ldots, 2p_{m*-1}\}.$$

In order to prove these assertions, we use the following statement, which is similar to Lemma 5.3:

Lemma 5.7. *Let V be a sufficiently small neighborhood of a point* $y \in \mathcal{R}_{m*} \setminus \bigcup_{0 \le i < p_{m*}} \partial J^*_{p_{m*},i}$ *and let* $S = \bigcup_{0 \le i < p_{m*}} f^i(V)$. *Then there exists* $j = j(V)$ *such that* $f^j(S) \supset \Phi^{(0)}_{m*}$.

It follows from Lemma 5.7 that $C^{(0)}_{m*} = \Phi^{(0)}_{m*}$ whenever $\mathcal{B}_{m*} \cap \Phi^{(0)}_{m*} = \varnothing$ (the set $\mathcal{B}_{m*}$ is defined above in constructing the phase diagram of the map f). If $\mathcal{B}_{m*} \cap \Phi^{(0)}_{m*} \neq \varnothing$, then it follows from Lemma 5.7 that $\Omega^{(0)}_{m*}$ is a nowhere dense set and $C^{(0)}_{m*}$ is a perfect nowhere dense set. By analogy with the case $m < m^*$, one can prove the other properties of the sets $\Omega^{(0)}_{m*}$ and $C^{(0)}_{m*}$. We only note that the equality $m = m^*$ implies that if $\Omega^{(0)}_{m*} \setminus C^{(0)}_{m*} \neq \varnothing$, then a neighborhood of the point c is a wandering interval of the map f. In this case, $\Omega^{(0)}_{m*} \setminus C^{(0)}_{m*} = \mathrm{NW}(f) \setminus \overline{\mathrm{Per}(f)}$.

The Case $m^* = \infty$. Here, $\Phi^*_\infty = \bigcap_{m>1} \Phi^*_m$, $\mathcal{B}_\infty = \operatorname{int} \Phi^*_\infty$, and $\mathcal{R}_\infty = \Phi^*_\infty \setminus \mathcal{B}_\infty$. The investigation of the limit behavior of trajectories of the map f on the set Φ^*_∞ is based on the use of the following lemma:

Lemma 5.8. *For any point* $y \in \mathcal{R}_\infty$ *and any its neighborhood U, there exist* $j \ge 0$ *and* $m \ge 1$ *such that*

$$f^{p_m}(\Phi^*_m) \subset f^j\Big(\bigcup_{0 \le i < p_m} f^i(U)\Big).$$

Proof. Let U be a neighborhood of the point $y \in \mathcal{R}_\infty$. Then $U \cap \partial \Phi^*_m \neq \varnothing$ for some $m \ge 1$. Therefore, $f^{p_m}(U) \cap f^{2p_m}(U) \neq \varnothing$ and $\bigcap_{i \ge 0} \bigcup_{j \ge i} f^j(U)$ is a cycle of intervals. If $c \notin f^j(U)$ for all $j \ge 0$, then this cycle of intervals does not contain the point

c and, consequently, the ω-limit set of any point of the interval U is a cycle. However, for the point $y \in U$, this is impossible because $\omega(y) \subset \mathcal{R}_\infty$ and $\mathcal{R}_\infty \cap \mathrm{Per}(f) = \varnothing$. Hence, $c \in f^j(U)$ for some $j \geq 0$.

On the other hand, $f^i(U) \cap \partial\Phi_m^* \neq \varnothing$ for any $i \geq 0$ because $f(\partial\Phi_m^*) \subset \partial\Phi_m^*$. Therefore, $f^{j+1}(\bigcup_{0 \leq i < p_m} f^i(U))$ contains $f^{p_m}(\Phi_m^*)$.

Lemma 5.8 yields the following properties of trajectories of the set Φ_∞^*.

Property 1. *For any point $y \in \mathcal{R}_\infty$, its domain of influence $Q(y, f)$ coincides with $\Phi_\infty^0 = \bigcap_{m \geq 1} f^{p_m}(\Phi_\infty^0)$.*

Proof. It follows from Lemma 5.8 that $\Phi_\infty^0 \subset Q(y, f)$. Note that the set $Q(y, f)$ is invariant. On the other hand, it is not difficult to show that Φ_∞^0 is the maximal invariant subset of the set Φ_∞^* because $\bigcap_{m \geq 1} f^{p_m}(\Phi_m^*) = \bigcap_{i \geq 1} f^i(\bigcap_{m \geq 1} \Phi_m^*)$.

Consider the set $\Omega_\infty^{(0)} = \Phi_\infty^{(0)} \cap \mathrm{NW}(f)$. Since each component of the set $\mathcal{B}_\infty$ is wandering, it follows from Property 1 that $\Omega_\infty^{(0)} = \Phi_\infty^* \cap \mathrm{NW}(f)$.

Property 2. *If $y \in \Omega_\infty^{(0)}$ and $y \neq f^i(c)$, $i = 1, 2, \ldots$, then $y \in \overline{\mathrm{Per}(f)}$.*

Proof. Under the conditions of Property 2, the point y is an internal point of the set $f^{p_m}(\Phi_m^*)$ for any $m \geq 1$. By Lemma 5.8, for any sufficiently small interval U that contains the point y, we have $U \subset f^j(U)$ for some $j \geq 0$. Hence, the map f^j possesses a fixed point in the interval U and $y \in \overline{\mathrm{Per}(f)}$.

Denote the set $\Omega_\infty^{(0)} \cap \overline{\mathrm{Per}(f)}$ by $C_\infty^{(0)}$.

Property 3. *If $\Omega_\infty^{(0)} \setminus C_\infty^{(0)} \neq \varnothing$, then $\Omega_\infty^{(0)} \setminus C_\infty^{(0)} = \{f^i(c),\ i \geq 1\}$.*

Proof. Assume that $\Omega_\infty^{(0)} \setminus C_\infty^{(0)} \neq \varnothing$. If $c \notin \mathcal{B}_\infty$, then $c \in \mathcal{R}_\infty$ and we have $c \in \Omega_\infty^{(0)}$ by Property 1 and $c \in \overline{\mathrm{Per}(f)}$ by Property 2. Hence, $f^i(c) \in \overline{\mathrm{Per}(f)}$ for any $i \geq 1$. Therefore, we must consider the case $c \in \mathcal{B}_\infty$.

If $f(c) \in \mathcal{B}_\infty$, then $f^i(c) \in \mathcal{B}_\infty$ for any $i \geq 1$ because the map f is monotone on any component of the set $\mathcal{B}_\infty$ that does not contain the point c. Hence, $f(c) \notin \mathcal{B}_\infty$. By Property 1, $f(c) \in \Omega_\infty^{(0)}$ and, therefore, $f^i(c) \in \Omega_\infty^{(0)}$ for any $i \geq 1$.

Let us show that $f^i(c) \notin \overline{\mathrm{Per}(f)}$ for $i \in \{1, 2, 3, \ldots\}$. Let $i \geq 1$ and $m > i$. Then $f^i(c)$ is the end of the interval $f^{p_m}(J_{p_m, i}^*)$. The equality $m^* = \infty$ implies that the point $f^i(c)$ does not lie on the boundary of the set Φ_m^*. Therefore, there exists a neighborhood U of the point $f^i(c)$ such that $f^{p_m}(U) \subset f^{p_m}(\Phi_m^*)$. Let U^+ be the part of the neighborhood U that lies in $f^{p_m}(J_{p_m, i}^*)$ and let U^- be the remaining part of this neighborhood. One can regard U^+ as the image of the component of the set $\mathcal{B}_\infty$, which contains the point c. Then $f^j(U^+) \cap U^+ = \varnothing$ for any $j \geq 1$, i.e., $U^+ \cap \mathrm{Per}(f) = \varnothing$. On the other hand, $U^- \cap \mathrm{Per}(f) = \varnothing$ because $f^{p_m}(U^-) \subset f^{p_m}(\Phi_m^*)$, $U^- \cap f^{p_m}(\Phi_m^*) = \varnothing$, and the set $f^{p_m}(\Phi_m^*)$ is invariant. This completes the proof.

Property 4. *For any point* $y \in \Phi_\infty^*$, *its* ω*-limit set coincides with the set* $C_\infty^{(0)}$.

Proof. Let $J_{p_m,i}^*$ be an interval from the cycle of intervals $A_{p_m}^*$. There are p_{m+j}/p_m intervals from $A_{p_{m+j}}^*$ in $J_{p_m,i}^*$. The utmost left and right intervals in this collection are called one-sided in the sense that all other intervals from $A_{p_{m+j}}^*$ lying in the interval $J_{p_m,i}^*$ are located on the one side of the indicated intervals. All other intervals are called two-sided. One can show that the intervals $J_{p_{m+j},k}^*$, $k = 1, 2, \ldots, 2p_m$, are one-sided intervals of the cycle of intervals $A_{p_{m+j}}^*$ in the intervals of $A_{p_m}^*$.

Let $J_{p_{m+j},s}^*$ be a one-sided interval in $J_{p_m,i}^*$. If, e.g., the interval $J_{p_m,i}^*$ contains no intervals of $A_{p_{m+j}}^*$ to the left of $J_{p_{m+j},s}^*$, then the trajectories of the points of the set Φ_∞^* cannot have limit points in $J_{p_m,i}^*$ to the left of $J_{p_{m+j},s}^*$ but any trajectory of this sort has limit points in $J_{p_m,i}^*$ to the right of $J_{p_{m+j},s}^*$. If the interval $J_{p_{m+j},s}^*$ is two-sided in $J_{p_m,i}^*$, then the trajectories of all points of the set Φ_∞^* has limit points in $J_{p_m,i}^*$ both to the left and to the right of $J_{p_{m+j},s}^*$.

Let $[a, b]$ be a component of the set Φ_∞^* $(a \leq b)$. Then there exists a sequence $\{i_m\}_{m=1}^\infty$ such that $i_m \in \{0, 1, \ldots, p_{m-1}\}$ and $[a, b] = \bigcap_{m\geq 1} J_{p_m,i_m}^*$. It follows from the inclusion $J_{p_{m+1},i_{m+1}}^* \subset J_{p_m,i_m}^*$ that $i_{m+1} = i_m + k_m p_m$, $k_m \in \{0, 1, \ldots, p_{m+1}/p_m - 1\}$, and $\{i_m\}_{m=1}^\infty$ is a nondecreasing sequence. Since $p_1 = 1$, we have $i_1 = 0$ and $i_{m+1} = k_1 p_1 + k_2 p_2 + \ldots + k_m p_m$ for $m > 1$, where $k_n \in \{0, 1, \ldots, p_{n+1}/p_n - 1\}$, $n = 1, 2, \ldots, m$. Hence, there exists a one-to-one correspondence between the family of components of the set Φ_∞^* and the family of infinite sequences of integer numbers of the form $(k_1, k_2, \ldots)$, where $k_n \in \{0, 1, \ldots, p_{n+1}/p_n - 1\}$.

Thus, let $[a, b] = \bigcap_{m\geq 1} J_{p_m,i_m}^*$. Assume that there exists $m \geq 1$ such that $J_{p_{m+j},i_{m+j}}^*$ is a one-sided interval in J_{p_m,i_m}^* for any $j \geq 1$. In this case, $i_{m+j} = 2p_m$ for any $j \geq 1$. This inequality implies the equality $k_n = 0$ for sufficiently large n. If, for any m, there exists $j = j(m)$ for which $J_{p_{m+j},i_{m+j}}^*$ is a two-sided interval in J_{p_m,i_m}^*, then it is possible to show that either $k_n = 0$ for any $n \geq 1$ or there are infinitely many nonzero elements in the sequence $(k_1, k_2, \ldots)$ that corresponds to the component $[a, b]$.

Let $[a, b] = \bigcap_{m\geq 1} J_{p_m,i_m}^*$ be a component of the set Φ_∞^* and let $(k_1, k_2, \ldots)$ be the corresponding sequence. Assume that either all k_n are equal to zero or the number of nonzero elements in this sequence is infinite. We fix $\varepsilon > 0$ and choose m such that $J^*_{p_m, i_m} \subset (a - \varepsilon, b + \varepsilon)$. By assumption, there exists $j \geq 1$ for which the interval $J^*_{p_{m+j}, i_{m+j}}$ is two-sided in $J^*_{p_m, i_m}$. Therefore, for any point $y \in \Phi^*_\infty$, we have $\omega_f(y) \cap (a - \varepsilon, a) \neq \varnothing$ and $\omega_f(y) \cap (b, b + \varepsilon) \neq \varnothing$. Hence, the points a and b belong to the set $\omega_f(y)$ because this set is closed.

If $k_m \neq \varnothing$ for some $m \geq 1$ and $k_{m+j} = 0$ for all $j \geq 1$, then the component $[a, b]$ contains the point $f^{i_{m+1}}(c)$, where $i_{m+1} = k_1 p_1 + k_2 p_2 + \ldots + k_m p_m \neq 0$. Thus, the interval $J^*_{p_{m+j}, i_{m+j}}$ is one-sided in $J^*_{p_m, i_m}$ because $i_{m+j} = i_m$ for any $j \geq 1$. Therefore, if $a \neq b$, then one end of the component $[a, b]$ lies in $\omega_f(y)$ and the other end does not belong to this set.

Following the proof of Property 2, we can show that $f^i(c) \notin \omega_f(y)$ for any $i \geq 0$ whenever $c \in \mathcal{B}_\infty$. This completes the proof of Property 4.

Consider the dynamics of the components of the set Φ^*_∞ under the map f in more details. As shown in the proof of Property 4, there exists a one-to-one correspondence between the components of the set Φ^*_∞ and the set of infinite sequences $(k_1, k_2, \ldots)$ with $k_i \in \{0, 1, \ldots, p_{i+1}/p_i - 1\}$. (Note that each of these sequences can be interpreted as a digital representation of a number from the interval $[0, 1]$ similar to its decimal representation.) In what follows, for the sake of brevity, the sequences $(k_1, k_2, k_3, \ldots)$ are written in the form $0.k_1 k_2 k_3 \ldots$ and each k_i is called the value of the ith digit.

We define the sum of two sequences (numbers) $K = 0.k_1 k_2 \ldots$ and $L = 0.l_1 l_2 \ldots$ as follows: The value of the ith digit in K is added to the value of the ith digit in L modulo p_{i+1}/p_i and the overflow unit is added to the next digit. Thus, if $p_{i+1}/p_i = 10$ for all $i \geq 1$, then $0.999\ldots + 0.100\ldots = 0.000\ldots$. It is easy to check that the family of sequences corresponding to components of the set Φ^*_∞ equipped with this operation of addition is an Abelian group. The action of the map f corresponds to the operation of adding the number $F = 0.100\ldots$. More precisely, if a component K of the set Φ^*_∞ corresponds to the number $K = 0.k_1 k_2 \ldots$, then the number $K + F$ corresponds to the component of Φ^*_∞ which contains $f(K)$ and the number $K - F$ corresponds to the preimage of the component K under the map f.

Denote the family of components of the set Φ^*_∞ by $\mathcal{K}$ and consider a map $F : \mathcal{K} \to \mathcal{K}$ defined as follows: For $K, L \in \mathcal{K}$, $F(K) = L$ if $f(K) \subset L$. (As mentioned above, $F(K) = K + F$, where $F = 0.100\ldots$.) The distance ρ between elements $K = 0.k_1 k_2 k_3 \ldots$ and $L = 0.l_1 l_2 l_3 \ldots$ of the set $\mathcal{K}$ is defined by the formula

$$\rho(K, L) = \sum_{i=1}^{\infty} \frac{|k_i - l_i|}{(r_i)^i}, \quad \text{where} \quad r_i = \frac{p_{i+1}}{p_i}.$$

By using the reasoning presented above, one can establish the following properties of the set $\mathcal{K}$ and the map F:

(i) the set $\mathcal{K}$ has the cardinality of continuum;

(ii) the map $F: \mathcal{K} \to \mathcal{K}$ is a homeomorphism of $\mathcal{K}$ onto itself;

(iii) the α-limit and ω-limit sets of any trajectory of the dynamical system generated by the map F on $\mathcal{K}$ coincide with $\mathcal{K}$, i.e., $\mathcal{K}$ is the minimal set of this dynamical system;

(iv) the map F possesses the mixing property on $\mathcal{K}$; more precisely, if U_ε is the ε-neighborhood of a point $K \in \mathcal{K}$, then one can indicate $j = j(\varepsilon)$ such that $F^j(U_\varepsilon) = \mathcal{K}$.

Let us now summarize the results obtained in this section.

For any unimodal map f, the set of its central motions admits the following representation:

$$\overline{\mathrm{Per}(f)} = P_0(f) \cup \bigcup_{m=1}^{m^*} C_m^{(0)},$$

where $m^* \leq \infty$ and

(a) $C_m^{(0)}$, $m = 1, 2, \ldots, m^*$, are nonempty closed invariant sets and $P_0(f)$ is an invariant subset of $\mathrm{Per}(f)$ (which may be empty);

(b) the sets $P_0(f)$ and $C_m^{(0)}$, $m = 1, 2, \ldots, m^*$, are mutually disjoint except, possibly, the sets $C_{m^*}^{(0)}$ and $C_{m^*-1}^{(0)}$ with $m^* < \infty$, which may have a common cycle;

(c) the map $f|_{C_m^{(0)}}$ possesses the mixing property; in particular, this map is transitive, i.e., one can indicate a point $y \in C_m^{(0)}$ such that $\omega_f(y) = C_m^{(0)}$;

(d) the set $C_{m^*}^{(0)}$ contains $\omega_f(c)$;

(e) for $m < m^*$, the set $C_m^{(0)}$ is either a cycle or a Cantor set; for $m^* < \infty$, the set $C_{m^*}^{(0)}$ is either a cycle, or a cycle of intervals, or a Cantor set; for $m^* = \infty$, the set $C_{m^*}^{(0)}$ is a Cantor set, which is the minimal set of f.

This decomposition is usually called the spectral decomposition of the set $\overline{\mathrm{Per}(f)}$. Generally speaking, by using the phase diagram, one can construct similar decompositions for any invariant set of the map f. Thus, for the investigation of the dynamics of a map, it might be useful to have the relevant decomposition of the set of its nonwandering points; in the case of unimodal maps, this decomposition slightly differs from the decomposition of the set $\overline{\mathrm{Per}(f)}$, namely,

$$\mathrm{NW}(f) = P_0(f) \cup \bigcup_{m=1}^{m^*} \Omega_m^{(0)},$$

where $m^* \le \infty$, and the sets $\Omega_m^{(0)}$ coincide, respectively, with the sets $C_m^{(0)}$ for all m except, possibly, either $\Omega_{m^*}^{(0)}$ or $\Omega_{m^*-1}^{(0)}$ (one of these sets may differ from the corresponding set in the decomposition of the set $\overline{\mathrm{Per}(f)}$ by the presence of a part of the trajectory or of the entire trajectory of the point $f(c)$ that does not belong to $\overline{\mathrm{Per}(f)}$).

The results established in this section imply the following important properties of the dynamics of unimodal maps:

(i) the sets $\mathrm{NW}(f)$ and $\overline{\mathrm{Per}(f)}$ may be different if and only if $c \notin \mathrm{NW}(f)$ and $f(c) \in \mathrm{NW}(f)$;

(ii) $\mathrm{NW}(f) \setminus \overline{\mathrm{Per}(f)}$ is the set of points from the set $\mathrm{NW}(f) \setminus \mathrm{Per}(f)$ isolated in $\mathrm{NW}(f)$;

(iii) $\mathrm{NW}(f) \setminus \overline{\mathrm{Per}(f)} \subset \{f^i(c)\}_{i=1}^{\infty}$;

(iv) $\mathrm{NW}(f|_{\mathrm{NW}(f)}) = \overline{\mathrm{Per}(f)}$.

Note that it follows from Property 1 that if $c \in \mathrm{NW}(f)$ or $f(c) \notin \mathrm{NW}(f)$, then $\mathrm{NW}(f) = \overline{\mathrm{Per}(f)}$.

Another problem, which might be important for the investigation of the behavior of trajectories of dynamical systems, is to describe the behavior of "typical" (generic) trajectories. In the most general case, a property of trajectories should be called generic if it is observed for the trajectories of the points of some set of the second category. For any unimodal map $f: I \to I$, generic trajectories possess one of the following three properties:

(a) after finitely many steps, a trajectory hits a cycle of intervals, where the map f possesses the mixing property;

(b) the trajectory is attracted by $\omega_f(c)$;

(c) the trajectory hits an open invariant set of the map f which does not contain the point c.

To prove this assertion, we show that the set $\mathfrak{X}$ of points of the interval I whose trajectories satisfy one of the conditions (a), (b), or (c) can be represented in the form of at most countable intersection of open dense subsets of the interval I.

Indeed, it follows from the results of this section that

$$I = \bigcup_{m=1}^{m^*} P(\mathcal{B}_m, f) \cup \bigcup_{m=1}^{m^*} P(C_m^{(0)}, f)$$

(recall that $P(A, f) = \{ x \in I : \omega_f(x) \subset A \}$ for $A \in 2^I$). For $m < m^*$, we can write

$$P(C_m^{(0)}, f) \subset \bigcup_{i \geq 0} f^{-i}(\mathcal{B}_m) \cup \bigcup_{i \geq 0} f^{-i}(C_m^{(0)}).$$

Note that $\bigcup_{i \geq 0} f^{-i}(C_m^{(0)})$ is a nowhere dense set because $C_m^{(0)}$ is a nowhere dense set. One can easily show that

$$\mathfrak{X} = I \setminus \left(\bigcup_{1 \leq m < m^*} \bigcup_{i \geq 0} f^{-i}(C_m^{(0)}) \right).$$

The assertion formulated above now becomes obvious.

3. Maps with Negative Schwarzian

The general form of phase diagrams of unimodal maps reflects dynamics possible for continuous maps. Thus, it is important to study the problem of realization of these possibilities for smooth maps. In particular, it is quite interesting to answer the following two questions: What maps are characterized by the "most simple" phase diagrams (i.e., by phase diagrams without wandering intervals and cycles of intervals that do not contain the point c)? Are there smooth maps characterized by the "most complicated" dynamics (i.e., maps for which the estimates for the number of classes $B_m^{(i)}$ and $\Gamma_m^{(i)}$ given in Section 2 are attained)? These and other similar questions are discussed in the present section.

It is clear that, in typical situations, each class of cycles of intervals $B_m^{(i)}$ is associated with an attracting cycle. In the following lemma, this assertion is formulated rigorously:

Lemma 5.9. *Assume that a map* $g : [a, b] \to [a, b]$ $(a \neq b)$ *is continuous and monotonically nondecreasing,* $g(a) = a$, *and* $g(b) = b$. *Then there exist a fixed point* $z \in [a, b]$ *of the map* g *and* $\varepsilon > 0$ *such that either* $z + \varepsilon < b$ *and* $g(x) \leq x$ *for all* $x \in (z, z + \varepsilon)$ *or* $z - \varepsilon > a$ *and* $g(x) \geq x$ *for all* $x \in (z - \varepsilon < z)$.

Proof. Assume that the assertion of the lemma does not hold for fixed points a and b. Then one can indicate points $x, y \in (a, b)$ such that $x < y$, $g(x) > x$, and $g(y) < y$. In this case, it follows from the continuity of the map g that it possesses one more fixed point in the interval (a, b). Hence, if the assertion of the lemma is violated for two fixed points of the map g, then the interval between these points contains a fixed point.

Thus, if the lemma is not true, then the set $\text{Fix}(g) = \{x \in [a, b] \mid g(x) = x\}$ is dense in $[a, b]$ and, consequently, $g(x) = x$ for all $x \in [a, b]$ but this contradicts the assumption of the lemma. Lemma 5.9 is proved.

Corollary 5.7. *If* $[c, d]$ *is a component of* $\text{Fix}(g)$, *then* $g(x) = x$ *for all* $x \in [c, d]$. *If* (c, d) *is a component of* $[a, b] \backslash \text{Fix}(g)$, *then* $g(c) = c$, $g(d) = d$, *and either* $g(x) > x$ *for all* $x \in (c, d)$ *or* $g(x) < x$ *for all* $x \in (c, d)$.

Note that the case of an orientation-reversing map g (i.e., the map g is nonincreasing) can easily be reduced to the case described above by passing to the map g^2.

Thus, the question about the number of classes $B_m^{(i)}$ in the phase diagram is closely related to the question about the number of attracting or semiattracting cycles of a map.

It is well known that unimodal maps defined by quadratic polynomials may have at most one attracting or semiattracting cycle. This property of quadratic maps was established as early as at the beginning of the century by Julia [1] and Fatou [1] when investigating rational endomorphisms of the Riemann sphere. It is important to clarify which property of quadratic maps is responsible for the restrictions imposed on the number of sinks. Julia and Fatou showed that the number of sinks of the maps under consideration is bounded by the number of critical points of these maps. However, by using Lemma 5.9, one can easily construct even a monotone map of the interval (of any smoothness) with any (finite or countable) number of attracting cycles.

Another important property of quadratic maps is their convexity. This property also cannot play a decisive role in this case because convexity is not invariant under iterations of maps. This observation is clarified by the following examples:

Example 5.1. Consider the orientation-reversing homeomorphism generated by the function $f(x) = \frac{1}{x}$, $x \in [\frac{1}{2}, 2]$. We have $f([\frac{1}{2}, 2]) = [\frac{1}{2}, 2]$ and, for $x \in [\frac{1}{2}, 2]$, $f'(x) = -\frac{1}{x^2} \le -\frac{1}{4}$ and $f''(x) = \frac{2}{x^3} \ge \frac{1}{4}$. On the other hand, $f^2(x) = x$ and, consequently, $(f^2)(x) = 0$ for any $x \in [\frac{1}{2}, 2]$.

Example 5.2. Consider the map $g_\varepsilon \in C^\infty([\frac{1}{2}, 2])$ generated by the equality

$$g_\varepsilon(x) = \begin{cases} 0, & 1/2 \le x \le 1, \\ \varepsilon e^{1/1-x} \sin \dfrac{\pi}{x-1}, & 1 < x \le 2. \end{cases}$$

It is not difficult to show that, for any $r \ge 0$, $\| g_\varepsilon \|_{C^r} \to 0$ as $\varepsilon \to 0$ and the set of zeros of the function $g_\varepsilon(x)$ coincides with the set

$$Z = \left[\frac{1}{2}, 1\right] \cup \left\{\frac{i+1}{i}\right\}_{i=1}^{\infty}.$$

Moreover, the function $g_\varepsilon(x)$ changes its sign on passing through each isolated point of the set Z ($g_\varepsilon'\left(\frac{i+1}{i}\right) \neq 0$ for all $i \geq 1$).

We choose $\varepsilon > 0$ such that $\| g_\varepsilon \|_{C^2} < \frac{1}{8}$ and consider the map $f_1(x) = f(x) + g_\varepsilon(x)$, where $f(x) = \frac{1}{x}$. Then

$$f_1 \in C^\infty([\tfrac{1}{2}, 2]),\ f_1([\tfrac{1}{2}, 2]) = [\tfrac{1}{2}, 2],\ \ f_1'(x) \leq \tfrac{1}{8},\ \text{and}\ \ f_1''(x) \geq \tfrac{1}{8}\ \text{ for } x \in [\tfrac{1}{2}, 2].$$

For the map f_1^2, we arrive at the equality

$$\operatorname{Fix}(f_1^2) = \{1\} \cup \left\{\frac{i}{i+1}, \frac{i+1}{i}\right\}_{i=1}^{\infty},$$

which can be established by direct computation. Thus, if $x \in [\frac{1}{2}, 1) \setminus \{\frac{i}{i+1}\}_{i=1}^{\infty}$, then

$$f_1(x) = f(x) \in (1, 2] \setminus \left\{\frac{i+1}{i}\right\}_{i=1}^{\infty}$$

and

$$f_1(f_1(x)) = f(f(x)) + g_\varepsilon(f(x)) = x + g_\varepsilon(f(x)) \neq x$$

because $g_\varepsilon(f(x)) \neq 0$. For any $i \geq 1$, we have $g_\varepsilon'\left(\frac{i+1}{i}\right) \neq 0$; therefore,

$$(f^2)'\left(\frac{i+1}{i}\right) = f'\left(\frac{i+1}{i}\right)\left(f\left(\frac{i+1}{i}\right) + g_\varepsilon'\left(\frac{i+1}{i}\right)\right) = 1 + f'\left(\frac{i+1}{i}\right) + g_\varepsilon'\left(\frac{i+1}{i}\right) \neq 1.$$

As follows from Corollary 5.7, the map f_1^2 has countably many attracting fixed points and countably many repelling fixed points. In particular, f_1 has countably many attracting cycles of period two (these cycles are formed by the pairs of points $\{\frac{i}{i+1}, \frac{i+1}{i}\}$ with odd $i \geq 1$).

Example 5.3. Let

$$g_{n,\varepsilon}(x) = \varepsilon(x-1) \prod_{i=1}^{n} \left(\frac{i}{i+1} - x\right)\left(\frac{i+1}{i} - x\right), \quad x \in \left[\frac{1}{2}, 2\right].$$

For any $n \geq 1$ and $r \geq 1$, $\| g_{n,\varepsilon} \|_{C^r} \to 0$ as $\varepsilon \to 0$. We fix $n \geq 1$ and choose $\varepsilon > 0$ such that $\| g_{n,\varepsilon} \|_{C^2} < \frac{1}{8}$. Consider the map $f_2(x) = f(x) + g_{n,\varepsilon}(x)$, where $f(x) = \frac{1}{x}$, $x \in [\frac{1}{2}, 2]$. Acting as in Example 5.2, we conclude that

$$\operatorname{Fix}(f_2^2) = \{1\} \cup \left\{ \frac{i}{i+1}, \frac{i+1}{i} \right\}_{i=1}^{\infty}$$

and the pairs of points $\frac{i}{i+1}$ and $\frac{i+1}{i}$ form 2-periodic attracting cycles of the map f_2 for odd i and repelling cycles for even i.

Note that the monotone map f_2 constructed in Example 5.3 is analytic. Hence, analyticity, as well as convexity, cannot guarantee the uniqueness of attracting cycle for unimodal maps.

On the other hand, unimodal maps from the family $f_{\alpha,\beta}: x \to \alpha x e^{-\beta x}$, $x \geq 0$, $\alpha > 0$, $\beta > 0$, are neither convex nor concave. Nevertheless, they have at most one sink just as quadratic maps (Jakobson [2]).

Maps from this family and quadratic maps are characterized by the following common property:

Their *Schwarzian*

$$Sf = \frac{f'''}{f'} - \frac{3}{2}\left[\frac{f''}{f'}\right]^2$$

(which is also called the *differential Schwarz invariant* or *Schwarzian derivative*) is negative in the entire domain of its definition. A remarkable property of the Schwarzian is the invariance of its sign under iterations of the map: Since

$$S(f \circ g) = Sf(g)(g') + Sg$$

(this equality can be verified by direct calculation), we have $Sf^n < 0$ (> 0) whenever $Sf < 0$ (> 0). Below, we show that just the negativity of the Schwarzian and the fact that the corresponding maps possess a unique critical point are responsible for the existence of at most one sink.

Before studying dynamical systems, we consider some properties of the maps whose Schwarzian preserves its sign.

Parallel with quadratic functions, there are many other primary functions with sign-preserving Schwarzian. Thus, it is negative for x^3, e^x, $\sin x$, and $\tan^{-1} x$ and positive for the corresponding inverse functions $\sqrt[3]{x}$, $\ln x$, $\sin^{-1} x$, and $\tan x$ (at all points where the functions and their Schwarzians are well defined). The indicated property of inverse functions is explained by the formula $Sf^{-1}(x) = -Sf(x)/[f'(x)]^2$.

It is not difficult to show that $Sf(x) = 0$ on an interval I if and only if $f(x)$ is a linear-fractional function on this interval.

Let $f \in C^3(I, I)$. Assume that the Schwarzian of the map f is well defined and preserves sign on an interval I (i.e., it is either always negative, or always positive, or always equal to zero). In this case, the map f has the following properties:

Property 1. *If $f'(x)Sf(x) < 0$, then the function $f'(x)$ has no local minima on an interval I; if $f'(x)Sf(x) > 0$, then it has no local maxima on this interval; if $f'(x)Sf(x) = 0$, then the function $f'(x)$ is monotone.*

Proof. Assume that $f'(x)$ possesses a local minimum at a point a. Then $f''(a) = 0$ and the condition of minimum implies the inequality $f'(a)\,Sf(a) = f'''(a) \geq 0$. The other assertions are proved similarly.

Property 2. *If $Sf(x) < 0$, then $\min\limits_{x \in I} |f'(x)| = \min\limits_{x \in \partial I} |f'(x)|$. If $Sf(x) > 0$, then $\max\limits_{x \in I} |f'(x)| = \max\limits_{x \in \partial I} |f'(x)|$.*

Property 3. *The function $f(x)$ has at most one point of inflection in I (i.e., at most one point where $f''(x) = 0$).*

These properties immediately follow from Property 1.

An exclusive place occupied by maps with negative Schwarzian in the collection of maps with sign-preserving Schwarzian is explained by the following assertion:

Proposition 5.1. *Let $f \in C^1(I, I)$ be a unimodal map and let $K(f) = \{ x \in I \mid f'(x) = 0 \}$. Suppose that $f \in C^3(I \backslash K(f))$ and the Schwarzian of the map f preserves its sign on the set $I \backslash K(f)$. Then $Sf(x) < 0$ for $x \in I \backslash K(f)$.*

Proof. It follows from the definition of Schwarzian that if $g(x+d) = af(x) + b$ for some $a \in \mathbb{R} \backslash \{0\}$, $b, d \in \mathbb{R}$, then $Sg(x+d) = Sf(x)$. Hence, without loss of generality, we can assume that $0 \in I$ and $c = 0$ is the maximum point of the map $f\colon I \to I$.

Let $I = [y, y']$. Then $f(0) > f(y')$. We choose a constant b such that the map $g(x) = f(x) + b$ satisfies the inequalities $g(0) > 0$ and $g(y') < 0$. Then there exists a unique point $z \in (0, y')$ for which $g(z) = 0$.

Consider the function g^2 on the interval $[0, z]$. It is easy to check that g^2 increases on $(0, z)$ and satisfies the equalities $(g^2)'(0) = (g^2)'(z) = 0$. If $Sg(x) \geq 0$, then, by Property 2 of the maps with sign-preserving Schwarzian, $g^2(x) \geq 0$ for all $x \in [0, z]$ but this contradicts the inequality $g^2(0) = g(g(0)) < g(0) = g^2(z)$.

Remark. The condition of unimodality of f in Proposition 5.1 is not essential:

It follows from the proof that it suffices to impose the condition that f is not a constant. One can also omit the inclusion $f(I) \subset I$.

According to the proof of Proposition 5.1, unimodal maps with positive and zero Schwarzian cannot be differentiable at the point c. Moreover, for these maps, both one-sided derivatives are not equal to zero at the point c.

In what follows, unimodal maps satisfying the conditions of Proposition 5.1 are referred to as SU-maps or S-unimodal maps.

To establish restrictions that should be imposed on the number of sinks for SU-maps, we consider some properties of periodic trajectories of maps with negative Schwarzian.

Lemma 5.10. *Let $f \in C^3(I, I)$, let $Sf(x) < 0$ for $x \in I \setminus K(f)$, where $K(f) = \{x \in I \mid f'(x) = 0\}$, and let $B = \{\beta_0, \beta_1, \dots, \beta_{n-1}\}$ be a cycle of the map f. Assume that $|\mu(B)| \le 1$, where $\mu(B) = f'(\beta_0) \cdot f'(\beta_1) \cdot \ldots \cdot f'(\beta_{n-1})$ is the multiplier of the cycle B. Then B is either an attracting cycle or a semiattracting cycle of the map f.*

Proof. It suffices to consider the case $|\mu(B)| = 1$. If $\mu(B) = 1$, then, for the map $g = f^n$, we have $g'(\beta_0) = \mu(B) = 1$. If, in this case, $g''(\beta_0) \neq 0$, then the cycle B is semiattracting. If $g''(\beta_0) = 0$, then it follows from the condition $Sf(x) < 0$ that $Sg(\beta_0) = g'''(\beta_0) < 0$ and, consequently, B is an attracting cycle.

If $\mu(B) = -1$, then we consider the map $g^2 = f^{2n}$. For this map, $(g^2)'(\beta_0) = (g'(\beta_0))^2 = 1$ and $(g^2)''(\beta_0) = g''(\beta_0)(g'(\beta_0))^2 + g''(\beta_0)g'(\beta_0) = 0$. The condition $Sf(x) < 0$ implies that $Sg^2(\beta_0) = (g^2)'''(\beta_0) < 0$ and, consequently, in this case, the cycle B is also attracting.

Lemma 5.11. *Assume that $f \in C^1(I, I)$, $f \in C^3(I \setminus K(f))$, and $Sf(x) < 0$ for $x \in I \setminus K(f)$. Let $B = \{\beta_0, \beta_1, \dots, \beta_{n-1}\}$ be an n-periodic attracting or semiattracting cycle of the map f and let $P_0(B)$ be its domain of immediate attraction. If $n > 2$, then $P_0(B) \cap K(f) \neq \varnothing$ and if $n \le 2$, then $P_0(B) \cap \{K(f) \cup \partial I\} \neq \varnothing$.*

Proof. As follows from the results established in Chapter 1, the set $P_0(B)$ consists of disjoint intervals $J_0, J_1, \dots, J_{n-1}$, which form an n-periodic cycle of intervals. If $n > 2$, then the indicated collection of intervals contains an interval J_i such that $J_i \cap \partial I = \varnothing$. Let J_0 be an interval of this sort and let a and b be its ends. Then the following three cases are possible for the map $g = f^n$:

(i) $g(a) = a$ and $g(b) = b$;

(ii) $g(a) = b$ and $g(b) = a$;

(iii) $g(a) = g(b)$.

Let us show that, in all cases, the interval J_0 contains the critical point of the map g. In case (iii), this is obvious. Case (ii) is reduced to case (i) if we consider the map g^2.

Let $g(a) = a$ and $g(b) = b$. Assume that $J_0 \cap c(g) = \varnothing$ and the cycle B is attracting. Then $\beta_0 \in (a, b)$, the inequality $g(x) > x$ holds for $x \in (a, \beta_0)$, and the inequality $g(x) < x$ holds for $x \in (\beta_0, b)$. Hence, by the law of mean, there exist points $z_1 \in (a, \beta_0)$ and $z_2 \in (\beta_0, b)$ such that $g'(z_1) = g'(z_2) = 1$. By Property 2 of maps with sign-preserving Schwarzian, we have $g'(x) \geq 1$ for all $x \in [z_1, z_2]$, which is impossible in view of the fact that $g(z_2) - g(z_1) < z_2 - z_1$.

If B is a semiattracting cycle, then either $\beta_0 = a$ or $\beta_0 = b$; moreover, we have $g'(\beta_0) = 1$. If $\beta_0 = a$, then $g'(z) = 1$ for some point $z \in (a, b)$. Hence, we again conclude that $g'(x) \geq 1$ for all $x \in [\beta_0, z]$, which is impossible because $g(x) < x$ for $x \in (\beta_0, b)$. The case $\beta_0 = b$ can be investigated similarly.

Thus, for $n > 2$, we have $J_0 \cap K(g) \neq \varnothing$. Let $c_1 \in J_0 \cap K(g)$. It is easy to show that

$$K(g) = \bigcup_{i=0}^{n-1} f^{-1}(K(f)).$$

Hence, $f^k(c_1) \in K(f)$ for some $k < n$, i.e., $f^k(J_0) \cap K(f) \neq \varnothing$ and $P_0(B) \cap K(f) \neq \varnothing$.

Similar reasoning is applicable in the case where $n \geq 2$. However, in this case, it is possible that the ends of intervals of the domain of immediate attraction are not fixed points of the map g (and an end of the interval I is attracted by the cycle B).

Corollary 5.8. *Let $f \in C^1(I, I)$ and let $f \in C^3(I \backslash K(f))$. Assume that $Sf(x) < 0$ for $x \in I \backslash K(f)$. Then the number of attracting and semiattracting cycles of the map f does not exceed the number of components of the set $K(f)$ plus two.*

The assertion of Corollary 5.8 immediately follows from the statement and proof of Lemma 5.11.

Example 5.4. Consider a map $f: I \to I$ defined by the formula (see Fig. 40):

$$f(x) = \begin{cases} \dfrac{\sqrt{2}}{4} \sin \pi \left(x - \dfrac{1}{4}\right) + \dfrac{1}{4}, & x \in \left[0, \dfrac{1}{2}\right], \\[2ex] \dfrac{\sqrt{2}}{4} \sin \pi \left(x - \dfrac{3}{4}\right) + \dfrac{2+\sqrt{2}}{8}, & x \in \left(\dfrac{1}{2}, 1\right]. \end{cases}$$

We have $f \in C^1(I, I)$ and $f \in C^3(I \backslash \{\frac{1}{2}\})$. The Schwarzian of the map f is negative everywhere except the point $\frac{1}{2}$ (where it is not defined) and f is a monotone function without critical points. Nevertheless, the map f has three attracting fixed points: 0, $\frac{1}{2}$, and 1.

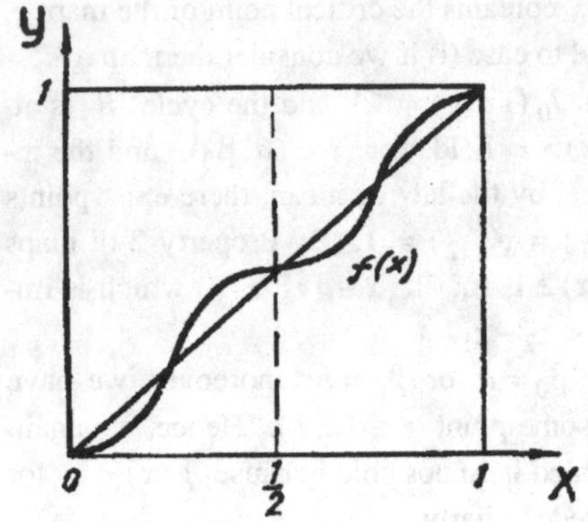

Fig. 40

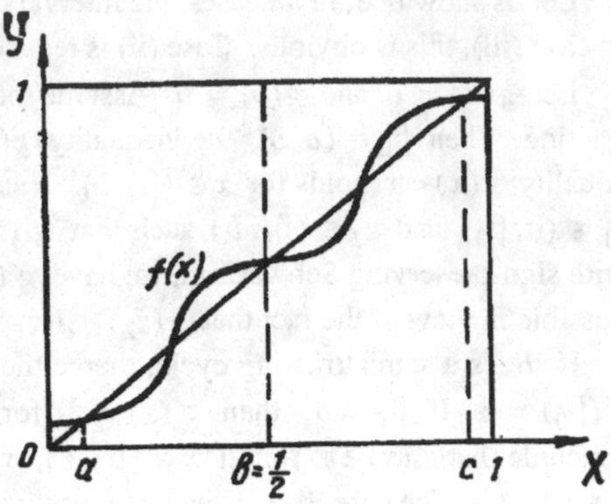

Fig. 41

Similarly, one can construct a homeomorphism $f: I \to I$ with negative Schwarzian which belongs to the class C^3 everywhere except countably many points (where $f \in C^1$) and has countably many attracting cycles.

Example 5.5. Let

$$f(x) = \begin{cases} \frac{1}{4} g\left(2x - \frac{1}{2}\right) + \frac{1}{4}, & x \in \left[0, \frac{1}{2}\right], \\ \frac{1}{4} g\left(2x - \frac{3}{2}\right) + \frac{3}{4}, & x \in \left(\frac{1}{2}, 1\right], \end{cases}$$

where

$$g(x) = 140 \int_0^x \left(\frac{1}{2} - x\right)^3 \left(\frac{1}{2} + x\right)^3 dx$$

(see Fig. 41).

The map f belongs to $C^3(I, I)$ and has three attracting points a, b, and c. It is not difficult to show that the Schwarzian Sf is negative everywhere except the points 0, $\frac{1}{2}$, and 1, where the first three derivatives vanish. Thus, the map $f: [a, c] \to [a, c]$ has exactly one critical point $x = \frac{1}{2}$ and three attracting fixed points.

For SU-maps, the assertion of Corollary 5.8 implies the following theorem:

Theorem 5.1. *Let $f: I \to I$ be an SU-map. Assume that $f'(x) \neq 0$ for $x \neq c$. Then f has at most two attracting or semiattracting cycles. Moreover, if there are two* *cycles of this sort, then one of these cycles is a fixed point attracting the trajectory of at least one of the ends of the interval I.*

Proof. Let $I = [a, b]$, let c be the maximum point, and let n be the period of an attracting or semiattracting cycle B of the map f. As follows from Lemma 5.11, we must consider only the cases $n = 2$ and $n = 1$. If $n = 2$, then $P_0(B) = J_0 \cup J_1$. If, in this case, $J_i \cap \partial I = \varnothing$ for some $i \in \{0, 1\}$, then it follows from the proof of Lemma 5.11 that $c \in P_0(B)$. Now assume that $a \in J_0$ and $b \in J_1$. In view of the unimodality of the map f, in this case, one can also prove that $c \in P_0(B)$. Hence, under the conditions of the theorem, $c \in P_0(B)$ for any $n \geq 2$.

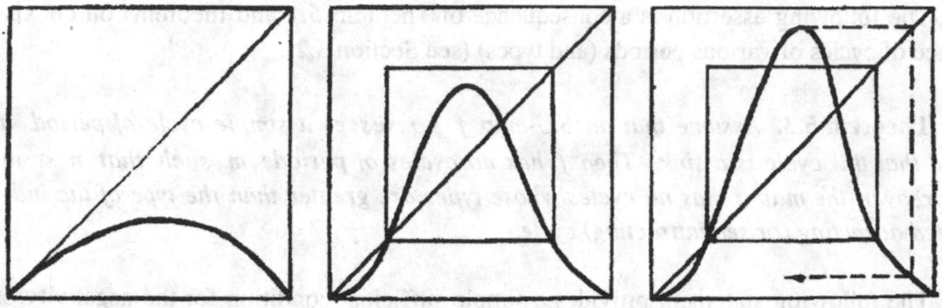

a **b** **c**

Fig. 42

For $n = 1$, let x^* denote a fixed point of the map f such that $a \notin P(x^*)$. In this case, by using Property 2 of maps with sign-preserving Schwarzian and the unimodality of the map f, we can also prove that $c \in P(x^*)$. This completes the proof of the theorem.

Corollary 5.9. *If, under the conditions of Theorem 5.1,* $f'(x) \notin [0, 1]$ *for* $x \in \partial I$, *then* f *has at most one attracting or semiattracting cycle.*

This statement follows from the proof of Theorem 5.1. Indeed, if $a \in P_0(x^*)$ for some attracting fixed point x^*, then Property 2 implies that $f'(a) < 1$ (provided that c is the maximum point). Simple analysis demonstrates (see Fig. 42) that if $f'(a) < 1$, then either the structure of the map f is quite simple or the investigation of this map can be reduced to the investigation of a map that satisfies the conditions of the corollary.

Theorem 5.2. *Let* $f: I \to I$ *be an SU-map. Then the set* $\mathrm{Fix}(f^n)$ *is finite for any* $n \geq 1$. *In particular, the set* $\mathrm{Per}(f)$ *is at most countable.*

Proof. Under the conditions of the theorem, the map f^n has finitely many intervals of monotonicity. If the set $\mathrm{Fix}\,(f^n)$ is infinite, then one can find an interval of monotonicity of the map f^n that contains infinitely many fixed points. Hence, this interval contains infinitely many points at which the derivative of the map f^n is equal to one. By virtue of Property 2, there exists an interval on which f^n coincides with the identity map but this contradicts, e.g., Lemma 5.10.

Remark. By using Theorem 5.2, one can collect a lot of information about the structure of the set $\mathrm{Per}\,(f)$ for SU-maps. Thus, it follows from Theorem 5.2 and results established in Sections 5.1 and 5.2 that, for SU-maps, the set $\mathrm{Per}\,(f)$ is closed if and only if $m^* < \infty$ and $p_{m+1}/p_m = 2$ for all $m < m^*$ in the phase diagram of the map f. In this case, the periods of periodic orbits of the map f are uniformly bounded.

The following assertion is a consequence of Theorem 5.1 and theorems on coexistence of cycles of various periods (and types) (see Section 3.2).

Theorem 5.3. *Assume that an SU-map f possesses a simple cycle of period n and that this cycle is a sink. Then f has no cycles of periods m such that $n \lhd m$. Moreover, the map f has no cycles whose types are greater than the type of the indicated attracting (or semiattracting) cycle.*

The following statement provides a simple sufficient condition for the negativity of Schwarzian in the case of polynomial maps.

Proposition 5.2. *If $f(x)$ is a polynomial whose degree is greater than one and all roots of the equation $f'(x) = 0$ are real, then $Sf(x) < 0$ for all x such that $f'(x) \neq 0$.*

Proof. Let $n+1$ be the degree of the polynomial $f(x)$, $n \geq 1$. Then, by the condition of Proposition 5.2, we can write

$$f'(x) = a_0 \prod_{i=1}^{n} (x - a_i),$$

where $a_0 \neq 0$ and $a_i \in \mathbb{R}$. Therefore,

$$Sf(x) = 2 \sum_{i=1}^{n-1} \sum_{j=2}^{i+1} \frac{1}{(x-a_i)(x-a_j)} - \frac{3}{2}\left(\sum_{i=1}^{n} \frac{1}{(x-a_i)}\right)^2 < 0$$

because

$$2 \sum_{i=1}^{n-1} \sum_{j=2}^{i+1} \frac{1}{(x-a_i)(x-a_j)} \leq \left(\sum_{i=1}^{n} \frac{1}{(x-a_i)}\right)^2.$$

In particular, the Schwarzian of a polynomial is negative whenever all its zeros are real. The well-known fact that the Schwarzians of quadratic maps are negative also follows from Proposition 5.2.

Note that polynomial maps have the following property, which is not based on the characteristics of the Schwarzian:

Let $f(x) = a_n x^n + a_{n-1} x^{n-1} + \ldots + a_0$, where $n \geq 2$ and $a_i \in \mathbb{R}$. Then the map $f\colon \mathbb{R} \to \mathbb{R}$ has at most $n-1$ attracting or semiattracting trajectories. More exactly, if $F(z) = a_n z^n + a_{n-1} z^{n-1} + \ldots + a_0$, where $z \in \mathbb{C}^1$, then the number of sinks of the real map $f\colon \mathbb{R} \to \mathbb{R}$ does not exceed the number of different complex roots of the equation $F'(z) = 0$.

Indeed, let $z = x + iy$. Then $F(z) = \varphi(x, y) + i\psi(x, y)$, where $\varphi\colon \mathbb{R}^2 \to \mathbb{R}$ and $\psi\colon \mathbb{R}^2 \to \mathbb{R}$ are such that $\varphi(x, 0) = f(x)$ and $\psi(x, 0) = 0$ for $x \in \mathbb{R}$. It follows from the Cauchy–Riemann conditions for complex functions that $\frac{\partial\varphi}{\partial x} = \frac{\partial\psi}{\partial y}$ and $\frac{\partial\varphi}{\partial y} = -\frac{\partial\psi}{\partial x}$. Hence, for $z = x + i \cdot 0$, we have $\frac{\partial\varphi}{\partial y} = -\frac{\partial\psi}{\partial x} = 0$ and $dF(z) = f'(x)dz$. Therefore, every attracting cycle of $f\colon \mathbb{R} \to \mathbb{R}$ is an attracting cycle of the map $F\colon \mathbb{C}^1 \to \mathbb{C}^1$ and every neutral cycle of the map $f\colon \mathbb{R} \to \mathbb{R}$ is a neutral rational cycle of the map $F\colon \mathbb{C}^1 \to \mathbb{C}^1$. As shown by Julia [1] and Fatou [1], the number of attracting and neutral rational cycles of the polynomial map $F\colon \mathbb{C}^1 \to \mathbb{C}^1$ does not exceed the number of its critical points, i.e., the number of different roots of the equation $F'(z) = 0$.

Note that this fact is not a consequence of Corollary 5.8. Thus, Singer [1] constructed an example of a unimodal map given by a polynomial of the fourth degree with two sinks (see also the remark at the end of this section).

Our interest to the class of maps with sign-preserving Schwarzian is not restricted to maps with negative Schwarzian. Thus, some problems are connected with the study of iterations of maps that consist of finitely many pieces of linear-fractional functions (see Aliev et al. [1]), i.e., of maps of the form $g(x) = \frac{\alpha x + \beta}{\gamma x + \delta}$. Their Schwarzian is equal to zero, and one can formulate the following analog of Theorem 5.1 for these maps:

Theorem 5.4. *If $f\colon I \to I$ is a unimodal map, $f \in C^3(I \setminus \{c\})$, and $Sf(x) = 0$ for $x \in I \setminus \{c\}$, then the set of all cycles of the map f that are not repelling is either empty, or consists of one attracting or semiattracting cycle, or is a cycle of closed intervals $B = \{J_0, J_1, \ldots, J_{n-1}\}$ of period $n \geq 1$ such that the point c is one of the ends of the interval J_0 and $f^n(x) = x$ for any point $x \in J_0$.*

Theorem 5.4 can be proved just as Theorem 5.1 by using Properties 1–3 of maps with sign-preserving Schwarzian.

For any unimodal map f with positive Schwarzian, the set $\mathrm{Fix}(f^n)$ is also finite for any $n \geq 1$. At the same time, unimodal maps with positive Schwarzian may have more than one attracting cycle. (Properties 1–3 do not impose any direct prohibition even on the existence of countably many attracting cycles of these maps.) Thus, let

$$g_\lambda(x) = \begin{cases} \lambda \sqrt[4]{\dfrac{x}{2}}, & 0 \le x < \dfrac{1}{2}, \\ \lambda\sqrt{1-x}, & \dfrac{1}{2} \le x \le 1, \quad \lambda \in (0, \sqrt{2}). \end{cases}$$

For any $x \in [0, 1] \setminus \{1/2\}$, we have $Sg_\lambda(x) > 0$. One can easily show that, for $\lambda < \frac{2}{\sqrt{3}}$, the map g_λ possesses an attracting fixed point (other than the fixed point $x = 0$). For $\frac{1+\sqrt{5}}{2\sqrt{2}} < \lambda < \sqrt{2}$, this map possesses two cycles of period two (attracting and repelling). Hence, for $\lambda \in \left(\frac{1+\sqrt{5}}{2\sqrt{2}}, \frac{2}{\sqrt{3}}\right)$, the map g_λ has both an attracting fixed point and an attracting cycle of period two.

Remark. The importance of the function

$$Sf(x) = \frac{f'''(x)}{f'(x)} - \frac{3}{2}\left[\frac{f''(x)}{f'(x)}\right]^2$$

was first noticed by Herman Schwarz [1] in the second half of the 19th century in connection with the investigation of conformal maps. Allwright [1] and Singer [1] almost simultaneously applied the notion of Schwarzian to the study of one-dimensional dynamical systems. Thus, Singer [1] gave the first proof of Theorem 5.1 and constructed the following example of a unimodal convex map with two sinks:

$$f(x) = 7.86x - 23.31x^2 + 28.75x^3 - 13.30x^4.$$

For this map, $\beta = 0.7263986\ldots$ is a sink of period one and $\beta_1 = 0.3217591\ldots$, $\beta_2 = 0.9309168\ldots$ is a sink of period two. It is clear that the Schwarzian of this map cannot preserve its sign (this is a consequence of Proposition 5.1).

It is worth noting that, in Chapter 8, the expression for the Schwarzian appears in a natural way as a characteristic of the period doubling bifurcation for periodic trajectories of smooth one-dimensional maps.

Note that the constancy of sign of the Schwarzian is a sufficient condition for the validity of Properties 1–3 used in the study of the dynamics of maps. There are more general conditions, which, in particular, do not require the existence of the third derivative of a map but their verification is more complicated. Some of these conditions are considered below.

Matsumoto [1] studied C^2-maps such that any their iteration has at most one inflection point in each interval of monotonicity of this iteration (cf. Property 3 presented above).

For a map $f \in C^1(I, I)$, one can use the properties of concavity, convexity, or linearity of the function $g(x) = |f'(x)|^{-1/2}$ instead of the negativity, positivity, or equality to zero of the Schwarzian, respectively.

Indeed, if $f \in C^3(I)$, then $g''(x) = \frac{1}{2} g(x) Sf(x)$ (this equality can be checked by direct computation).

If the existence of derivatives of a map f is not required, then one can use the so-called hypergraphic property, which is defined as follows: Let $f \in C^0(I)$ be a monotone function defined on the interval I and let $x_1 < x_2 < x_3 < x_4$ be points of this interval. Consider a function

$$G(x_1, x_2, x_3, x_4) = \frac{(x_4 - x_1)(x_3 - x_2)}{(x_4 - x_3)(x_2 - x_1)}.$$

A function $f(x)$ is called hypergraphic on I if the quantity

$$\mathcal{R}_f(x_1, x_2, x_3, x_4) = G(x_1, x_2, x_3, x_4) - G(f(x_1), f(x_2), f(x_3), f(x_4))$$

has the same sign for any set of four points $x_1 < x_2 < x_3 < x_4$ from the interval I. If $f \in C^3$, then the negativity, positivity, and equality to zero of the quantity $\mathcal{R}_f(x_1, x_2, x_3, x_4)$ on the interval I are, respectively, equivalent to the negativity, positivity, and equality to zero of the Schwarzian $Sf(x)$ of the function f on the interval I. This fact can be proved by using the relation

$$G(f(x + x_1 t), f(x + x_2 t), f(x + x_3 t), f(x + x_4 t))$$

$$= G(x_1, x_2, x_3, x_4)\left(1 + \frac{t^2}{6}(x_1 - x_2)(x_3 - x_4) Sf(x) + O(t^3)\right).$$

This observation enables us to reformulate Properties 1–3 and Theorem 5.1 for a broader class of unimodal maps.

In the next section, we shall prove that unimodal maps from the class C^2 with nondegenerate critical point cannot have wandering intervals. This result and Theorem 5.1 imply the following assertion:

Theorem 5.5. *Let f be an SU-map, let $f'(x) \neq 0$ for $x \neq c$, and let $f''(c) \neq 0$. Then the phase diagram of the map f consists only of central nodes and, possibly, of cycles of intervals $B_1^{(1)}$ of period one and $B_{m^*}^{(1)}$ of period p_{m^*} (in this case, $m^* < \infty$) corresponding to the domains of immediate attraction of attracting or semiattracting cycles of the map.*

Corollary 5.10. *If a map $f: I \to I$ satisfies the conditions of Theorem 5.5, then* $\mathrm{NW}(f) = \overline{\mathrm{Per}(f)}$.

This assertion follows from Theorem 5.5 and the results established in Section 5.2.

Note that the corresponding statements can also be formulated for the spectral decomposition of the set $\mathrm{NW}(f)$ of maps of this sort.

4. Maps with Nondegenerate Critical Point

Let $f: I \to I$ be a map from the class $C^1(I, I)$ and let $K(f) = \{x \in I : f'(x) = 0\}$ be the set of its critical points. A point $c \in K(f)$ is called nondegenerate if there exists a neighborhood U of the point c in the interval I such that $f \in C^2(U)$ and $f''(c) \neq 0$. Similarly, a point $c \in K(f)$ is called nonflat if there exist a neighborhood U of the point c in the interval I and $r > 1$ such that $f \in C^r(U)$ and

$$\frac{d^r f}{dx^r}(c) \neq 0.$$

In this section, we describe the dynamics of maps from the class $C^2(I, I)$ with nonflat critical points.

To formulate our principal results, we introduce the following notation:

For a unimodal map $f: I \to I$, let $\mathcal{B}_0(f)$ denote the set of points from the classes of cycles of interval $B_m^{(i)}$, i.e.,

$$\mathcal{B}_0(f) = \bigcup_{m=1}^{m^*} \bigcup_{i=1}^{j_m} B_m^{(i)}.$$

The domain of attraction of $\mathcal{B}_0(f)$ is denoted by $\mathcal{B}(f)$. It follows from the results of Section 5.2 that $\mathcal{B}(f) = \bigcup_{i \geq 0} f^{-i}(\mathcal{B}_0(f))$. An interval U is called a wandering interval of the map f if $U \not\subset \mathcal{B}(f)$ and $f^i(U) \cap f^j(U) = \varnothing$ for $i \neq j$. The set of points that belong to wandering intervals is denoted by $\Gamma(f)$. It is clear that the intervals from the classes $\Gamma_m^{(i)}$ belong to the set $\Gamma(f)$.

It should be noted that significant results in this field were obtained after the appearance of the Russian version of this book. Thus, the results obtained by Martens, de Melo, and van Strien and by Blokh and Lyubich made it possible to formulate the following two theorems, which establish the fact that the dynamics of smooth maps is, in a certain sense, similar to the dynamics of maps with negative Schwarzian derivative:

Theorem 5.6. *Let f be a C^∞-map with nonflat critical points. Then $\Gamma(f) = \varnothing$, i.e., the map f has no wandering intervals.*

Theorem 5.7. *Let f be a C^∞-map with nonflat critical points. Then the set $\mathcal{B}_0(f)$ consists of finitely many intervals, i.e., the periods of attracting or semiattracting periodic points of the map f are uniformly bounded.*

For phase diagrams of unimodal maps, these theorems imply the following assertion:

Corollary 5.11. *The phase diagrams of C^∞-unimodal maps with nonflat critical points satisfy the relations*

$$\sum_{m=1}^{m^*} j_m < \infty \quad \text{and} \quad \sum_{m=1}^{m^*} k_m = 0.$$

For the original ideas of the proof of Theorems 5.6 and 5.7, we refer the reader to the works by de Melo and van Strien [1] and van Strien [2, 3]. A detailed presentation of the dynamics of smooth maps of an interval can be found in the book by de Melo and van Strien [2].

The following two theorems demonstrate that the conditions of nonflatness in Theorems 5.6 and 5.7 cannot be omitted.

Theorem 5.8. *There exists a unimodal C^∞-map of type 2^∞ with three critical points (two critical points that are not extrema are flat) such that* $\mathrm{NW}(f) \setminus \overline{\mathrm{Per}\, f} \neq \varnothing$.

Proof. We fix $\varepsilon \in (0, 1)$ and choose a sequence $\beta_1 > \beta_2 > \ldots > \beta_i > \ldots > 0$ satisfying the condition

$$\sum_{i=1}^{\infty} \beta_i < \varepsilon.$$

For example, let $\beta_i = \varepsilon 2^{-i-1}$.

Assume that an interval $I_1 = I_1^1 = [x_0, y_0] \subset I$ is such that $\operatorname{mes} I_1^1 = \varepsilon$. We define $I_2 = I_{10}^2 \cup I_{11}^2 \subset I_1^1$, where I_{10}^2 and I_{11}^2 are closed disjoint intervals such that one end of the first interval coincides with the point x_0 and an end of the second interval coincides with the point y_0. The lengths of the intervals I_{10}^2 and I_{11}^2 are, respectively, $(\varepsilon - \beta_1)(\delta_1 + 1)^{-1}$ and $(\varepsilon - \beta_1)\delta_1(\delta_1 + 1)^{-1}$, $\delta_1 > 1$. Denote $U_1 = I_1 \setminus I_2$.

The sets I_m with $m > 2$ are defined by induction. Assume that we have already constructed the set $I_{m-1} = \bigcup_\alpha I_\alpha^{m-1}$, where $\alpha = \alpha_1 \alpha_2 \ldots \alpha_{2^{m-2}}$ is a sequence of zeros and ones such that, for any $i = 1, 2, 4, \ldots, 2^{m-3}$, we have either $\alpha_1 \ldots \alpha_i = \alpha_{i+1} \ldots \alpha_{2i}$ or $\alpha_1 \ldots \alpha_i = \overline{\alpha}_{i+1} \ldots \overline{\alpha}_{2i}$, where $\overline{\alpha}_i = 1 - \alpha_i$. In this case, the set $U_{m-2} = \bigcup_\alpha U_\alpha^{m-2}$, where $U_\alpha^{m-2} \subset I_\alpha^{m-2}$ are open intervals, is also defined. Open intervals $U_\alpha^{m-1} \subset I_\alpha^{m-1}$ are defined so that their lengths are equal to $\gamma_{m-1} \cdot \beta_{m-1}$, where

$$\gamma_{m-1} = \operatorname{mes} I_\alpha^{m-1}\left(\varepsilon - \sum_{i=1}^{m-2}\beta_i\right)^{-1},$$

and the intervals from the set $I_\alpha^{m-1} \setminus U_\alpha^{m-1}$ satisfy the conditions

(i) $I_\beta^m \cup I_{\beta'}^m = I_\alpha^{m-1} \setminus U_\alpha^{m-1}$, where $\beta = \alpha\alpha$, $\beta' = \alpha\overline{\alpha}$, and the interval $I_{\beta'}^m$ is located to the right (left) of U_α^{m-1} if $\alpha_1 \ldots \alpha_{2^{m-3}} = \overline{\alpha}_{2^{m-3}+1} \ldots \overline{\alpha}_{2^{m-2}}$ ($\alpha_1 \ldots \alpha_{2^{m-3}} = \alpha_{2^{m-3}+1} \ldots \alpha_{2^{m-2}}$);

(ii) $$\operatorname{mes} I_{\beta'}^m = \left(\varepsilon - \sum_{i=1}^{m-1}\beta_i\right)\gamma_{m-1}\delta_{m-1}(\delta_{m-1}+1)^{-1},$$

$$\operatorname{mes} I_\beta^m = \left(\varepsilon - \sum_{i=1}^{m-1}\beta_i\right)\gamma_{m-1}(\delta_{m-1}+1)^{-1}, \quad \delta_{m-1} > 1.$$

Also let δ_m, $m \geq 1$, be such that

$$\prod_{m=1}^{\infty}\delta_m(\delta_m+1)^{-1} \neq 0.$$

For example, let $\delta_m = 2^m$. It follows from the construction that

$$I_\infty = \bigcap_{m=1}^{\infty}\bigcup_{\alpha} I_\alpha^m$$

contains countably many intervals (denoted by J_i) and $I_\infty \setminus \bigcup_i J_i$ is a Cantor set.

A map $f(I_1) = I_1$ satisfying the conditions of the theorem is determined as follows: First, we take two sequences x_i, y_i $i \geq 1$, of points from I such that

$$x_{2i+1} = x_{2i} + \left(\varepsilon - \sum_{j=1}^{2i+2}\beta_j\right)\prod_{j=1}^{2i+1}\delta_j(\delta_j+1)^{-1}(\delta_{2i+2}+1)^{-1},$$

$$x_{2i+2} = x_{2i+1} + \beta_{2i+2}\prod_{j=1}^{2i+1}\delta_j(\delta_j+1)^{-1},$$

$$y_{2i+1} = y_{2i} - \left(\varepsilon - \sum_{j=1}^{2i+1}\delta_j(\delta_j+1)^{-1}\right)(\delta_{2i+1}+1)^{-1},$$

$$y_{2i+2} = y_{2i+1} - \beta_{2i+1} \prod_{j=1}^{2i} \delta_j(\delta_j + 1)^{-1},$$

$$y_1 = y_0 - (\varepsilon - \beta_1)(\delta_1 + 1)^{-1}.$$

The sequences x_i, y_i $i \geq 1$, are formed by the ends of the intervals U_α^m such that $\alpha_1 \ldots \alpha_j = \overline{\alpha}_{j+1} \ldots \overline{\alpha}_{2j}$ for all $j = 1, 2, 4, \ldots, 2^{m-2}$.

At the points x_i, y_i $i \geq 0$, the map f is defined as follows:

$$f(x_{2i-1}) = y_0 - \left(\varepsilon - \sum_{j=1}^{2i} \beta_j\right) \prod_{j=1}^{2i} (\delta_j + 1)^{-1} - \beta_{2i} \prod_{j=1}^{2i-1} (\delta_j + 1)^{-1},$$

$$f(x_{2i}) = y_0 - \left(\varepsilon - \sum_{j=1}^{2i+1} \beta_j\right) \prod_{j=1}^{2i+1} (\delta_j + 1)^{-1},$$

$$f(y_{2i}) = y_0 - \left(\varepsilon - \sum_{j=1}^{2i} \beta_j\right) \prod_{j=1}^{2i} (\delta_j + 1)^{-1},$$

$$f(y_{2i+1}) = y_0 - \left(\varepsilon - \sum_{j=1}^{2i+1} \beta_j\right) \prod_{j=1}^{2i+1} (\delta_j + 1)^{-1} - \beta_{2i+1} \prod_{j=1}^{2i-1} (\delta_j + 1)^{-1},$$

$$f(y_1) = y_0 - (\varepsilon - \beta_1)(1 + \delta_1)^{-1} - \beta_1, \quad f(x_0) = y_1, \quad f(y_0) = x_0.$$

In the intervals (x_{2i}, x_{2i+1}) and (y_{2i+1}, y_{2i}), we set $f(x)$ to be equal to

$$f(x_{2i}) + \delta_{2i+2} \prod_{j=1}^{2i+1} \delta_j^{-1}(x - x_{2i}), \quad x \in (x_{2i}, x_{2i+1});$$

$$f(y_{2i}) + \delta_{2i+1} \prod_{j=1}^{2i} \delta_j^{-1}(y_{2i} - x), \quad x \in (y_{2i+1}, y_{2i}).$$

We extend the map $f: I_1 \to I_1$ by continuity to the intervals (x_{2i+1}, x_{2i+2}) and (y_{2i+2}, y_{2i+1}) as monotone functions and to the interval $(\lim_{i\to\infty} x_i, \lim_{i\to\infty} y_i)$ as a segment of the straight line $y = y_0$. It is not difficult to show that the map thus constructed is a map of type 2^∞ with "flat" extremum (the interval where the map f attains its maximum value).

For the problem under consideration, we define the map f on the intervals $[0, x_0)$ and $(y_0, 1]$ as C^∞-smooth monotone functions. In the interval $(\lim_{i\to\infty} x_i, \lim_{i\to\infty} y_i)$, the "flat" extremum is replaced by a unimodal C^∞-function whose derivatives at the points $\lim_{i\to\infty} x_i$ and $\lim_{i\to\infty} y_i$ are equal to zero. Thus, we defined a unimodal map $f: I \to I$. This map is of type 2^∞ if the trajectory of the interval $J = [y_0, f(c)]$ lies in $I \setminus I_\infty$.

Note that the trajectory of J belongs to the set $\bigcup_{m=1}^{\infty} U_m$. Moreover, it "cuts off" the left and right ends of each interval U_α^m from this set. In our construction of a C^∞-map f, we do not change its values on the intervals $[x_{2i}, x_{2i+1}]$ and $[y_{2i+1}, y_{2i}]$ and assume that each interval from the trajectory of J constitutes a third part of the interval U_α^m such that $J \subset U_\alpha^m$.

Denote the derivative of the function $f|_{[x_{2i}, x_{2i+1}]}$ by δ_i. Let δ_i^+ denote the "averaged derivative" of f on the interval from the trajectory of J whose end coincides with the point x_{2i+1} (this interval is denoted by (x_{2i+1}, x'_{2i+1})), let δ_{i+2}^- denote the "averaged derivative" of f on the interval from the trajectory of J whose end coincides with the point x_{2i+2} (this interval is denoted by (x'_{2i+2}, x_{2i+2})), and let δ_i^* denote the "averaged derivative" of f on the interval (x_{2i+1}, x_{2i+2}) (the "averaged derivative" is defined as the ratio of the length of the image of an interval under the map f to the length of the original interval). One can easily show that

$$\delta_i^+ = \prod_{j=1}^{i-2} \delta_j^{-1}, \quad \delta_i' = \prod_{j=1}^{i-1} \delta_j^{-1}, \quad \delta_{i+2} = \frac{\beta_{i+1}}{\beta_i} \prod_{j=1}^{i+1} \delta_j^{-1} (\delta_{i+2} + 1)^{-1},$$

$$\delta_i^* = \left(\varepsilon - \sum_{j=1}^{i-1} \beta_j\right)(1 + \beta_i + \beta_{i+1}(\delta_i + 1)^{-1})\beta_i^{-1} \prod_{j=1}^{i-1} \delta_j$$

and, hence, $\delta_i^* > \delta_i^+ > \delta'_{i+2} > \delta_{i+2}^-$.

We extend the rectilinear segment

$$y = f(x_{2i+2}) + \delta_{2i+4} \prod_{j=1}^{2i+3} \delta_j^{-1} (x - x_{2i+2}), \quad x \in [x_{2i+2}, x_{2i+3}],$$

to the interval (x'_{2i+2}, x_{2i+2}) so that the length of the projection of its extended part onto the abscissa is equal to $\frac{1}{4} |x_{2i+2} - x'_{2i+2}|$. The rectilinear segment

$$y = f(x_{2i}) + \delta_{2i+2} \prod_{j=1}^{2i+1} \delta^{-1} (x - x_{2i}), \quad x \in [x_{2i}, x_{2i+1}],$$

is extended to the interval (x_{2i+1}, x'_{2i+1}) so that the length of the projection of its extended part onto the abscissa is $\frac{1}{4} | x'_{2i+1} - x_{2i+1} |$. Finally, the rectilinear segment

$$y = f(x_{2i+1}) + \frac{f(x_{2i+2}) - f(x'_{2i+1})}{(x'_{2i+1} - x_{2i+1})}(x - x_{2i+1}), \quad x \in [x'_{2i+1}, x'_{2i+2}],$$

is extended to the intervals (x_{2i+1}, x'_{2i+1}) and (x'_{2i+2}, x_{2i+2}) so that the length of the projections of its extended parts onto the abscissa are, respectively, $\frac{1}{4}|x'_{2i+1} - x_{2i+1}|$ and $\frac{1}{4}|x_{2i+2} - x'_{2i+2}|$.

Further, we connect the values of f at the points

$$x_{2i+1} + \frac{x'_{2i+1} - x_{2i+1}}{4} \quad \text{and} \quad x_{2i+1} - \frac{x'_{2i+1} - x_{2i+1}}{4}$$

and

$$x_{2i+2} + \frac{x_{2i+2} - x'_{2i+2}}{4} \quad \text{and} \quad x_{2i+2} - \frac{x_{2i+2} - x'_{2i+2}}{4}$$

by rectilinear segments.

Let $a_0 = \delta_i$ and

$$a_1 = \left(f\left(x_{2i+1} - \frac{x'_{2i+1} - x_{2i+1}}{4} \right) - f\left(x_{2i+1} + \frac{x'_{2i+1} - x_{2i+1}}{4} \right) \right) \left(\frac{x'_{2i+1} - x_{2i+1}}{2} \right)^{-1}.$$

By using the function

$$g^*(x) = f(x_{2i+1}) + a_1(x - x_{2i+1}) + (a_0 - a_1)\left(\frac{x'_{2i+1} - x_{2i+1}}{2} \right) \varphi\left(\frac{2(x - x_{2i+1})}{x'_{2i+1} - x_{2i+1}} \right),$$

where

$$\varphi(x) = \int_0^x \left(1 - \frac{\int_0^y g(x)dx}{\int_0^1 g(x)dx} \right) dy, \quad g(x) = \exp\{ -[x(1-x)] \}^{-1},$$

we connect the points

$$\{x_{2i+1}, f(x_{2i+1})\} \quad \text{and} \quad \left\{ \frac{x'_{2i+1} + x_{2i+1}}{2}, f\left(\frac{x_{2i+1} + x'_{2i+1}}{2} \right) \right\}$$

by C^∞-smooth curves. In the same way, we connect the points

$$\left\{ \frac{x_{2i+1} + x'_{2i+1}}{2}, f\left(\frac{x_{2i+1} + x'_{2i+1}}{2} \right) \right\} \quad \text{and} \quad \{ x'_{2i+1}, f(x'_{2i+1}) \},$$

$$\{x'_{2i+2}, f(x'_{2i+2})\} \quad \text{and} \quad \left\{\frac{x_{2i+2}+x'_{2i+2}}{2}, f\left(\frac{x_{2i+2}+x'_{2i+2}}{2}\right)\right\},$$

and

$$\left\{\frac{x_{2i+2}+x'_{2i+2}}{2}, f\left(\frac{x_{2i+2}+x'_{2i+2}}{2}\right)\right\} \quad \text{and} \quad \{x_{2i+2}, f(x_{2i+2})\}.$$

Since

$$\sup_{x\in(x_{2i+1},\, x_{2i+2})} |f^k(x)| \le C_k \delta_i (x_{2i+2}-x_{2i+1})^{1-k} \sup_{x\in[0,1]} |\varphi^{(k)}(x)|, \quad k > 1,$$

where C_k is a number that depends only on k, we have

$$\lim_{j\to\infty} \sup_{x\in(x_{2i+1},\, x_{2i+2})} |f^{(k)}(x)| \le 0$$

for any $k \ge 1$.

By repeating the same procedure for the intervals (y_{2i+2}, y_{2i+1}), we obtain a map $f \in C^2(I, I)$ of type 2^∞ with wandering interval.

Now assume that J is the maximal wandering interval of the map f. In this case, it is easy to check that $f^n(c) \in \mathrm{NW}(f)$ and $f^n(c) \notin \overline{\mathrm{Per}\, f}$ for any $n \ge 1$. Hence,

$$\mathrm{NW}(f) \setminus \overline{\mathrm{Per}\, f} = \bigcup_{n=1}^{\infty} \{f^n(c)\}.$$

Theorem 5.9. *There exists a unimodal C^∞-map with single flat critical point, countably many sinks, and wandering interval.*

Proof. Here, we present the proof of a simpler assertion (which can be regarded as a modification of a result established by Sharkovsky and Ivanov [1]), namely, we construct an example of a unimodal C^∞-map with countably many critical points, countably many sinks, and wandering interval.

First, we construct a map with attracting cycles of arbitrarily large periods.

1. We choose an arbitrary number $\lambda > 1$ and a sequence $x_0 = \lambda/(1+\lambda^2) > x_1 > x_2 > \ldots > x_n > x_{n+1} > \ldots \to 1/2$ and set

$$f(x) = \begin{cases} \lambda x, & x \in [0, x_0], \\ \lambda(1-x), & x \in [1-x_0, 1]. \end{cases}$$

Let $N_1 < N_2 < \ldots$ be a sequence of natural numbers specified below. We define the quantities $f_n = 1 - x_n / \lambda^{N_n}$ and $h_n = f_{n+1} - f_n$ and the functions

$$g_n(x) = \begin{cases} \exp\left\{-1/(x-x_n)^2(x-x_{n+1})^2\right\}, & x \in (x_n, x_{n+1}), \\ 0, & x = x_n, \end{cases}$$

and

$$J_n = \int\limits_{x_n}^{x_{n+1}} g_n(x)dx, \quad n = 1, 2, \ldots .$$

On the interval $[x_0, 1/2]$, we set

$$f(x) = \begin{cases} f_n + \dfrac{h_n}{J_n}\displaystyle\int_{x_n}^{x} g_n(x)dx, & x \in [x_n, x_{n+1}],\ n = 1, 2, \ldots, \\ 0, & x = 1. \end{cases}$$

It remains to extend the definition of $f(x)$ to the remaining part of the interval to obtain a unimodal C^∞-map of the interval into itself. Thus, on $[1/2, 1 - x_0]$, the map $f(x)$ can be defined as an arbitrary monotonically decreasing function from the class C^∞ satisfying the conditions $f(1/2) = 1$, $f^{(m)}(1/2) = 0$ for $m \geq 1$, $f(1 - x_0) = \lambda x_0$, $f'(1 - x_0) = -\lambda$, and $f^{(m)}(1 - x_0) = 0$ for $m > 1$. The function $f(x)$ thus defined is continuous in the entire interval $I = [0, 1]$ and has continuous derivatives of all orders everywhere except, possibly, the point $x = 1/2$. Each point x_n, $n \geq 0$, is a periodic attracting point of period $N_n + 1$ for the map $f: I \to I$, i.e.,

$$f^{N_n+1}(x_n) = f^{N_n}(f_n) = \lambda^{N_n}(1 - f_n) = x_n, \quad \frac{d}{dx} f^{N_n+1}(x)\big|_{x = x_n} = 0.$$

Let us show that, for a properly chosen sequence $\{N_n, n = 1, 2, \ldots\}$, the derivatives $f^{(m)}(x)$, $m > 0$, exist for $x = 1/2$ and are equal to zero. For this purpose, it suffices to prove that $\lim\limits_{n\to\infty} \max\limits_{x \in I_n} |f^{(m)}(x)| = 0$ for $m > 0$; here, $I_n = [x_n, x_{n+1}]$.

Consider the interval (x_n, x_{n+1}). By successive differentiation, we obtain $f'(x) = (h_n/J_n)\, g_n(x)$ and $f^{(m)}(x) = (h_n/J_n)(Q_{n,m}(x)/ P_n^{L_m}(x))$, where $Q_{n,m}$ is a polynomial with continuous coefficients that depend only on m, x_n, and x_{n+1}, $P_n(x) = (x - x_n)(x - x_{n-1})$, and $L_{k+1} \leq 2L_k$, $k = 2, 3, 4, \ldots$. Since $x_n \to 1/2$, we have

$$\max_{x \in I_n} |Q_{n,m}(x)| \leq C_1(m),$$

where $C_1(m)$ is a constant that depends only on m.

Denote $\alpha_n = x_{n-1} - x_n$. Then

$$J_n = \int_{x_n}^{x_{n+1}} \exp\{-1/(x-x_n)^2(x-x_{n+1})^2\}\,dx$$

$$= \alpha_n \int_{-1/2}^{1/2} \exp\left\{-1\Big/\left(x+\frac{1}{2}\right)^2\left(x-\frac{1}{2}\right)^2\alpha_n^4\right\}dx.$$

By using the Hölder inequality, we conclude that

$$J_n \geq \alpha_n\left(\int_{-1/2}^{1/2} \exp\left\{-1\Big/\left(x+\frac{1}{2}\right)^2\left(x-\frac{1}{2}\right)^2\right\}dx\right)^{1/\alpha_n^4} = \alpha_n\kappa^{1/\alpha_n^4},$$

where

$$\kappa = \int_{-1/2}^{1/2} \exp\left\{-1\Big/\left(x+\frac{1}{2}\right)^2\left(x-\frac{1}{2}\right)^2\right\}dx.$$

Therefore,

$$\max_{x\in I_n}|f^{(m)}(x)| \leq C_1(m)(h_n/J_n)\max_{x\in I_n}|g_n(x)/P^{L_m}(x)|.$$

Since $\lim\limits_{n\to\infty}[\exp\{-\frac{1}{t}\}/t^k] = 0,\ k>0,$ we can write

$$\lim_{n\to\infty}\max_{x\in I_n}|g_n(x)/P_n^{L_m}(x)| = 0.$$

Further,

$$h_n/J_n \leq (1/\lambda)^{N_n}(x_{n+1}-x_n/\lambda^{N_{n+1}-N_n})/\alpha_n\kappa^{\alpha_n^{-4}} \leq 1/\lambda^{N_n}\alpha_n\kappa^{\alpha_n^{-4}} \leq 1$$

if $\lambda^{N_n}\alpha_n\kappa^{\alpha_n^{-4}} \geq 1$, i.e., if

$$N_n \geq -(\ln\alpha_n + \alpha_n^{-4}\ln\kappa)/\ln\lambda. \tag{5.1}$$

Hence, if inequality (5.1) is true, then

$$\lim_{n\to\infty}\max_{x\in I_n}|f^{(m)}(x)| = 0$$

for any sequence x_n, $n \geq 1$. Thus, by setting $\lambda = 1/\kappa$ and $(x_{n+1} - x_n) = 1/n^2$, we conclude that $N_n \geq n^8 + 2\ln n/\ln\lambda$ for sufficiently large n. This means that, for the indicated sequence of x_n, it suffices to set $N_n = n^8 + n$.

2. To construct a unimodal C^∞-map with an wandering interval, we choose an arbitrary number $\lambda > 1$ and sequences J_n and z_n, $n = 1, 2, \ldots$, such that $\frac{\lambda}{1+\lambda^2} > y_1 > z_1 > y_2 > z_2 > \ldots > y_n > z_n > y_{n+1} > z_{n+1} > \ldots \to 1/2$. We set $\alpha_n = z_{n+1}/\lambda^{N'_n}$ and $\beta_n = y_{n+1}/\lambda^{N'_n}$, where N'_n, $n \geq 1$, is a sequence of natural numbers.

Note that, in the construction realized above, one can set $N_{2k-1} = N_{2k}$, $k \geq 1$, in addition to inequality (5.1). Further, if we set $x_{2n-1} = y_n$, $x_{2n} = z_n$, $f_{2n-1} = \alpha_n$, $f_{2n} = \beta_n$, $h_n = f_{n+1} - f_n$, and $N_n = N^1_{[(n+1)/2]}$, $n \geq 1$, then, for any sequence x_n there exists a sequence N^1_n such that the function $f(x)$ constructed as in the previous case belongs to the class $C^\infty[0, 1]$. In this case, the interval (y_1, z_1) (as well as any other interval (y_n, z_n), $n > 1$) is a maximal wandering interval.

6. METRIC ASPECTS OF DYNAMICS

The phase space of dynamical systems under consideration, i.e., the interval I, is endowed with Lebesgue measure. It is thus useful to establish some properties of dynamical systems that are typical with respect to this measure, i.e., properties exhibited by trajectories covering sets of full measure.

1. Measure of the Set of Lyapunov Stable Trajectories

By using the phase diagram constructed and studied in Sections 5.1 and 5.2, we have already described the limiting behavior of trajectories of unimodal maps. Let us now describe the set of points of an interval that generate Lyapunov stable trajectories (recall that the trajectory of a point $x \in I$ of the map $f: I \to I$ is called Lyapunov stable if, for any $\varepsilon > 0$, there exists $\delta > 0$ such that the inclusion $f^i(y) \in (f^i(x) - \varepsilon, f^i(x) + \varepsilon)$ holds for any $y \in I \cap (x - \delta, x + \delta)$ and all $i \geq 0$).

Let $f: I \to I$ be a unimodal map and let $\mathfrak{L}(f)$ be the set of the points of the interval I whose trajectories are Lyapunov stable. If $x \in \Gamma(f)$, i.e., if the point x belongs to a wandering interval U, then $x \in \mathfrak{L}(f)$ because $|f^i(U)| \to 0$ as $i \to \infty$. Hence, $\Gamma(f) \subset \mathfrak{L}(f)$.

Assume that $x \in \mathcal{B}(f)$, i.e., there exist a cycle of intervals $B_m^{(i)} = \{ J_0, J_1, \ldots, J_{n-1} \}$ which does not contain the point of extremum of f and a number $k < \infty$ such that $f^k(x) \in J_0$. If $f^k(x) \notin \mathrm{Per}(f)$, then the point $f^k(x)$ belongs to the domain of immediate attraction of some point of the set $J_0 \cap \mathrm{Fix}(f^{2n})$. In this case, $x \in \mathfrak{L}(f)$. Now let $f^k(x) \in \mathrm{Per}(f)$. Then $f^k(x) \in J_0 \cap \mathrm{Fix}(f^{2n})$. We set $\beta = f^k(x)$ and assume that there exists $\varepsilon > 0$ such that either $f^{2n}(y) < y$ for all $y \in (\beta - \varepsilon, \beta)$ or $f^{2n}(y) > y$ for all $y \in (\beta, \beta + \varepsilon)$. In this case, $\beta \notin \mathfrak{L}(f)$ and, hence, $x \notin \mathfrak{L}(f)$. One can easily show that $x \notin \mathfrak{L}(f)$ in all other cases. The set of all periodic points of $\mathcal{B}(f)$ whose trajectories are not Lyapunov stable is denoted by $\Lambda_0 = \Lambda_0(f)$. The set Λ_0 is invariant and consists of at most countably many cycles because, for each of these cycles, one can indicate its half neighborhood that does not contain periodic points of the map.

The results established in Section 5.2 imply that

$$I = \mathcal{B}(f) \cup \Gamma(f) \bigcup_{m \le m^*} \bigcup_{i \ge 0} f^{-i}(C_m^{(0)})$$

for $m^* < \infty$ and

$$I = \mathcal{B}(f) \cup \Gamma(f) \cup P(C_\infty^{(0)}, f) \bigcup_{m \le \infty} \bigcup_{i \ge 0} f^{-i}(C_m^{(0)})$$

for $m^* = \infty$. (Recall that the elements of the spectral decomposition of the set $\overline{\mathrm{Per}(f)}$ are denoted by $C_m^{(0)}$, and $P(C_\infty^{(0)}, f)$ denotes the domain of attraction of the set $C_\infty^{(0)}$.) Therefore,

$$I \setminus \mathfrak{L}(f) \subset \bigcup_{\substack{m < \infty \\ m \le m^*}} \bigcup_{i \ge 0} f^{-i}(C_m^{(0)}) \bigcup \bigcup_{i \ge 0} f^{-i}(\Lambda_0) \cup \Lambda_\infty,$$

where $\Lambda_\infty = \varnothing$ if $m^* < \infty$ and Λ_∞ is a subset of the set $P(C_\infty^{(0)}, f)$ if $m^* = \infty$ (it can be shown that Λ_∞ is empty if the set Φ_∞^* is nowhere dense and nonempty if Φ_∞^* contains intervals). Note that trajectories from the set $\bigcup_{i \ge 0} f^{-i}(C_m^{(0)})$ are not Lyapunov stable for $p_{m+1}/p_m > 2$ because, in this case, $C_m^{(0)}$ is a Cantor set and the map $f|_{C_m^{(0)}}$ possesses the mixing property.

The representation of the set $I \setminus \mathfrak{L}(f)$ constructed above demonstrates that the answer to the question as to whether Lyapunov stability is a generic property of trajectories of a given unimodal map f essentially depends on the Lebesgue measure of elements of the spectral decomposition of the set $\overline{\mathrm{Per}(f)}$. The following theorem is a simple consequence of this observation:

Theorem 6.1. *Let $f: I \to I$ be a unimodal map. Assume that f has no cycles whose periods are not a power of two and has no wandering intervals. Then the set of points of the interval I whose trajectories are not Lyapunov stable is a set of Lebesgue measure zero.*

Proof. Under the conditions of the theorem, in the phase diagram of the map f, we have $p_{m+1}/p_m = 2$ for any $m < m^*$ (here, we use the same notation as in Section 5.2). Hence, for any $m < m^*$, the set $C_m^{(0)}$ is a cycle or a pair of cycles, i.e., mes $C_m^{(0)} = 0$.

Consider the set $C_{m^*}^{(0)}$ for $m^* < \infty$. Since the map f has no cycles whose periods are not a power of two, the set $C_{m^*}^{(0)}$ can be neither a cycle of intervals nor a Cantor set. Therefore, $C_{m^*}^{(0)}$ is a periodic trajectory and mes $C_{m^*}^{(0)} = 0$. Thus, mes $(I \setminus \mathfrak{L}(f)) = 0$ whenever $m^* < \infty$.

If $m^* = \infty$, then the fact that wandering intervals are absent implies that the set $\Phi^*_\infty = \bigcup_{m\geq 1}\Phi^*_m$ is nowhere dense (see Section 5.2). Hence, the Lebesgue measure of each component of the set Φ^*_m tends to zero as m increases. On the other hand, if $x \in P(C^{(0)}_\infty, f)$, then, for any $m < \infty$, there exists a neighborhood $U = U(m)$ of the point x such that $f^k(U) \subset \Phi^*_m$ for some $k < \infty$. Therefore, in this case, we have $x \in \mathfrak{L}(f)$, and the proof of Theorem 6.1 is completed.

If the map is not "simple", i.e., if it has cycles whose period is not a power of two, then both the situation where Lyapunov stability is a typical property of trajectories of a given map and the situation where this is not true are possible. Simple examples of maps of both kinds are presented below.

For maps with negative Schwarzian, the following assertion is true:

Theorem 6.2. *Let $f \in C^3(I, I)$ be a unimodal map such that its critical point c is not flat and $Sf(x) < 0$ for $x \in I\setminus\{c\}$. Then the inequality* $\operatorname{mes}\mathfrak{L}(f) < \operatorname{mes} I$ *holds if and only if $m^* < \infty$ and $C^{(0)}_{m^*}$ is a cycle of intervals.*

The proof of Theorem 6.2 immediately follows from the properties of the spectral decomposition of the set $NW(f)$ for unimodal maps with negative Schwarzian, Theorem 5.5, and Theorem 6.3 formulated below. Note that if $\operatorname{mes}\mathfrak{L}(f) < \operatorname{mes} I$ and both fixed points of the map f are repelling, then $\operatorname{mes}\mathfrak{L}(f) = 0$.

Example 6.1. Consider the map $g(x) = 1 - 2|x - \frac{1}{2}|$, $x \in [0, 1]$, encountered somewhat earlier. The map g is unimodal and consists of two linear pieces. Moreover, it is expanding, i.e., $|g'(x)| = 2$ for $x \in [0, 1]\setminus\{\frac{1}{2}\}$. It is easy to check that g possesses the mixing property on the interval $[0, 1]$. Therefore, $\overline{\operatorname{Per}(g)} = [0, 1]$ and $\mathfrak{L}(g) = \varnothing$, i.e., the map g has no Lyapunov stable trajectories.

The point $x^* = \frac{2}{3}$ is a repelling fixed point of the map g. We fix arbitrary $\varepsilon < \frac{1}{12}$ and replace the right branch of the function $g(x)$ by a piecewise linear function $\tilde{g}(x)$ such that

$$\tilde{g}(x^* - 2\varepsilon) = x^* + 2\varepsilon, \quad \tilde{g}(x^* + 2\varepsilon) = x^* - 2\varepsilon,$$

$$\tilde{g}(x^* - \varepsilon) = x^* + \frac{\varepsilon}{2}, \quad \text{and} \quad \tilde{g}(x^* + \varepsilon) = x^* - \frac{\varepsilon}{2} \quad \text{(see Fig. 43).}$$

For $x \in [0, \frac{1}{2}]$, we set $\tilde{g}(x) = g(x)$. The function $\tilde{g}(x)$ obtained as a result is unimodal, the point x^* is an attracting fixed point of the map $\tilde{g}$, and the interval $(x^* - 2\varepsilon, x^* + 2\varepsilon)$ is the domain of immediate attraction of the point x^*. We have

$$\overline{\operatorname{Per}(\tilde{g})} = \{x^*\} \cup C^{(0)}_1,$$

where $C_1^{(0)} = I \setminus \bigcup_{i\geq 0} \tilde{g}^{-i}((x^* - 2\varepsilon,\ x^* + 2\varepsilon))$ is a Cantor set. By direct calculation, one can show that the Lebesgue measure of the set $\bigcup_{i\geq 0} \tilde{g}^{-i}((x^* - 2\varepsilon,\ x^* + 2\varepsilon))$ is equal to the measure of the entire interval $[0, 1]$, i.e., $\text{mes}\ C_1^{(0)} = 0$ (independently of the choice of ε). Hence, for the map $\tilde{g}$, we have $\text{mes}\ \mathfrak{B}(\tilde{g}) = \text{mes}([0, 1])$ because $\mathfrak{B}(\tilde{g}) = \bigcup_{i\geq 0} \tilde{g}^{-i}((x^* - 2\varepsilon,\ x^* + 2\varepsilon))$.

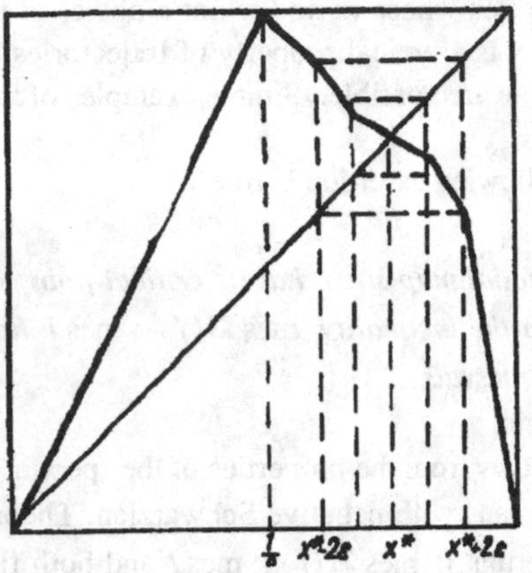

Fig. 43

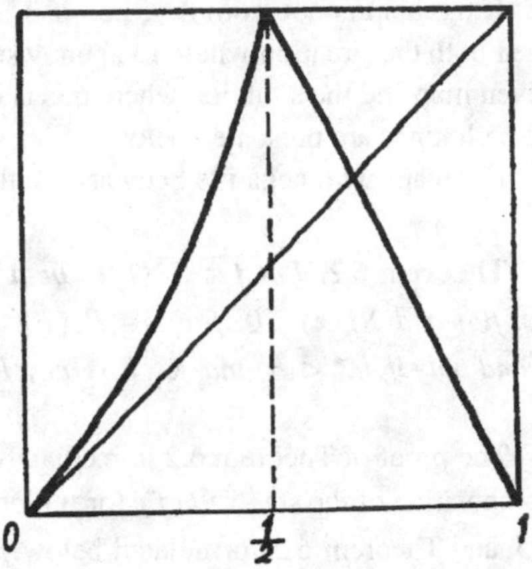

Fig. 44

Thus, from the metric point of view, the limiting behavior of a "typical" trajectory of a unimodal map $f: I \to I$ is determined by the structure of a certain subset $\mathcal{A}(f)$ of the set $\overline{\text{Per}(f)}$ and the dynamics of the map $f|_{\mathcal{A}(f)}$. It is natural to choose the smallest possible $\mathcal{A}(f)$. Thus, for the map $\tilde{g}$ constructed in Example 6.1, we must take $\mathcal{A}(\tilde{g}) = \{x^*\}$. The exact definitions are presented below.

Recall some definitions from Chapter 1. The probabilistic limit set $\mathcal{M}(f)$ of a map $f: I \to I$ (according to Milnor) is defined as the smallest closed set that contains the ω-limit sets of trajectories of almost all points of the interval I (with respect to the Lebesgue measure). It is clear that, for any unimodal map f, the set $\mathcal{M}(f)$ is an invariant subset of $\overline{\text{Per}(f)}$. On the other hand, if, for any measurable set $\Lambda \subset I$ such that $f(\Lambda) \subset \Lambda$, we define $\mu(\Lambda) = \text{mes}\,\{x \in I: \omega_f(x) \subset \Lambda\}$, then $\mathcal{M}(f)$ is the smallest closed invariant subset of $\overline{\text{Per}(f)}$ such that $\mu(\mathcal{M}(f)) = \text{mes}\, I$.

For any set $M \subset I$, let

$$p(x, M) = \lim_{n\to\infty} \inf \frac{1}{n} \sum_{k=0}^{n-1} \chi_M(f^k(x)),$$

where $x \in I$ and χ_M is the indicator of the set M. We say that M is the center of attraction (of almost all trajectories) of the map f if, for any neighborhood U of the set M, the equality $p(x, U) = 1$ holds for almost all points $x \in I$ with respect to the Lebesgue measure. The minimal center of attraction of almost all trajectories $\mathcal{A}(f)$ is defined as the center of attraction that contains no other centers of attraction.

It is not difficult to show that $\mathcal{A}(f) \subset \mathcal{M}(f)$ and these sets are distinct even in the case of unimodal maps. Thus, for the map

$$x \to g(x) = \begin{cases} x + 2^m x^{m+1}, & x \in [0, 1/2], \\ 2(1-x), & x \in [1/2, 1], \end{cases}$$

where $m \geq 1$ (see Fig. 44), we have $\mathcal{M}(g) = \overline{\operatorname{Per}(g)} = [0, 1]$ and $\mathcal{A}(g) = 0$. In the neighborhood of the nonhyperbolic fixed point 0, the motion of trajectories is significantly decelerated and the time $\tau(\varepsilon)$ of expansion of the ε-neighborhood of the point 0 to its 2ε-neighborhood infinitely increases as $\varepsilon \to 0$, i.e., $\tau(\varepsilon) \sim \varepsilon^{-m}/m$ as $\varepsilon \to 0$. One can also construct an analytic map which possesses the indicated property.

As shown above, metric properties of a given unimodal map f depend on the metric properties of the elements of the spectral decomposition of the set $\overline{\operatorname{Per}(f)}$. Numerous important results were obtained in this direction for smooth maps.

As usual, for a map $f \in C^r(I, I)$, a closed set Λ such that $f(\Lambda) \subset \Lambda$ is called hyperbolic if there exist $\lambda > 1$ and $C > 0$ such that either $|(f^n)'(y)| \geq C\lambda^n$ or $|(f^n)'(y)| \leq \lambda^{-n}/C$ for any point $y \in \Lambda$ and all $n \geq 0$.

Theorem 6.3. *Assume that $f \in C^2(I, I)$, Λ is a closed set, $f(\Lambda) \subset \Lambda$, and $\Lambda \cap K(f) = \varnothing$, where $K(f) = \{x \in I : f'(x) = 0\}$. Then there exists $N < \infty$ such that all periodic orbits in Λ whose periods are greater than N are hyperbolic and repelling. Moreover, if Λ does not contain any nonhyperbolic periodic orbits, then Λ is a hyperbolic set, and if, in addition, Λ does not contain attracting periodic orbits, then* mes $\Lambda = 0$.

This theorem was first proved by Mañé [1]. Another proof was suggested by van Strien [3].

2. Conditions for the Existence of Absolutely Continuous Invariant Measures

We study the asymptotic behavior of trajectories. This type of behavior can be efficiently described, e.g., for maps preserving a measure μ (such that $\mu(f^{-1}(A)) = \mu(A)$ for any measurable set A). It is well known that the support of a measure of this sort must belong to the set $\overline{\mathrm{Per}(f)}$.

In order to exclude trivial cases, we require that $\mu(\mathrm{Per}(f)) = 0$. The best possibility is to guarantee the absolute continuity of a measure μ with respect to the Lebesgue measure, i.e., to require that the condition $\mathrm{mes}\, A = 0$ imply the equality $\mu(A) = 0$.

There are some general results concerning the existence of absolutely continuous invariant measures for nonsingular maps of an interval. (Recall that a map f is called nonsingular if the equality $\mathrm{mes}\, A = 0$ yields the equality $\mathrm{mes}\, f^{-1}(A) = 0$ for any measurable set A.)

Theorem 6.4 (Foguel [1]). *Let $f: I \to I$ be a nonsingular map. Then the following assertions are equivalent:*

(i) There exists an invariant measure of the map f absolutely continuous with respect to the Lebesgue measure;

(ii) there exists $\varepsilon < 1$ such that the condition $\mathrm{mes}(A) < \varepsilon$ *implies the inequality* $\limsup_{n\to\infty} \mathrm{mes}(f^{-n}(A)) \leq \frac{1}{2}\, \mathrm{mes}\, I$ *for all* $n \geq 0$;

(iii) there exists $\varepsilon < 1$ such that the condition $\mathrm{mes}(A) < \varepsilon$ *implies the inequality*

$$\limsup_{n\to\infty} \left(\frac{1}{n} \sum_{k=0}^{n-1} \mathrm{mes}\,(f^{-k}(A)) \right) \leq \frac{1}{2}\, \mathrm{mes}\, I.$$

The absolute continuity of a given measure with respect to the Lebesgue measure can be established by using the Radon–Nikodym theorem:

A probability measure μ is absolutely continuous with respect to the Lebesgue measure if and only if there exists an L^1-function $p(x)$ such that

$$\mu(A) = \int_A p(x)\, dx$$

for any measurable set A.

By using this representation of absolutely continuous measures, one can prove the existence of invariant measures absolutely continuous with respect to the Lebesgue measure for expanding maps of an interval from the class C^2.

Theorem 6.5 (Lasota and Yorke [1]). *Let $f: [-1, 1] \to [-1, 1]$ be such that*

(i) there are points $-1 = c_0 < c_1 < \ldots < c_n = 1$ *such that* $f \in C^2((c_i, c_{i+1}))$, $i = 0, 1, \ldots, n-1$;

(ii) $|f'| > 1$ *on* $[-1, 1] \setminus \{c_1, \ldots, c_{n-1}\}$.

Then f possesses an invariant measure absolutely continuous with respect to the Lebesgue measure.

The proof of Theorem 6.5 is based on the use of the Frobenius–Perron operator. This operator realizes a transformation in the set of densities of measures corresponding to the transformations of the Lebesgue measure under iterations of the map f. By applying the Frobenius–Perron operator to the original density $p_0(x)$, we obtain a density $p_n(x)$ given by a function of bounded variation. The conditions of smoothness imposed on the map f enable us to conclude that the functions $p_n(x)$ converge to a limit function $p(x)$ which is a function of bounded variation and, hence, an L^1-function. Note that, in the conditions of Theorem 6.5, C^2-smoothness can be replaced by $C^{1+\varepsilon}$-smoothness.

By using Theorem 6.4, one can show that a smooth map similar to the map displayed in Fig. 44 has no finite invariant measure absolutely continuous with respect to the Lebesgue measure. This example clarifies the importance of the requirement of hyperbolicity of periodic trajectories in the conditions of theorems establishing the existence of absolutely continuous measures (see Theorem 6.7 below).

Consider a simple unimodal map $f: I \to I$, i.e., a unimodal map with topological entropy equal to zero. In this case, in the phase diagram of the map f, we have $p_{m+1}/p_m = 2$ for any $m < m^*$. If $m^* < \infty$, then $\mathrm{NW}(f) = \mathrm{Per}(f)$ and, moreover, the periods of all points of the set $\mathrm{Per}(f)$ are uniformly bounded. Therefore, it remains to consider the case $m^* = \infty$ where $\overline{\mathrm{Per}(f)} = \mathrm{Per}(f) \cup C_\infty^{(0)}$ ($C_\infty^{(0)}$ is a Cantor set). The following assertion is true (see also Collet and Eckmann [1] and Misiurewicz [2]):

Theorem 6.6. *Let* $f: I \to I$ *be a unimodal map in the phase diagram of which* $m^* = \infty$ *and* $p_{m+1}/p_m = 2$ *for all* $m < m^*$. *Then there exists a unique invariant probability measure* μ *on* $\overline{\mathrm{Per}(f)}$ *equal to zero on any subset of the set* $\mathrm{Per}(f)$. *The following equality holds for any point* $y \in P(C_\infty^{(0)}, f)$ *and any continuous function* $g(x)$ *defined on I:*

$$\lim_{n\to\infty} \frac{1}{n} \sum_{k=0}^{n-1} g(f^k(y)) = \int g \, d\mu.$$

The proof of Theorem 6.6 is split into two parts: first, we construct the measure μ and then study its properties. The construction of the measure μ depends on the form of the phase diagram of the map f. It follows from the results established in Section 5.2 that, under the conditions of the theorem, $C_\infty^{(0)} \subset \bigcap_{m\geq 1} \Phi_m^*$, where

$$\Phi_m^* = \bigcup_{i=0}^{p_m - 1} J_m^{(i)}$$

and the intervals $J_m^{(0)}, J_m^{(1)}, \ldots, J_m^{(p_m-1)}$ form a cycle of period p_m (in the case under consideration, $p_m = 2^m$). The measure of each interval $J_m^{(0)}, J_m^{(1)}, \ldots, J_m^{(2^m-1)}$ is assumed to be equal to 2^{-m}. For a more detailed proof, see Misiurewicz [2].

If f is a convex unimodal map with a single nondegenerate critical point, then $\operatorname{mes} C_\infty^{(0)} = 0$. This fact and some other possibilities connected with the Lebesgue measure of sets of type $C_\infty^{(0)}$ are discussed at the end of this section.

If the topological entropy of the map $f: I \to I$ is positive, then we have the following sufficient condition for the existence of an invariant measure absolutely continuous with respect to the Lebesgue measure:

Theorem 6.7. *Let $f \in C^2(I, I)$ and let all critical points of the map f be nonflat. Assume that f has no attracting and nonhyperbolic periodic trajectories and*

$$K(f) \cap \overline{\bigcup_{i \geq 1} f^i(K(f))} = \varnothing.$$

Then f possesses an invariant measure absolutely continuous with respect to the Lebesgue measure.

Note that if a unimodal map f satisfies the conditions of Theorem 6.7, then, in its phase diagram, $m^* < \infty$ and $C_{m^*}^{(0)}$ is a cycle of intervals (the invariant measure is concentrated just on this cycle of intervals). Guckenheimer [3] conjectured that, at least for unimodal maps with negative Schwarzian, an invariant measure absolutely continuous with respect to the Lebesgue measure exists if and only if $m^* < \infty$ and $C_{m^*}^{(0)}$ is a cycle of intervals. The following example suggested by Johnson demonstrates that this is not true. It is clear that, for the map constructed in this example, the condition

$$K(f) \cap \overline{\bigcup_{i \geq 1} f^i(K(f))} = \varnothing$$

cannot be satisfied.

Theorem 6.8 (Johnson [1]). *For the family of maps $f_\lambda: x \to \lambda x(1-x)$, one can indicate a value of the parameter λ_∞ such that the set $C_{m^*}^{(0)}(f_{\lambda_\infty})$ in the phase diagram of f_{λ_∞} is a cycle of intervals and any finite invariant measure of f_{λ_∞} is singular.*

In [1], by using the lemmas formulated below, Johnson proved that, for the map $f_{\lambda_\infty}: I \to I$, one can indicate a sequence of sets $\{G_{m_i}\}_{i=1}$ such that $f_{\lambda_\infty}^{m_i}(G_m) \subset I \setminus G_{m_i}$ and

$\text{mes}(I \backslash G_m) \to 0$ as $i \to \infty$. Therefore, any finite invariant measure of the indicated map is not absolutely continuous with respect to the Lebesgue measure.

Let $\beta \in \text{Per}(f_\lambda)$ and let $\beta' \neq \beta$ be such that $f(\beta') = f(\beta)$. The closed interval $\langle \beta, \beta' \rangle$ with ends at β and β' is denoted by J_λ. If $f_\lambda^n(J_\lambda) \subset J_\lambda$ for some $n \geq 1$, then we say that J_λ is a periodic interval (of period n if n is the least possible number with the indicated property). If $f_\lambda^n(J_\lambda) = J_\lambda$, then the interval J_λ is called strictly periodic.

Let $[\lambda_a, \lambda_b]$ be the maximal interval such that the map f_λ possesses a periodic interval J_λ of period n_0 for any $\lambda \in [\lambda_a, \lambda_b]$.

Lemma 6.1 (Guckenheimer [2]). *For any* $\sigma > 0$, *there exists* $\delta > 0$ *such that, for any* $\lambda \in (\lambda_b, \lambda_b + \delta)$, *one can indicate a set* P_λ *and* $m \in \mathbb{N}$ *for which* $\text{mes}(I \backslash P_\lambda) < \sigma$ *and* $f_\lambda^m P_\lambda \subset \bigcup_{0<i\leq n_0} f_\lambda^i(J_\lambda)$.

Lemma 6.2. *For any* $\varepsilon > 0$ *and* $\sigma > 0$, *there exist* $\lambda' \in (\lambda_b, \lambda_b + \delta)$ *and a strictly periodic interval* $J_{\lambda'}$ *such that* $\text{mes}(\text{orb}\, J_{\lambda'}) < \sigma$.

As λ_∞, we take the limit of the sequence $\{\lambda_i\}$ formed by the values of the parameter equal to λ_b for periodic intervals J_{λ_i} of periods n_i, where $n_i \to \infty$ as $i \to \infty$. In this case, by Lemma 6.2, the intervals J_{λ_i} can be chosen so that $(\bigcup_{0<j\leq n_i} f^j(J_{n_i})) \to 0$ as $i \to \infty$ and the set G_i is chosen so that $I \backslash G_i = (I \backslash P_{\lambda_i}) \cup \bigcup_{0<j\leq n_i} f^j(J_{\lambda_\infty})$, where P_{λ_i} are prescribed by Lemma 6.1.

For maps with negative Schwarzian, we have the following theorem, which characterizes the probabilistic limit set $A(f)$:

Theorem 6.9 (Blokh and Lyubich [1]). *Let* $f \in C^3(I, I)$ *be a unimodal map and let* $Sf(x) < 0$ *for* $x \in I \backslash \{c\}$. *Then, for almost all points of the interval* I *with respect to the Lebesgue measure, only one of the following three possibilities is realized:*

(i) $\omega_f(x)$ *is an attracting or semiattracting cycle;*

(ii) $\omega_f(x)$ *is a cycle of intervals (and coincides with the set* $C_{m^*}^{(0)}$*);*

(iii) $\omega_f(x) = \omega_f(c) \subset C_{m^*}^{(0)}$ *(in this case,* $c \in \omega_f(c)$*).*

The proof of Theorem 6.9 follows from the estimates established in the previous chapter and the proof of Theorem 6.7 (see van Strien [3]).

3. Measure of Repellers and Attractors

The results of this section can be regarded as a supplement to the results established in Section 6.2. The following assertion demonstrates that the condition of continuity of the second derivative in Theorems 6.3 and 6.7 is fundamental:

Theorem 6.10. *There exists a unimodal map from the class* C^1 *which possesses a repeller* K^* *in the form of an invariant hyperbolic Cantor set of positive Lebesgue measure; moreover, the Lebesgue measure is invariant on* K^*.

Proof. We fix an arbitrary number $\varepsilon \in (0, 1)$ and a sequence of numbers $\beta_0 > \beta_1 > \beta_2 > \ldots$ such that

$$\text{(i)} \quad \sum_{i=0}^{\infty} \beta_i = \beta < \varepsilon \quad \text{and} \quad \text{(ii)} \quad \lim_{i\to\infty} \frac{\beta_{i+1}}{\beta_i} = 1.$$

Thus, we can take

$$\beta_i = \frac{\varepsilon}{(i+10)^2}, \quad i = 1, 2, \ldots.$$

First, we construct a set $K^* \subset I$ homeomorphic to the standard Cantor set $\hat{K}$ and such that $\operatorname{mes} K^* = 1 - \beta$.

In the ternary notation, the Cantor set $\hat{K}$ takes the form $\{0.i_1 i_2 \ldots,$ where $i_s = 0$ or $2,\ s = 1, 2, \ldots\}$. Denote by $\hat{U}_{i_1 \ldots i_k}$ the intervals $(0.i_1 \ldots i_k 022 \ldots, 0.i_1 \ldots i_k 200 \ldots)$ "removed" at the $(k+1)$th step, $k \geq 1$ (for each $k \geq 1$, there are 2^k intervals of this sort).

Let h be a homeomorphism from I into I. Denote $h_{i_1 i_2 \ldots} = h(0.i_1 i_2 \ldots)$, $U = (h_{022\ldots}, h_{200\ldots})$, and $U_{i_1 \ldots i_k} = h(\hat{U}_{i_1 \ldots i_k})$. Assume that $h(I) = I$ and

(i) the interval U is equidistant from 0 and 1 and $\operatorname{mes} U = \beta_0$;

(ii) the intervals $U_{i_1 \ldots i_k}$, $k \geq 1$, are equidistant from the points $h_{i_1 \ldots i_k 00 \ldots}$ and $h_{i_1 \ldots i_k 22 \ldots}$, respectively, and $\operatorname{mes} U_{i_1 \ldots i_k} = \frac{\beta_k}{2^k}$.

The set $K^* = h(\hat{K})$ is homeomorphic to the Cantor set. Indeed,

$$K^* = (I \setminus U) \setminus \bigcup_{\substack{i_s = 0;2 \\ 1 \leq s \leq k,\ k \geq 1}} U_{i_1 \ldots i_s \ldots i_k}, \qquad \operatorname{mes} K^* = 1 - \beta > 1 - \varepsilon.$$

The map $f\colon R \to R$ is constructed as follows:

(1) $f(x) = 2x$ for $x \le 0$;

(2) for $0 < x \le \frac{1-\beta_0}{2}$, the function $f(x)$ is monotone continuous and such that

(a) $f(h_{i_1 i_2 i_3 \ldots}) = h_{i_2 i_3 i_4 \ldots}$ (hence, $f(U_{i_1 i_2 \ldots i_k}) = U_{i_2 \ldots i_k}$ for $k > 1$, $f(U_0) = U$);

(b) on the intervals $U_{i_1 \ldots i_k}$, the function $f(x)$ is defined as an arbitrary function from the class C^1 satisfying the conditions

(b′) $f'(x) \ge 2$;

(b″) $\lim_{x \to \partial U_{i_1 \ldots i_k}} f'(x) = 2$ (this condition can be satisfied because, for $k > 1$, we have

$$\frac{\operatorname{mes} U_{i_2 \ldots i_k}}{\operatorname{mes} U_{i_1 i_2 \ldots i_k}} = 2\,\frac{\beta_{k-1}}{\beta_k} > 2$$

for any $i_1, i_2, \ldots, i_k$ and

$$\frac{\operatorname{mes} U}{\operatorname{mes} U_0} = 2\,\frac{\beta_0}{\beta_1} > 2);$$

(b‴) $\sup_{x \in U_{i_1 i_2 \ldots i_k}} f'(x) \to 2$ as $k \to \infty$ (this condition can also be satisfied because $\beta_{k-1}/\beta_k \to 1$ as $k \to \infty$);

(3) for

$$\frac{1-\beta_0}{2} < x \le \frac{1}{2},$$

the function $f(x)$ is defined as an arbitrary function from the class C^1 satisfying the conditions

(a) $f(x) > 1$;

(b) $\lim\limits_{x \to \frac{1-\beta_0}{2}} f'(x) = 2$;

(c) $f'(\frac{1}{2}) = 0$;

(4) $f(x) = f(1-x)$ for $x \geq \dfrac{1}{2}$.

It follows from the construction of the function f that

$$\{x \in I: f^i(x) \in I,\ i = 1, 2, \ldots\} = (I \setminus U) \setminus \bigcup_{\substack{i_s = 0,2 \\ 1 \leq s \leq k,\ k = 1,2,3,\ldots}} U_{i_1 \ldots i_s \ldots i_k} = K^*$$

and mes $K^* > 1 - \varepsilon$. It remains to show that f is a function from the class C^1. For this purpose, it suffices to check that $f'(x)$ exists for $x \in K^*$ and is equal to 2 for $x \leq 1/2$ and to -2 for $x > 1/2$. Indeed, if this is true, then $f'(x)$ is continuous in K^* (by virtue of (2b″) and (2b‴)) and, hence, in the entire interval I.

We introduce the following notation: $\alpha_1 = 1 + \beta_0$, $\alpha_i = 1 - \beta_0 - \ldots - \beta_{i-2} + \beta_{i-1}$, $i = 2, 3, \ldots$. Then $\alpha_i \to 1 - \beta$ as $i \to \infty$ and, hence, $\alpha_{i+1}/\alpha_i \to 1$. By the construction of the set K^*, we have

$$h_{i_1 i_2 i_3 \ldots} = i_1 \frac{\alpha_1}{2^2} + i_2 \frac{\alpha_2}{2^3} + i_3 \frac{\alpha_3}{2^4} + \ldots .$$

Therefore, if $x' = h_{i_1' i_2' \ldots} \in K^*$ and $i_s' = i_s$ for $s = 1, 2, 3, \ldots, m$, then

$$\frac{f(x') - f(h_{i_1 i_2 \ldots})}{x' - h_{i_1 i_2 \ldots}} = 2 \frac{(i'_{m+1} - i_{m+1})\alpha_m + (i'_{m+2} - i_{m+2})\dfrac{\alpha_{m+1}}{2} + \ldots}{(i'_{m+1} - i_{m+1})\alpha_{m+1} + (i'_{m+2} - i_{m+2})\dfrac{\alpha_{m+2}}{2} + \ldots}.$$

The smaller the difference $|x' - h_{i_1 i_2 \ldots}|$, the closer this ratio to two. At the same time, if $x' \in U_{i_1' \ldots i_k'}$, i.e., $h_{i_1' \ldots i_k' 022 \ldots} < x' < h_{i_1' \ldots i_k' 200 \ldots}$ and, for definiteness, $x' > h_{i_1 i_2 \ldots}$, then

$$\frac{f(h_{i_1' \ldots i_k' 022 \ldots}) - f(h_{i_1 i_2 \ldots})}{h_{i_1' \ldots i_k' 022 \ldots} - h_{i_1 i_2 \ldots}} < \frac{f(x') - f(h_{i_1 i_2 \ldots})}{x' - h_{i_1 i_2 \ldots}} < \frac{f(h_{i_1' \ldots i_k' 200 \ldots}) - f(h_{i_1 i_2 \ldots})}{h_{i_1' \ldots i_k' 200 \ldots} - h_{i_1 i_2 \ldots}}.$$

In this case, the smaller the difference $|x' - h_{i_1 i_2 \ldots}|$, the smaller the quantity

$$\big|\, |x' - h_{i_1 i_2 \ldots}|^{-1} \, |f(x') - f(h_{i_1 i_2 \ldots})| - 2 \,\big|.$$

Thus, $f'(h_{i_1 i_2 \ldots}) = 2$ whenever $h_{i_1 i_2 \ldots} < 1/2$.

It follows from the construction of the map f that K^* is a hyperbolic invariant set of positive Lebesgue measure.

Let $A_{i_1 i_2 \ldots i_k}$ denote the maximal closed interval such that

$$U_{i_1 i_2 \ldots i_k} \subset A_{i i_2 \ldots i_k} \subset \Lambda \bigcup_{\substack{l=1 \\ i_s=0;2 \\ 1 \le s \le l}}^{k-1} U_{i_1 \ldots i_s \ldots i_l}$$

Then

$$2 \operatorname{mes}(A_{i_1 i_2 \ldots i_k} \cap K^*) = \operatorname{mes}(A_{i_2 i_3 \ldots i_k} \cap K^*)$$

and, hence, the Lebesgue measure defined on K^* is invariant under the map f.

By using Theorem 6.3, one can estimate the measures of the sets $C_m^{(0)}$ with $m < m^*$ in the spectral decomposition of the set $\overline{\operatorname{Per} f}$. The set $C_{m^*}^{(0)}$ contains the critical point and, therefore, cannot satisfy the conditions of the theorem. The following theorem determines the measure of the set $C_{m^*}^{(0)}$ with $m^* = \infty$ for maps with negative Schwarzian.

Theorem 6.11 (Guckenheimer [3]). *Let f be an S-unimodal map with the following properties:*

(i) it possesses a unique nonflat critical point (the point of extremum);

(ii) f is a map of type 2^∞.

Then the Lebesgue measure of the quasiattractor of the map f is equal to zero.

Proof. We prove this assertion for maps symmetric with respect to their point of extremum. For the sake of convenience, we assume that the point of extremum of the map f is located at the origin and $f(0) > 0$. Denote the points $f^j(0)$ by c_j. It follows from the conditions of the theorem that, for any $n > 0$, there are 2^{n-1} mutually disjoint intervals $\mathcal{J}_{n,j} = [c_j, c_{2^{n-1}+j}]$. Each of these intervals contains a single (repelling) point of period 2^{n-1}. Denote the left fixed repelling point of f by p_{-1}, the periodic point of period 2^n closest to the point of extremum by p_n, and the points $f^j(p_n)$ by $p_{n,j}$. For any n, we have $\mathcal{J}_{n+1,j} \cup \mathcal{J}_{n+1,2^{n-1}+j} \subset \mathcal{J}_n$ and $K = \bigcap_{n>0} (\bigcup_{j \ge 0} \mathcal{J}_{n,j})$ is a quasiattractor (see Barkovsky and Levin [1] and Misiurewicz [2]). Let q_n denote the first point to the right of c_1 such that $f^{2^n-1}(q_n) = -p_{n-1}$. Note that $f^{2^n}|_{(p_{n,1}, q_n)}$ is a homeomorphism.

Let $|\mathcal{J}|$ denote the length of the interval $\mathcal{J}$. To prove the theorem, it suffices to show that there exists $\alpha < 1$ such that the inequality $(|\mathcal{J}_{n+1,j}| + |\mathcal{J}_{n+1,j+2^n}|) / |\mathcal{J}_{n,j}| < \alpha$ holds for any $n > 0$. Indeed, in this case,

$$\sum_{j=1}^{2^n-1} |\mathcal{J}_{n,j}| < \alpha^n |\mathcal{J}_{1,1}|$$

and, hence, mes $K = 0$.

The proof is split into several steps.

Step I. $|Df^{2^n}(q_n)| > 1$.

We proceed by induction. Since f is symmetric, we have $q_0 = -p_{-1}$ and, therefore, $|f'(q_0)| = |f'(p_{-1})| > 1$. Suppose that $|Df^{2^n}(q_n)| > 1$. Then

$$|Df^{2^{n+1}}(q_n)| = |Df^{2^n}(q_n)||Df^{2^n}(p_{n,1})| > 1.$$

Note that q_{n+1} lies in the interval $(p_{n+1,1}, q_{n+1})$ and $f^{2^{n+1}}$ is monotone in this interval. Since all iterations of f have negative Schwarzians, $|Df^{2^{n+1}}(q_{n+1})|$ is greater than the minimum of $|Df^{2^{n+1}}(q_n)|$ and $|Df^{2^{n+1}}(p_{n+1,1})|$. Since all periodic orbits are repelling, we have $|Df^{2^{n+1}}(p_{n+1,1})| > 1$ and, hence, $|Df^{2^{n+1}}(q_{n+1})| > 1$.

Note that this inequality remains true for all n under weaker conditions than the symmetry of f. Thus, it suffices to require that $|f'(q_0)| > 1$.

Step II. $p_n / c_{2^n} < 0.71$ for all sufficiently large n.

Since the point of extremum is nonflat, the function f on the interval $[0, p_{n-1}]$ can be approximated (as $n \to \infty$) with any desired degree of accuracy by a function of the form $a - bx^2$. This enables us to conclude that

$$\frac{|c_1 - p_{n,1}|}{|c_1 - c_{2^n+1}|} \approx \left(\frac{p_n}{c_{2^n}}\right)^2.$$

Moreover,

$$|c_1 - c_{2^n+1}| = |p_{n,1} - c_{2^n+1}| + |c_1 - p_{n,1}| \quad \text{and} \quad |c_1 - p_{n,1}| < |p_{n,1} - c_{2^n+1}|$$

because $|Df^{2^n}(x)| > 1$ for any $x \in (p_{n,1}, c_1) \subset (p_{n,1}, q_1)$.

Consequently,

$$\frac{|c_1 - p_{n,1}|}{|c_1 - c_{2^n+1}|} < \frac{1}{2}$$

and, hence,

$$\left|\frac{p_n}{c_{2^n}}\right| < \frac{1}{\sqrt{2}} + \varepsilon < 0.71$$

for large n if we set $\varepsilon = 0.002$.

Step III. $\left|\dfrac{p_n}{c_{2^n}}\right| > \dfrac{1}{3}$.

Since f is symmetric with respect to the origin, the map f^{2^n} is symmetric in the interval $[p_{n-1}, -p_{n-1}]$. Since $Sf < 0$, the map f^{2^n} is expanding on $[p_n, -p_{n-1}]$. The inclusion $c_{2^{n+1}} \in (0, -p_n)$ implies the inequality $|c_{2^n} - p_n| < |p_n - c_{2^{n+1}}| < 2|p_n|$. Hence, $|c_{2^n}| = |p_n| + |c_{2^n} - p_n| < 3|p_n|$ and $|p_n/c_{2^n}| > 1/3$.

Lemma 6.3. *Let h be a C^3-diffeomorphism on* $[0, 1]$ *such that* $Sh < 0$, $h(0) = 0$, *and* $h(1) = 1$. *Then the inequalities*

$$\left|\frac{h''(x)}{(h'(x))^2}\right| < \frac{2}{\delta} \quad \textit{and} \quad \left|\frac{h'(x)}{h'(y)}\right| < \exp\left\{\frac{2}{\delta}\right\}$$

hold for any x and y from the interval $h^{-1}(\delta, 1-\delta)$.

A similar assertion was proved by van Strien [3] and we refer the reader to this paper for the proof.

The following statement is an immediate consequence of Lemma 6.3:

Step IV. There exists $\varepsilon > 0$ such that $|p_n / p_{n-1}| > \varepsilon$ for all n.

Step V. There exists a constant $\beta > 0$ such that

$$\frac{|c_{2^{n+1}} - p_{n-1}|}{|c_{2^{n+2}} - c_{2^{n+1}}|} > \beta \quad \text{and} \quad \frac{|p_n - c_{2^{n+2}}|}{|c_{2^{n+2}} - c_{2^{n+1}}|} > \beta.$$

If $p_{n-1} < 0$, then the points are ordered as follows: $p_{n-1} < -p_n < c_{2^{n+1}} < 0 < c_{2^{n+2}} < p_n < c_{2^n} < -p_{n-1}$. By using Step II, we obtain $|c_{2^{n+1}}/p_{n-1}| < |p_n / c_{2^n}| < 0.71$ and, therefore, $|c_{2^{n+1}} / p_n| < 0.71$. Since $|p_n / p_{n-1}| > \varepsilon > 0$ for any n, the results of Steps II and III imply that the quantities $|c_{2^{n+1}}/c_{2^n}|$ are also separated from zero for any n. This proves the existence of the constant β.

All preliminary steps of the proof of Theorem 6.11 are now completed, and we can make the following conclusions:

(i) $[c_{3\cdot 2^n}, c_{2^{n+2}}] \supset [p_n, c_{2^{n+2}}]$;

(ii) $|c_{2^n+2} - c_{2^n+1}|/|c_{2^n+1} - c_{2^n}|$ are separated from zero (see Step IV);

(iii) $|c_{2^n+2} - p_n|/|c_{2^n+2} - c_{2^n+1}|$ are separated from zero (see Step V).

Thus, a constant γ defined as the minimum of the ratio of the length of the "removed" interval to the length of the original interval exists and is positive. Hence,

$$\sum_{j=1}^{2^n} |\mathcal{I}_{n+1,j}| < \alpha \sum_{j=1}^{2^{n-1}} |\mathcal{I}_{n,j}|,$$

where $\alpha = 1 - \gamma$.

Theorem 6.12. *There exists a unimodal C^∞-map with flat extremum, which possesses a quasiattractor of positive Lebesgue measure.*

Proof. The corresponding example was suggested by Misiurewicz [4]. However, the map constructed in that example is characterized by a property that seems to be nontypical of smooth unimodal maps with nonflat extremum, namely, the multiplicator

$$\overline{\lim_{n\to\infty}} \sup_{x\in K} \left| \frac{df^n(x)}{dx} \right|$$

of the quasiattractor K of this map is unbounded.

An example presented below is free of this shortage. At the same time, the smoothness of the map at the point of extremum is not higher than $C^r, r \geq 0$ (see Kolyada [1]). It is worth noting that, in this example, one can also show that any invariant measure is singular.

We fix $\varepsilon \in (0, 1)$ and $\varepsilon_0 \in (0, \varepsilon)$ and take a sequence of numbers $\beta_1 > \beta_2 > \ldots > \beta_i > \ldots > 0$ such that

$$\sum_{i=1}^{\infty} \beta_i = \beta < \varepsilon - \varepsilon_0.$$

We construct a Cantor set

$$I_\infty = \bigcap_{m=1}^{\infty} I_m \subset I$$

such that $\operatorname{mes} I_\infty > 1 - \varepsilon$ as follows:

Let $m = 1$. We choose an arbitrary interval $[x_0, y_0] \subset \operatorname{int} I$ of length $1 - \varepsilon_0$ and set $I_1 = I_1^1 = [x_0, y_0]$. For $m = 2$, we define a set $I_2 = I_{10}^2 \cup I_{11}^2 \subset I_1^1$, where I_{10}^2 and I_{11}^2 are closed disjoint intervals, $x_0 \in \partial I_{10}^2$, $y_0 \in \partial I_{11}^2$, $\operatorname{mes} I_{10}^2 = (1 - \varepsilon_0 - \beta_1)\delta(\delta + 1)^{-1}$, and $\operatorname{mes} I_{11}^2 = (1 - \varepsilon_0 - \beta_1)\delta(\delta + 1)^{-1}$, where $\delta > 1$. Let $U_1 = I_1 \setminus I_2$. For $m > 2$, the set I_m is constructed recursively. Assume that we have already constructed the set $I_{m-1} = \bigcup_\alpha I_\alpha^{m-1}$, where $\alpha = \alpha_1 \alpha_2 \ldots \alpha_{2^{m-2}}$ is a sequence of 2^{m-2} zeros and ones such that, for any $i \in \{1, 2, 4, \ldots, 2^{m-3}\}$, either $\alpha_1 \ldots \alpha_i = \alpha_{i+1} \ldots \alpha_{2i}$ or $\alpha_1 \ldots \alpha_i = \overline{\alpha}_{i+1} \ldots \overline{\alpha}_{2i}$, where $\overline{\alpha}_i = 1 - \alpha_i$. Then the set $U_{m-2} = \bigcup_\alpha U_\alpha^{m-2}$, where $U_\alpha^{m-2} \subset I_\alpha^{m-2}$ is an open interval, is also well defined.

Let us now construct the set $U_{m-1} = \bigcup_\alpha U_\alpha^{m-1}$. We choose an open interval $U_\alpha^{m-1} \subset I_\alpha^{m-1}$ such that $\operatorname{mes} U_\alpha^{m-1} = \gamma_{m-1} \beta_{m-1}$, where

$$\gamma_{m-1} = \operatorname{mes} I_\alpha^{m-1} \left(1 - \varepsilon_0 - \sum_{i=1}^{m-2} \beta_i \right)^{-1},$$

and the intervals from the set $I_\alpha^{m-1} \setminus U_\alpha^{m-1}$ have the following properties:

(i) $I_{\beta'}^m \cup I_\beta^m = I_\alpha^{m-1} \setminus U_\alpha^{m-1}$, where $\beta' = \alpha\overline{\alpha}$, $\beta = \alpha\alpha$, and the interval $I_{\beta'}^m$ is located to the right (left) of U_α^{m-1} if $\alpha_1 \ldots \alpha_{2^{m-3}} = \overline{\alpha}_{2^{m-3}+1} \ldots \overline{\alpha}_{2^{m-2}}$ $(\alpha_1 \ldots \alpha_{2^{m-3}} = \alpha_{2^{m-3}+1} \ldots \alpha_{2^{m-2}})$;

(ii) $$\operatorname{mes} I_{\beta'}^m = \left(1 - \varepsilon_0 - \sum_{i=1}^{m-1} \beta_i \right) \gamma_{m-1} \delta(\delta + 1)^{-1},$$

$$\operatorname{mes} I_\beta^m = \left(1 - \varepsilon_0 - \sum_{i=1}^{m-1} \beta_i \right) \gamma_{m-1} \delta(\delta + 1)^{-1}.$$

Thus, we have constructed the set $I_\infty = \bigcup_{m=1}^\infty \bigcup_\alpha I_\alpha^m$. Since

$$\sup_\alpha \operatorname{mes} (I_\infty \cap I_\alpha^m) = \left(1 - \varepsilon_0 - \sum_{i=1}^{m-1} \beta_i \right) \left(\frac{\delta}{\delta + 1} \right)^{m-1},$$

the set I_∞ does not contain intervals. Hence, by construction, it is a Cantor set of positive Lebesgue measure: $\operatorname{mes} I_\infty = 1 - \varepsilon_0 - \beta > 1 - \varepsilon$.

We now construct a map $f: I \to I$ of type 2^∞ whose quasiattractor coincides with I_∞. For this purpose, we choose two sequences $\{x_i\}_{i=1}^\infty$ and $\{y_i\}_{i=1}^\infty$, where x_i and y_i

are the ends of the intervals U_α^m such that $\alpha_1 \ldots \alpha_j = \overline{\alpha}_{j+1} \ldots \overline{\alpha}_{2j}$ for all $j = 1, 2, 4, \ldots$. Then $x_0 < x_i < \ldots < x_{2i} < \ldots$ and $y_0 > y_1 > \ldots > y_{2i} > \ldots$ are such that

$$x_{2i+1} = x_{2i} + \left(1 - \varepsilon_0 - \sum_{j=1}^{2i+2} \beta_j\right) \delta^{2i+1} (1+\delta)^{-(2i+2)},$$

$$x_{2i+2} = x_{2i+1} + \beta_{2i+2} \delta^{2i+1} (1+\delta)^{-(2i+1)},$$

$$y_{2i+2} = y_{2i+1} - \beta_{2i+1} \delta^{2i} (1+\delta)^{-2i},$$

$$y_1 = y_0 - (1 - \varepsilon_0 - \beta_1)(1+\delta)^{-1},$$

$$y_{2i+1} = y_{2i} - \left(1 - \varepsilon_0 - \sum_{j=1}^{2i+1} \beta_j\right) \delta^{2i} (1+\delta)^{-(2i+1)},$$

where $i = 1, 2, \ldots$ and x_0 and y_0 are the points used in the construction of the set I_∞. At the points $x_i, y_i, i = 0, 1, 2, \ldots$, we define the values of the map f as follows:

$$f(x_{2i-1}) = y_0 - \left(1 - \varepsilon_0 - \sum_{j=1}^{2i} \beta_j\right)(\delta+1)^{-2i} - \beta_{2i}(1+\delta)^{-(2i+1)},$$

$$f(x_{2i}) = y_0 - \left(1 - \varepsilon_0 - \sum_{j=1}^{2i+1} \beta_j\right)(\delta+1)^{-(2i+1)},$$

$$f(y_{2i}) = y_0 - \left(1 - \varepsilon_0 - \sum_{j=1}^{2i} \beta_j\right)(\delta+1)^{-2i},$$

$$f(y_{2i+1}) = y_0 - \left(1 - \varepsilon_0 - \sum_{j=1}^{2i+1} \beta_j\right)(\delta+1)^{-(2i+1)} - \beta_{2i+1}(\delta+1)^{-2i},$$

$$f(y_1) = y_0 - (1 - \varepsilon_0 - \beta_1)(1+\delta)^{-1} - \beta_4, \quad i = 1, 2, \ldots,$$

$$f(x_0) = y_1, \quad f(y_0) = x_0.$$

In the intervals $[x_{2i}, x_{2i+1}]$ and $[y_{2i+1}, y_{2i}]$, the map f is defined as follows: $f(x_{2i}) + \delta^{-2i}(x - x_{2i})$, $x \in [x_{2i}, x_{2i+1}]$, and $f(y_{2i}) + \delta^{1-2i}(y_{2i} - x)$, $x \in [y_{2i+1}, y_{2i}]$. Let us extend the definition of the map to the remaining intervals (x_{2i+1}, x_{2i+2}) and (y_{2i+2}, y_{2i+1}). Denote the interval (x_{2i+1}, x_{2i}) by K_i and the interval (y_{2i+2}, y_{2i+1}) by K_i'.

To paste the relevant parts of the map in these intervals, we determine the coordinates of the points of intersection of the straight lines $y = f(x_{2i}) + \delta^{-2i}(x - x_{2i})$ and $y = f(x_{2i+2}) + \delta^{-2i-2}(x - x_{2i+2})$, namely,

$$x = x_{2i+2} + \frac{\delta^2}{\delta^2 - 1}[(f(x_{2i+2}) - f(x_{2i}))\delta^{2i} - (x_{2i+2} - x_{2i})],$$

$$y = f(x_{2i+2}) + \frac{\delta^{-2i}}{\delta^2 - 1}[(f(x_{2i+2}) - f(x_{2i}))\delta^{2i} - (x_{2i+2} - x_{2i})],$$

and of the straight lines $y = f(y_{2i}) + \delta^{1-2i}(y_{2i} - x)$ and $y = f(y_{2i+2}) + \delta^{-2i-1}(y_{2i+2} - x_{2i})$, i.e.,

$$x = y_{2i+2} + \frac{\delta^2}{\delta^2 - 1}[(f(y_{2i+2}) - f(y_{2i}))\delta^{2i-1} - (y_{2i} - y_{2i+2})],$$

$$y = f(y_{2i+2}) + \frac{\delta^{-2i}}{\delta^2 - 1}[(f(y_{2i+2}) - f(y_{2i}))\delta^{2i-1} - (y_{2i} - y_{2i+2})].$$

Let us now compute the quantities

$$\operatorname{sign}[(f(x_{2i+2}) - f(x_{2i}))\delta^{2i} - (x_{2i+2} - x_{2i})]$$

and

$$\operatorname{sign}[(f(y_{2i+2}) - f(y_{2i}))\delta^{2i-1} - (y_{2i} - y_{2i+2})].$$

We have

$$\operatorname{sign}\left[\frac{\delta^{2i}}{(\delta+1)^{2i+3}}\left(\left(1 - \varepsilon_0 - \sum_{j=1}^{2i+1}\beta_j\right)\delta - \beta_{2i+2}(\delta^3 + \delta - 1) + \beta_{i+3}\right)\right] > 0$$

for large i (in particular, for $\delta = 2$, we have $\beta_i = \frac{1-\varepsilon_0}{(i+1)^2}$ for any $i = 1, 2, \ldots$),

$$\operatorname{sign}\left[\frac{\delta^{2i-1}}{(\delta+1)^{2i+2}}\left(\left(1 - \varepsilon_0 - \sum_{j=1}^{2i}\beta_j\right)\delta - \beta_{2i+1}(\delta^3 + \delta - 1) + \beta_{2i+2}\right)\right] > 0$$

for large i (in particular, for $\delta = 2$, we have $\beta_i = \frac{1-\varepsilon_0}{(i+1)^2}$ for any $i = 2, 3, \ldots$). For $i = 1$, we can write

$$\operatorname{sign}\left[f(y_2) - f(y_0)\right)\delta^{-1} - (y_0 - y_2)\right] < 0,$$

$y = y_2 + \Delta y$, and $\Delta y \le \frac{\beta_1}{2}$. Hence, only in the last case, it is possible to define f on the interval $[y_3, y_0]$ as a convex function.

Consider a segment $[a, b]$ and straight lines $y = \alpha(x-a) + A$ and $y = \beta(x-b) + B$ such that $\frac{B-A}{b-a} > \alpha > \beta > 0$. Then the coordinate x of the point of intersection of these straight lines belongs to the segment $(b, \varepsilon]$, $\varepsilon > b$. We denote $\frac{B-A}{b-a}$ by γ and $2\gamma - \alpha$ by γ' and construct a straight line $y = \gamma'(x - a_0) + b_0$ such that the coordinate x of the point of intersection of this line with the straight line $y = \beta(x-b) + B$ satisfies the condition $1 < \frac{a_0 - x}{x - a} < k = \text{const}$. In view of the fact that $\frac{a_0 - x}{x - a} = \frac{\gamma - \beta}{\gamma - \alpha}$, this is possible only in the case where $\gamma > \frac{k\alpha - \beta}{k-1}$. Since $\alpha = \delta^2\beta$, for the construction of the required example, one must check the inequality $\gamma > \frac{k\delta^2 - \beta}{k-1}\beta$. Indeed, in the intervals (x_{2i-1}, x_{2i}), $i = 1, 2, 3, \ldots$, we have

$$\gamma = \frac{1 - \varepsilon_0 - \sum_{j=1}^{2i}\beta_j}{\beta_{2i}(\delta+1)\delta^{2i-1}} + \frac{1}{\delta^{2i-1}} - \frac{1 - \varepsilon_0 - \sum_{j=1}^{2i+1}\beta_j}{\beta_{2i}(\delta+1)^2\delta^{2i-1}}, \qquad \beta = \delta^{-2i}$$

and, hence, $\gamma > \frac{k\delta^2 - \beta}{k-1}\beta$ for large i and all $k > 1$. (Note that, for $\delta = 2$, $k = 6$ and $\delta = \frac{3}{2}$, $k = 2$, this inequality holds for any $i = 1, 2, \ldots$.) Similar reasoning is applicable to the intervals (y_{2i}, y_{2i-1}), $i = 2, 3, \ldots$.

Denote the point of intersection of the lines $y = \beta(x-b) + B$ and $y = \gamma'(x - a_0) + b_0$ by $\{a_1, b_1\}$. In the segment (a, a_0), we paste these lines by the function

$$g_0(x) = b_0 + \gamma'(x - a_0) + (\gamma' - \alpha)(a_1 - a)\varphi\left(\frac{x - a_0}{x_0 - a}\right)$$

$$= b_0 + (2\gamma - \alpha)(x - a_0) + 2(\gamma - \alpha)(a_0 - a)\varphi\left(\frac{x - a_0}{a_0 - a}\right),$$

in the segment $(a_0, 2a_1 - b]$, for this purpose, we use the function

$$y_1(x) = \gamma'(x - a_0) + b_0 = (2\gamma - \alpha)(x - a_0) + b_0,$$

and in the segment $(2a_1 - b, b)$, these lines are pasted by the function

$$g_2(x) = b_0 + (2\gamma - \alpha)(x - a_0) + 2(b - a_1)(2\gamma - \alpha - \beta)\,\varphi\left(\frac{x - 2a_1 + b}{2(b - a_1)}\right).$$

Since

$$\sup_{x\in[a,b]} |g_j^{(r)}(x)| = \max\{(2\gamma-\alpha-\beta)[2(b-a_1)]^{1-r}\sup_{x\in[0,1]}|\varphi^{(r)}(x)|,$$

$$2(\gamma-\alpha)(a_0-a)^{1-r}\sup_{x\in[0,1]}|\varphi^{(r)}(x)|\},\quad j=0,1,2,\quad r=2,3,\ldots,$$

we conclude that $g_j \in C^\infty([a, b])$.

By applying this construction to the segments (x_{2i-1}, x_{2i}) and (y_{2i+2}, y_{2i+1}), $i = 1, 2,$ one can easily show that, for any $r \geq 0$, there exists δ such that

$$\lim_{i\to\infty}\sup_{x\in(x_{2i-1},x_{2i})}|g_i^{(r)}(x)| = 0 \quad\text{and}\quad \lim_{i\to\infty}\sup_{x\in[y_{2i+2},y_{2i+1})}|g_i^{(r)}(x)| = 0.$$

The map constructed as a result belongs to the class C^r, $r \geq 0$. By connecting the point 0 with x_0 and the point y_0 with 1 by monotone C^∞-functions (under the relevant sewing conditions for the derivatives at the points x_0 and y_0), we obtain the required map.

Let us now discuss in brief the example suggested by Misiurewicz [4]. We use the notation introduced in the proof of Theorem 6.12 with certain modifications. Thus, the sequences $x_0 > x_1 > x_2 > \ldots$ and $y_0 < y_1 < y_2 < \ldots$ are defined as

$$x_0 = 0,\quad y_0 = 1,\quad x_{2i+1}-x_{2i} = \frac{1}{(2i+3)^2},\quad x_{2i+2}-x_{2i+1} = \frac{1}{(2i+2)(2i+3)^2},$$

$$y_{2i}-y_{2i+1} = \frac{1}{(2i+2)^2},\quad y_{2i+1}-y_{2i+2} = \frac{1}{(2i+1)(2i+2)^2},\quad i\geq 0.$$

It is not difficult to show that

$$\lim_{i\to\infty} x_i = \lim_{i\to\infty} y_i \overset{\text{def}}{=} c.$$

Let us construct the map f. For this purpose, we set

$$f(x) = 1,\quad f(x_{2i}) = 1 - \frac{1}{(2i+2)(2i+2)!},\quad f(x_{2i+1}) = 1 - \frac{1}{(2i+2)(2i+3)!},$$

$$f(y_{2i}) = 1 - \frac{1}{(2i+1)(2i+1)!},\quad f(y_{2i+1}) = 1 - \frac{1}{(2i+1)(2i+2)!}.$$

Then

$$0 = f(y_0) < f(y_1) < f(x_0) < f(x_1) < f(y_2) < f(y_3) < \ldots$$

and

$$\lim_{i\to\infty} f(x_i) = \lim_{i\to\infty} f(y_i) = f(c) = 1.$$

After this, the map f is constructed as in the previous example: First, in the intervals $[x_{2i}, x_{2i+1}]$ and $[y_{2i+1}, y_{2i}]$, it is defined as a linear function and then, in the remaining intervals, the relevant linear segments are C^∞-smoothly pasted. The map $f \in C^\infty(I, I)$ obtained as a result of this procedure possesses a quasiattractor of positive Lebesgue measure equal to

$$\prod_{n=1}^{\infty}\left(1-\left(\frac{1}{n+1}\right)^2\right).$$

7. LOCAL STABILITY OF INVARIANT SETS. STRUCTURAL STABILITY OF UNIMODAL MAPS

1. Stability of Simple Invariant Sets

1.1. Stability of Periodic Trajectories. Let $f: I \to I$ be a continuous map and let $B = \{\beta_0, \beta_1, \ldots, \beta_{n-1}\}$ be its cycle of period $n \geq 1$. One can distinguish between two types of stability of the cycle B, namely, between stability under perturbations of the initial data and stability under perturbations of the map. First, we consider the first type of stability.

Recall some definitions. A cycle B is called asymptotically stable or attracting if there exists a neighborhood U of this cycle such that

$$\bigcap_{i \geq 0} f^i(U) = B.$$

A cycle B is called repelling if there exists a neighborhood U of this cycle such that, for each point $x \in U \backslash B$, one can indicate $i \geq 0$ for which $f^i(x) \notin U$.

A cycle B is called semiattracting if there exists a neighborhood U of B such that, for any point $\beta_j \in B$, one can indicate its half neighborhood U'_j such that if $x \in \bigcap_{0 \leq j < n} U'_j$, then $f^i(x) \notin U$ for some $i \geq 0$ and the other half neighborhoods U''_j of the points β_j satisfy the equality

$$\bigcap_{i \geq 0} f^i\left(\overline{\bigcup_{0 \leq j < n} U''_j}\right) = B$$

As indicated in Chapter 1, these definitions do not exhaust all possibilities in the behavior of trajectories.

Theorem 7.1. *An n-periodic* $(n \geq 1)$ *cycle* $B = \{\beta_0, \beta_1, \ldots, \beta_{n-1}\}$ *of a continuous map* $f: I \to I$ *is attracting if and only if, for any point* x *from some neigh-*

borhood U_0 *of the point* β_0, *the inequality* $f^{2n}(x) > x$ *holds for* $x < \beta_0$ *and the inequality* $f^{2n}(x) < x$ *holds whenever* $x > \beta_0$.

A cycle B *is repelling if and only if, for any point* x *of some neighborhood* U_0 *of the point* β_0, $f^{2n}(x) \notin [x, \beta_0]$ *for* $x < \beta_0$ *and* $f^{2n}(x) \notin [\beta_0, x]$ *whenever* $x > \beta_0$.

A cycle B *is semiattracting if and only if, for any point* x *of some neighborhood* U_0 *of the point* β_0, *either* $f^{2n}(x) \in (x, \beta_0]$ *for* $x < \beta_0$ *and* $f^{2n}(x) > x$ *for* $x > \beta_0$ *or, vice versa,* $f^{2n}(x) < x$ *for* $x < \beta_0$ *and* $f^{2n}(x) \in [\beta_0, x)$ *for* $x > \beta_0$.

Theorem 7.1 can be proved by the direct investigation of the behavior of trajectories of the map f^{2n} in a neighborhood of its fixed point β_0 under the conditions of the theorem.

Consider the case where the map f is smooth in more details. Let $f \in C^r(I, I)$, $r \geq 1$, and let $B = \{\beta_0, \beta_1, \ldots, \beta_{n-1}\}$ be a cycle of the map f of period n. For $k \in \{1, 2\}$ and $i \in \{1, 2, \ldots, r\}$, we define the quantities

$$\mu_k^{(i)}(B) = \frac{d^i}{dx^i} f^{kn}(\beta_0) = (f^k)'(\beta_0)(f^k)'(\beta_1) \ldots (f^k)'(\beta_{n-1}).$$

The quantity $\mu(B) = \mu_1^{(1)}(B)$ is called the multiplier of the cycle B. The theorem below establishes the relationship between the values of $\mu_k^{(i)}(B)$ and the type of stability of the cycle B.

Theorem 7.2. *Let* $f \in C^r(I, I)$, $r \geq 1$, *and let* $B = \{\beta_0, \beta_1, \ldots, \beta_{n-1}\}$ *be a cycle of* f *with period* $n \geq 1$. *If* $|\mu(B)| > 1$, *then* B *is repelling.*

Suppose that $\mu(B) = 1$ *and there is* $s > 1$ $(s \leq r)$ *such that* $\mu_1^{(s)}(B) \neq 0$ *but* $\mu_1^{(i)}(B) = 0$ *for* $1 < i < s$. *If* s *is even, then* B *is a semiattracting cycle; if* s *is odd, then the cycle* B *is attracting if* $\mu_1^{(s)}(B) < 0$ *and repelling if* $\mu_1^{(s)}(B) > 0$.

Suppose that $\mu(B) = -1$ *and there exists* $s > 1$ $(s \leq r)$ *such that* $\mu_2^{(s)}(B) \neq 0$ *but* $\mu_2^{(i)}(B) = 0$ *for* $1 < i < s$. *Then* s *is odd and the cycle* B *is attracting whenever* $\mu_2^{(s)}(B) < 0$ *and repelling if* $\mu_2^{(s)}(B) > 0$.

Theorem 7.2 is proved by the direct verification of validity of the conditions of Theorem 7.1 under the conditions of Theorem 7.2. Here, we restrict ourselves to the proof of the following statement: If $\mu(B) = -1$, then $\mu_2^{(s)}(B) = 0$ for even s. Indeed, if $s = 2$, then

$$\mu_2^{(2)}(B) = (f^n)''(f^n(\beta_0))((f^n)'(\beta_0))^2 + (f^n)'(f^n(\beta_0))((f^n)''(\beta_0) = 0.$$

If s even and $\mu_2^{(i)}(B) = 0$ for $1 < i < s$, then

$$\mu_2^{(s)}(B) = \frac{d^s}{dx^s} f^n(f^n(\beta_0))((f^n)'(\beta_0))^s + (f^n)'(f^n(\beta_0)) \frac{d^s}{dx^s} f^n(\beta_0) = 0.$$

Hence, in this case, s must be odd and the cycle B cannot be semiattracting.

Parallel with the concept of asymptotic stability, one can also use the concept of Lyapunov stability.

Definition. A cycle $B = \{\beta_0, \beta_1, \ldots, \beta_{n-1}\}$ of period $n \geq 1$ of a map $f \in C^0(I, I)$ is called Lyapunov stable if, for any neighborhood U of B, there exists a neighborhood V of B, $V \subseteq U$, such that $f^i(V) \subset U$ for all $i > 0$.

It is clear that any attracting cycle is Lyapunov stable and any repelling or semiattracting cycle is not Lyapunov stable. It follows from Theorem 7.1 that if a cycle B of period n is Lyapunov stable but not attracting, then the points of this cycle are not isolated in the set of periodic points of period n or $2n$. Hence, if this cycle B is a cycle of a map $f \in C^r(I, I)$, $r \geq 1$, then either $\mu(B) = 1$ and $\mu_1^{(i)}(B) = 0$ for $1 < i \leq r$ or $\mu(B) = -1$ and $\mu_2^{(i)}(B) = 0$ for $1 < i \leq r$.

Thus, it follows from Theorem 7.2 that if a map $f: I \to I$ is analytic and $f(x) \neq x$ at least at one point $x \in I$, then any cycle of this map is either attracting, or repelling, or semiattracting. Note that this is not true even for maps from the class C^∞ because, for these maps, the set of periodic points of the same fixed period can be infinite.

Consider the problem of stability of periodic trajectories under perturbations of the map f.

Definition. We say that a cycle B of period $n \geq 1$ of a continuous map $f: I \to I$ survives under C^0-perturbations of the map f if, for any neighborhood U of the cycle B there exists a neighborhood $\mathfrak{U}$ of the map f in $C^0(I, I)$ such that any map $\tilde{f} \in \mathfrak{U}$ possesses a cycle of period n lying in the neighborhood U.

Theorem 7.3. *A cycle* $B = \{\beta_0, \beta_1, \ldots, \beta_{n-1}\}$ *of period* $n \geq 1$ *of a continuous map f survives under* C^0*-perturbations of the map f if and only if, for any neighborhood* U_0 *of the point* β_0, *there exist* $x_1, x_2 \in U_0$ *such that* $(f^n(x_1) - x_1)(f^n(x_2) - x_2) < 0$.

Proof. Suppose that $(f^n(x_1) - x_1)(f^n(x_2) - x_2) \geq 0$ for some neighborhood U_0 of the point β_0 and all points $x_1, x_2 \in U_0$. Without loss of generality, we can assume that $f^n(x) \leq x$ for $x \in U_0$. If $n = 1$, then, for any $\varepsilon > 0$, the map $\tilde{f} = f - \varepsilon$ has no fixed points in U_0. If $n > 1$, then we choose a neighborhood U_{n-1} of the point β_{n-1} such that $f(U_{n-1}) \subset U_0$. Let U'_{n-1} be a neighborhood of the point β_{n-1} which lies in

U_{n-1} together with its closure and let $\varphi(x)$ be a continuous function taking values from the interval [0, 1], equal to zero outside U_{n-1}, and equal to one inside U'_{n-1}. Then, for all sufficiently small $\varepsilon > 0$, the map $\tilde{\tilde{f}} = f - \varepsilon\varphi$ has no periodic points of period n in a certain neighborhood U'_0 of the point β_0 because $\tilde{\tilde{f}}^n(x) \leq x - \varepsilon$ for $x \in U'_0$. The other statements of Theorem 7.3 are obvious.

Corollary 7.1. *If a cycle is attracting or repelling, then it survives under C^0-perturbations of the map f.*

Note that the proof of Theorem 7.3 implies the following assertion: If a cycle does not survive under C^0-perturbations of the map, then it does not survive under C^r-perturbations of the map. Indeed, the function $\varphi(x)$ used in the proof of Theorem 7.3 can be taken even from the class C^∞.

If a cycle B does not survive under C^r-perturbations of the map f, $r \geq 1$, then, by virtue of Theorem 7.2 and Corollary 7.1, we can write $\mu(B) = 1$ and either there exists an even number $s \leq r$ such that $\mu_1^{(s)}(B) \neq 0$ but $\mu_1^{(i)}(B) = 0$ for $0 < i < s$ or $\mu_1^{(i)}(B) = 0$ for $1 < i \leq r$.

It is worth noting that the survival of cycles under perturbations of a map is not connected with the preservation of the structure of a dynamical system in the neighborhood of a cycle (i.e., with the behavior of trajectories): The behavior of trajectories of a perturbed map in the neighborhood of a cycle may significantly differ from the behavior of the original map in the neighborhood of the original cycle even if this cycle survives (for example, a cycle may change the type of stability). For this reason, we introduce the following definition:

Definition. A map $f \in C^r(I, I)$ is called C^r-structurally stable in the neighborhood of its cycle B if there exist a neighborhood U of the cycle B and a neighborhood $\mathcal{U}$ of the map f in $C^r(I, I)$ such that, for any $\tilde{f} \in \mathcal{U}$, one can indicate a homeomorphism $h = h(\tilde{f})$ of the interval I onto itself for which $\tilde{f} \circ h|_U = h \circ f|_U$.

The homeomorphism h translates trajectories (or parts of trajectories) of the map f lying in U into trajectories (or their parts) of the map $\tilde{f}$ and preserves the mutual arrangement of the points of these trajectories. This remark immediately implies the following assertion:

Theorem 7.4. *A map $f \in C^r(I, I)$, $r \geq 1$, is C^r-structurally stable in a neighborhood of its cycle B if and only if $|\mu(B)| \neq 1$ and $\mu(B) \neq 0$.*

Note that the concept of C^0-structural stability is meaningless because there are no

C^0-structurally stable maps: Indeed, for any point $x_0 \in I$, we can modify the map f to guarantee that $\tilde{f}(y) = \text{const}$ for all points in a certain neighborhood of x_0. If $f \neq \text{const}$ in this neighborhood, then the dynamics of trajectories of the map f undergoes significant changes near the indicated point. In all other cases, one can also easily construct the required C^0-perturbation of the map f.

1.2. Stability of Cycles of Intervals. By analogy with the stability of periodic trajectories, we now consider the problem of stability of cycles of intervals. Let $A = \{ I_0, I_1, \ldots , I_{n-1} \}$ be a cycle of intervals of period n of the map $f \in C^0(I, I)$. Without loss of generality, we can assume that the intervals I_i are closed. In order not to introduce new notation, we denote the set $\bigcup_{0 \le i < n} I_i$ also by A if this does not lead to misunderstanding. Finally, any open set that contains the set A is called a neighborhood of the cycle of intervals A.

By analogy with the general definitions of attractor, repeller, and quasiattractor, we introduce the corresponding definitions for cycles of intervals in order to characterize the behavior of trajectories in the neighborhood of a cycle of intervals.

Definition. We say that a cycle of intervals $A = \{ I_0, I_1, \ldots , I_{n-1} \}$ of a map $f \in C^0(I, I)$ is an attractor if one can indicate a neighborhood U of A such that $\bigcap_{i \ge 0} f^i(U) \subseteq A$.

A cycle of intervals A is called a repeller if there exists a neighborhood U of A such that, for any $x \in U \backslash A$, one can find $i = i(x)$ for which the point $f^i(x)$ does not belong to the set U.

A cycle of intervals A is called a quasiattractor if, for any its neighborhood U, there exists a neighborhood U' of A such that $f^i(U') \subset U$ for all $i \ge 0$.

For cycles of intervals, one can formulate an analog of Theorem 7.1.

Let $A = \{ I_1, I_2, \ldots , I_{n-1} \}$ be a cycle of intervals of a map $f \in C^0$ and let $I(A)$ be the component of the set $\bigcup_{0 \le i < n} I_i$ which contains I_0. Note that $I(A) = I_0$ whenever the intervals I_i are mutually disjoint; otherwise, n is even and $I(A) = I_0 \cup I_{n/2}$.

Theorem 7.5. *For a map $f \in C^0(I, I)$, let $A = \{ I_0, I_1, \ldots , I_{n-1} \}$ be a cycle of intervals of period n and let $I(A) = [a, b]$. The cycle of intervals A is an attractor if and only if there exists a neighborhood $\mathcal{U}$ of the interval $I(A)$ such that $f^{2n}(x) \notin [x, b]$ if $x \in \mathcal{U}$ and $x < a$ and $f^{2n}(x) \notin [a, x]$ if $x \in \mathcal{U}$ and $x > b$.*

The proof Theorem 7.5 is similar to the proof of Theorem 7.1.

Corollary 7.2. *Let $I_0 = [a_0, b_0]$. If $a_0 < f^n(a_0) < b_0$ and $a_n < f^n(b_0) < b_0$, then the cycle of intervals $A = \{ I_0, I_1, \ldots , I_{n-1} \}$ is an attractor.*

It is obvious that if a cycle of intervals A is an attractor, then it satisfies all conditions in the definition of quasiattractor. If a cycle of intervals A is a quasiattractor but not an attractor, then Theorem 7.5 implies that at least one end of the interval $I(A)$ is not isolated in the set of periodic points of period n or $2n$ and, consequently, either $a \in \mathrm{Per}(f)$ or $b \in \mathrm{Per}(f)$.

Let us now consider the problem of preservation of cycles of intervals under perturbations of a map.

Definition. We say that a cycle of intervals $A = \{ I_0, I_1, \ldots , I_{n-1} \}$ of period n of a map $f \in C^0(I, I)$ is preserved under C^0-perturbations of this map if, for any $\varepsilon > 0$, one can indicate a neighborhood $\mathcal{U} = \mathcal{U}(\varepsilon)$ of the map f in $C^0(I, I)$ such that any map $\tilde{f} \in \mathcal{U}$ possesses a cycle of intervals $\tilde{A}$ of period n and the Hausdorff distance between the sets A and $\tilde{A}$ is less than ε.

We say that a cycle of intervals A does not vanish under C^0-perturbations of the map f if, for any neighborhood U of A, there exists a neighborhood $\mathcal{U}$ of the map f in $C^0(I, I)$ such that any map $\tilde{f} \in \mathcal{U}$ has a cycle of intervals $\tilde{A}$ of period n and U is a neighborhood of this cycle.

As follows directly from this definition, a cycle of intervals preserved under perturbations of the map does not vanish in the indicated sense. It is also easy to show that attractors are preserved under C^0-perturbations.

We say that a cycle of intervals A of period n of a map f is maximal if the map f has no cycle of intervals $\tilde{A}$ of period n such that $A \subset \tilde{A}$ and $A \neq \tilde{A}$. In what follows, we restrict ourselves to the clarification of conditions under which maximal cycles of intervals of unimodal maps are preserved or do not vanish.

Let $A = \{ I_0, I_1, \ldots , I_{n-1} \}$ be a maximal cycle of intervals of period n for a map $f \in C^0(I, I)$. Suppose that the map $f^n|_{I_0}$ is monotone. Consider the interval $I(A) = [a, b]$ introduced above. Obviously, $f^n(I(A)) \subset I(A)$. Let s be the least positive integer of the form $n, 2n, 3n, \ldots$ for which $f^s|_{I_0}$ is nondecreasing. It is clear that s is equal either to n or to $2n$. It follows from the maximality of A that $f^s(a) = a$ and $f^s(b) = b$. Moreover, the invariant interval $[a, b]$ of f^s must be a repeller, i.e., the inequalities $f^s(x) < x$ for $x \in (a - \varepsilon, a)$ and $f^s(x) > x$ for $x \in (b, b + \varepsilon)$ must hold for some sufficiently small $\varepsilon > 0$.

Theorem 7.6. *Assume that a cycle of intervals A of a unimodal map $f \in C^0(I, I)$ does not contain the point of extremum. Then A does not vanish under C^0-perturba-*

tions of the map f if and only if the interval $I(A) = [a, b]$ contains points x_1 and x_2 such that $x_1 < x_2$, $f^s(x_1) > x_1$, and $f^s(x_2) < x_2$.

The cycle of intervals A is preserved under C^0-perturbations of the map f if and only if, for any $\varepsilon > 0$, one can indicate points $x_1 \in (a, a + \varepsilon)$ and $x_2 \in (b - \varepsilon, b)$ such that $f^s(x_1) > x_1$ and $f^s(x_2) < x_2$.

The proof of this theorem is similar the proof of Theorem 7.3.

Now assume that a cycle of intervals $A = \{I_0, I_1, \ldots, I_{n-1}\}$ of a unimodal map f contains its point of extremum c and is maximal. Let $I_0 = [a_0, b_0]$. Then it follows from the results of Chapter 5 that the map $f^n|_{[a_0, b_0]}$, where n is the period of A, is unimodal and either $f^n(a_0) = a_0$ and $f^n(b_0) = a_0$ or $f^n(a_0) = b_0$ and $f^n(b_0) = b_0$ (with obvious exceptions $n = 1$ and $n = 2$).

For the interval $I(A) = [a, b]$ defined above, there exists a unique number $s \geq 1$ such that $f^s(I(A)) \subset I(A)$ and $f^s|_{I(A)}$ is unimodal. It is clear that $s = n$ if the intervals of the cycle A are mutually disjoint and $s = n/2$ whenever $I_0 \cap I_{n/2} \neq \varnothing$.

Let $\langle f^s(c), f^{2s}(c)\rangle$ be the interval with ends at $f^s(c)$ and $f^{2s}(c)$. Denote this interval by $[a_1, b_1]$. Then $[a_1, b_1] \subset I(A)$ and the following theorem is true:

Theorem 7.7. *For a unimodal map $f \in C^0(I, I)$, let A be a cycle of intervals of period n that contains the point c. If there are points $x_1 \in (a, a_1)$ and $x_2 \in [b_1, b]$ such that $f^s(x_1) > x_1$ and $f^s(x_2) < x_2$, then the cycle of intervals A does not vanish under C^0-perturbations of the map f. Moreover, any unimodal map $\tilde{f}$ sufficiently close to f in $C^0(I, I)$ has a cycle of intervals of period n that contains the point of extremum of the map $\tilde{f}$.*

The cycle of intervals A is preserved under C^0-perturbations of the map f if and only if, for any $\varepsilon > 0$, there are points $x_1 \in (a, a + \varepsilon)$ and $x_2 \in (b - \varepsilon, b)$ such that $x_1, x_2 \notin [a_1, b_1]$, $f^s(x_1) > x_1$, and $f^s(x_2) < x_2$.

Proof. Without loss of generality, we can assume that the point c of the map $f^s|_{I(A)}$ is a point of maximum.

First, we consider the case $s = n/2$. If there are no points x_1 and x_2 indicated in the conditions of Theorem 7.7, then $f^s(c) = b$ and, as in the proof of Theorem 7.3, one can construct a small continuous perturbation of the map f such that the resulting perturbed map $\tilde{f}$ has no cycles of intervals of period n that contain the point c.

If $s = n$ and there are no suitable points x_1 and x_2, then either $f^s(c) = b$ or $f^s(x) \leq x$ for $x \in (a - \varepsilon, f^{2s}(c)]$ with some $\varepsilon > 0$. It is clear that, in both cases, the cycle of intervals A disappears under small C^0-perturbations of the map f.

Corollary 7.3. *A cycle of intervals A of period n of a unimodal map f survives* *under C^0-perturbations of the map if and only if, for any $\varepsilon > 0$, the map f possesses a cycle of intervals $\tilde{A}$ of period n such that*

(a) *the Hausdorff distance between the sets A and $\tilde{A}$ does not exceed ε;*

(b) *there exists a neighborhood U of the cycle of intervals $\tilde{A}$ which lies in A;*

(c) *the cycle of intervals $\tilde{A}$ is an attractor.*

This corollary is a consequence of the assertions and proof of Theorem 7.7.

As in the case of periodic trajectories, it follows from the proof of Theorem 7.7 that if a cycle of intervals A of a unimodal map $f \in C^r(I, I)$, $r \geq 1$, is not preserved under C^0-perturbations of the map f, then it is not preserved under C^r-perturbations of the map f even if the perturbed map $\tilde{f}$ remains in the class of unimodal maps. This observation is used in what follows.

Generally speaking, the problem of structural stability of the map f in the neighborhood of a cycle of intervals A under perturbations of the map f is not simpler than the problem of structural stability of the map f in the entire interval I. Therefore, we consider this problem in Section 3.

2. Stability of the Phase Diagram

2.1. Classification of Cycles of Intervals and Their Coexistence. In Chapter 3, we used the classification of cycles in terms of permutations to study the coexistence of periodic trajectories of continuous maps. Similar classification can be applied to the investigation of cycles of intervals.

Let $A = \{I_0, I_1, \ldots, I_{n-1}\}$ be a cycle of intervals of period n of a map $f \in C^0(I, I)$. This cycle of intervals is associated with a permutation

$$\pi(A) = \begin{pmatrix} 1 & 2 & \ldots & n \\ t_1 & t_2 & \ldots & t_n \end{pmatrix}$$

as follows:

(a) the intervals I_i, $i = 0, 1, \ldots, n-1$, are renumbered in the order of their location in the real line; as a result, we obtain an ordered collection of intervals $\tilde{A} = \{\tilde{I}_1, \tilde{I}_2, \ldots, \tilde{I}_n\}$;

(b) we set $t_i = j$ if $f(\tilde{I}_i) \subset \tilde{I}_j$, $i = 1, 2, \ldots, n$; the permutation $\pi(A)$ obtained as a result is called the type of the cycle of intervals A.

If a permutation

$$\pi = \begin{pmatrix} 1 & 2 & \ldots & n \\ t_1 & t_2 & \ldots & t_n \end{pmatrix}$$

is the type of a cycle of intervals of a continuous map, then the set $\{1, 2, \ldots, n\}$ is the minimal set of the map π of this set onto itself, i.e., it contains no proper invariant subsets. Permutations of this sort are called cyclic permutations. They were studied in Chapter 3. For any cyclic permutation π, one can easily construct a continuous map f: $I \to I$ which possesses a cycle of intervals A whose type $\pi(A)$ coincides with a given permutation π.

In Chapter 3, for continuous maps, we established several theorems on the coexistence of periodic trajectories of various periods and types. The following statement demonstrates that, for cycles of intervals, the situation is somewhat different because, unlike periodic orbits, cycles of intervals consist of nondegenerate intervals.

Proposition 7.1. *For any cyclic permutation*

$$\pi = \begin{pmatrix} 1 & 2 & \ldots & n \\ t_1 & t_2 & \ldots & t_n \end{pmatrix},$$

there exists a continuous map $f\colon \mathbb{R} \to \mathbb{R}$ *which has a cycle of intervals of type* π *but has no other cycles of intervals.*

Proof. Consider a permutation

$$\pi = \begin{pmatrix} 1 & 2 & \ldots & n \\ t_1 & t_2 & \ldots & t_n \end{pmatrix}.$$

For $i = 1, 2, \ldots, n$, we define $I_i = [4i-2, 4i]$. The map $f\colon \mathbb{R} \to \mathbb{R}$ is first defined at points with integer coordinates $j \in \{1, 2, \ldots, 4(n-1)+5\}$ as follows: If $j = 4i-2$ or $j = 4i$, then $f(j) = 4t_i - 2$; if $j = 4i-1$, then $f(j) = 4t_i$; at all other points, we set $f(j) = 0$. Then we extend f to the components of $\mathbb{R} \setminus \{1, 2, \ldots, 4n+1\}$ by linearity. As a result, we obtain the required piecewise linear map $f\colon \mathbb{R} \to \mathbb{R}$. This map is expanding because its derivative is greater than two at all points of its domain of definition. Hence, the trajectory of an arbitrary interval U either eventually hits one of the intervals I_i, $i \in \{1, 2, \ldots, n\}$ or covers the point of extremum of the map f which does not belong to these intervals. In the second case, $0 \in f^k(U)$ for some k. Since the intervals

$I_1, I_2, \ldots, I_n$ form a cycle of intervals and $f^m(0) \to -\infty$ as $m \to \infty$, this completes the proof of Proposition 7.1.

Nevertheless, under certain additional restrictions, the fact that a continuous map has cycles of intervals of a given type implies that it also has cycles of intervals of some other types. (The exact formulations are presented below.)

Let

$$\pi^{(1)} = \begin{pmatrix} 1 & 2 & \ldots & n \\ t_1 & t_2 & \ldots & t_n \end{pmatrix}$$

be a cyclic permutation. We say that a cyclic permutation

$$\pi^{(2)} = \begin{pmatrix} 1 & 2 & \ldots & k \\ s_1 & s_2 & \ldots & s_k \end{pmatrix}$$

divides the permutation $\pi^{(1)}$ if there exists $m \geq 1$ such that $n = m \cdot k$ and, for any $j \in \{1, 2, \ldots, k\}$, the map $\pi^{(1)}$ maps the set $\{mj - m + 1, mj - m + 2, \ldots, mj\}$ onto the set $\{ms_j - m + 1, ms_j - m + 2, \ldots, ms_j\}$.

It is clear from the definition that any permutation divides itself and that the permutation

$$\pi_1 = \begin{pmatrix} 1 \\ 1 \end{pmatrix}$$

divides any other permutation. A nontrivial example is given by the permutations

$$\pi_6 = \begin{pmatrix} 1 & 2 & 3 & 4 & 5 & 6 \\ 4 & 6 & 5 & 3 & 2 & 1 \end{pmatrix} \quad \text{and} \quad \pi_2 = \begin{pmatrix} 1 & 2 \\ 2 & 1 \end{pmatrix}.$$

It follows from the definition that if a permutation $\pi^{(3)}$ divides a permutation $\pi^{(2)}$ and the permutation $\pi^{(2)}$ divides the permutation $\pi^{(1)}$, then $\pi^{(3)}$ divides $\pi^{(1)}$.

Proposition 7.2. *Let $f \in C^0(I, I)$. Assume that the map f has cycles of intervals $A = \{I_0, I_1, \ldots, I_{n-1}\}$ and $\tilde{A} = \{\tilde{I}_0, \tilde{I}_1, \ldots, \tilde{I}_{k-1}\}$ of periods n and k, respectively, such that*

$$\bigcup_{0 \leq i < n} I_i \subset \bigcup_{0 \leq i < k} \tilde{I}_i.$$

Then the permutation $\pi(\tilde{A})$ divides the permutation $\pi(A)$.

Proof. Proposition 7.2 is a consequence of the definition of cycles of intervals.

Proposition 7.3. *Let* $A = \{I_0, I_1, \ldots, I_{n-1}\}$ *be a cycle of intervals of period* n *of a map* $f \in C^0(I, I)$. *Assume that the map* f *is monotone in any component of the set* $I/\bigcup_{0\le i<n} I_i$. *Then, for any permutation* π' *which divides the permutation* $\pi(A)$, *there exists a cycle of intervals* $A' = \{I'_0, I'_1, \ldots, I'_{k-1}\}$ *of the map* f *such that* $\pi(A') = \pi'$ *and*

$$\bigcup_{0\le i<n} I_i \subset \bigcup_{0\le i<k} \tilde{I}_i.$$

Proof. Let the conditions of Proposition 7.3 be satisfied and let k be the length of the permutation π'. Then $n = k \cdot m$ for some $m \ge 1$. We enumerate the intervals of the cycle A in the order of their location in the real line. As a result, we obtain an ordered collection of intervals $\tilde{A} = \{\tilde{I}_1, \tilde{I}_2, \ldots, \tilde{I}_n\}$. Let

$$\pi' = \begin{pmatrix} 1 & 2 & \ldots & k \\ t_1 & t_2 & \ldots & t_k \end{pmatrix}.$$

Then, under the conditions of the proposition, for $j = 1, 2, \ldots, k$, the intervals of $\tilde{A}$ with indices $jm - m + 1, jm - m + 2, \ldots, jm$ are mapped into the intervals with indices $t_j m - m + 1, t_j m - m + 2, \ldots, t_j m$, respectively. For $j = 0, 1, \ldots, k-1$, let $\tilde{I}_j$ be the smallest interval that contains the intervals of the set $\tilde{A}$ with indices $jm + 1, jm + 2, \ldots, jm + m$. Since f is monotone in components of the set $I/\bigcup_{0\le i<n} I_i$, we have $f(\tilde{I}_j) \subset \tilde{I}_{t_j}$ for $j \in \{0, 1, \ldots, k-1\}$ and the intervals $\tilde{I}_j$ form a cycle of intervals A' of period k such that $\pi(A') = \pi'$ and

$$\bigcup_{0\le i<n} I_i \subset \bigcup_{0\le i<k} \tilde{I}_i.$$

Note that Proposition 7.3 gives information about the nonlocal behavior of maps, which is used in what follows.

The following statement establishes conditions for the coexistence of periods of cycles of intervals and periods of periodic trajectories of continuous maps.

Proposition 7.4. *Let* $A = \{I_0, I_1, \ldots, I_{n-1}\}$ *be an* n*-periodic cycle of intervals of a map* $f \in C^0(I, I)$. *Then the map* f *possesses a periodic trajectory of period* s, *where* $s = n$ *if the intervals of the cycle* A *are mutually disjoint and* $s = n/2$ *if this is not true.*

Proof. Under the conditions of the proposition, we have $f^n(I_0) \subset I_0$. Hence, the map f^n possesses a fixed point β_0 in the interval I_0. If the intervals of the cycle A are mutually disjoint, then β_0 is an n-periodic point of the map f. Otherwise, it is not difficult to show that n is even and β_0 is an internal point of the interval $I_0 \cup I_{n/2} = I(A)$. Thus, the period of the trajectory of the point β_0 under the map f is not less than $n/2$. Clearly, in this case, the period of β_0 is equal either to n or to $n/2$. It follows from the results of Chapter 3 that, in both cases, the map f possesses a periodic trajectory of period $n/2$.

2.2. Conditions for the Preservation of Central Vertices. As shown in Chapter 5, the central vertices of the phase diagram of a unimodal map (i.e., vertices corresponding to the cycles of intervals that contain the point of extremum of a given map) are linearly ordered and their number is at most countable. In Chapter 5, central vertices of the phase diagram were denoted by $A^*_{p_m}$, $m \leq m^*$. They were identified with maximal cycles of intervals of period p_m covering the point of extremum. In this section, we formulate conditions under which central vertices do not disappear under C^0-perturbations of the map. These conditions, together with results established in Section 5.2, enable us to make some conclusions about the structural stability of unimodal maps, i.e., about the nonlocal behavior of dynamical systems.

In this section, we denote central vertices of the phase diagram of a unimodal map f by $A^*_{p_m}(f)$ and their number by $m^*(f)$ (recall that $m^*(f) \leq \infty$). The following assertion establishes the relationship between the behavior of trajectories for unimodal maps whose phase diagrams are characterized by central vertices of the same types.

Proposition 7.5. *If the equality* $\pi(A^*_{p_m}(f)) = \pi(A^*_{p_n}(f))$ *holds for unimodal maps* f *and* g *for some* $m \leq m^*(f)$ *and* $n \leq m^*(g)$, *then* $m = n$ *and*

$$\pi(A^*_{p_k}(f)) = \pi(A^*_{p_k}(g))$$

for any $k \leq m$.

Proof. Proposition 7.5 immediately follows from Propositions 7.3 and 7.4 and from the construction of phase diagrams in Chapter 5.

Proposition 7.6. *Let* f *be a unimodal map. Then, for any* $m < m^*(f)$, *there exists* $\varepsilon = \varepsilon(m) > 0$ *such that* $m^*(g) \geq m$ *for any unimodal map* g *with* $\|f-g\|_{C^0} < \varepsilon$ *and* $\pi(A^*_{p_n}(g)) = \pi(A^*_{p_n}(f))$ *for all* $n \leq m$.

Proof. If we assume that a cycle of intervals $A^*_{p_m}(f)$ vanishes under C^0-perturbations of the map f, then it follows from the proof of Theorem 7.7 that $m = m^*(f)$ but this is impossible by the condition of the proposition. Hence, the conditions of the first statement of Theorem 7.7 are satisfied. By virtue of Theorem 7.7, one can indicate $\varepsilon > 0$ such that any unimodal map g with $\|f - g\|_{C^0} < \varepsilon$ has a cycle of intervals A of period p_m which contains the extremum point of the map g. It is clear that, in this case, $\pi(A) = \pi(A^*_{p_m}(f))$. The required assertion now follows from Proposition 7.5.

Proposition 7.7. *Let f be a unimodal map. If the point c is not periodic and lies in the domain of attraction of an attracting cycle, then $m^*(g) = m^*(f)$ and*

$$\pi(A^*_{p_{m^*}}(g)) = \pi(A^*_{p_{m^*}}(f))$$

for any unimodal map g with sufficiently small $\|f - g\|_{C^0}$.

Proof. It follows from the results established in Chapter 5 that $m^*(f) < \infty$ under the conditions of Proposition 7.7. By Theorem 7.7, any unimodal map g sufficiently close to the map f in the metric of the space $C^0(I, I)$ has a cycle of intervals of period $p_{m^*(f)}$ that contains the point of extremum of the map g. Hence, $m^*(g) \geq m^*(f)$.

Assume that the trajectory of the point of extremum c of the map f is attracted by the trajectory of a periodic point β. Denote the period of the point β by k. According to Theorem 7.1, there is a neighborhood U of the point β such that $f^k(U) \subset U$ for $x \in U$, $f^{2k}(x) > x$ if $x < \beta$, and $f^{2k}(x) < x$ if $x > \beta$. Let U be the largest neighborhood of the point β with the indicated property. Then the domain of attraction of the trajectory of the point β coincides with the set $\bigcup_{i \geq 0} f^{-i}(U)$. The trajectory of the interval U forms a cycle of intervals, which is denoted by B.

Hence, under the conditions of the proposition, there exists $j \geq 0$ such that $f^j(c) \in U$. By Theorem 7.6, the cycle of intervals B is preserved under sufficiently small C^0-perturbations of the map f. In this case, if the perturbed map g is unimodal, then the first j iterations of its point of extremum $\tilde{c}$ are slightly different from the first j iterations of the point c of the map f. Hence, the point $g^j(\tilde{c})$ also belongs to a cycle of intervals which does not contain the point of extremum of the map g. This means that, under the conditions of the proposition, we have $m^*(g) = m^*(f)$ and

$$\pi(A^*_{p_{m^*(g)}}(g)) = \pi(A^*_{p_{m^*(f)}}(f)).$$

Let us now make several remarks. Let g be a unimodal map sufficiently close to a unimodal map f in $C^0(I, I)$. If $c \in \mathrm{Per}(f)$, then, by virtue of Proposition 7.6, we have $m^*(g) \geq m^*(f) - 1$. One can easily construct an example of g such that $m^*(g) > m^*(f)$; moreover, for any $k \geq 1$, one can find a map g such that $m^*(g) \geq m^*(f) + k$. On the other hand, it is not difficult to show that, for smooth unimodal maps f and g sufficiently close in $C^1(I, I)$, we have $m^*(f) \leq m^*(g) \leq m^*(f) + 1$.

If $\omega(c)$ is not a cycle and there is a neighborhood U of the point c such that $f^i(U) \cap f^j(U) = \varnothing$ for all $i \neq j$, then it follows from the proof of Theorem 7.7 that the cycle of intervals $A^*_{p_{m^*(f)}}(f)$ (with $m^*(f) < \infty$) does not vanish under C^0-perturbations of the map f. Consequently, $m^*(g) \geq m^*(f)$. It is not clear whether the equality $m^*(g) = m^*(f)$ is true under these conditions for smooth unimodal maps f and g sufficiently close in $C^r(I, I)$, $r \geq 1$. The same question remains open for $m^*(f) = \infty$.

3. Structural Stability and Ω-stability of Maps

In this section, we study the problem of stability of the dynamical structure of dynamical systems. In order to compare the dynamics of various systems, we use the concept of topological equivalence introduced in Chapter 1.

We recall the corresponding definition. Maps $f: I \to I$ and $g: I \to I$ are called topologically conjugate if there exists a homeomorphism $h: I \to I$ such that $g \circ h = h \circ f$ in I.

It follows from this definition that if maps f and g are topologically conjugate, then the homeomorphism h transforms trajectories of the map f into trajectories of the map g. This means that topologically conjugate maps generate topologically equivalent dynamical systems.

By using this relation of equivalence of maps, one can introduce all necessary characteristics of the stability of the structure of trajectories in $C^r(I, I)$, $r \geq 0$.

Let $\Lambda: C^r(I, I) \mapsto 2^I$ be a map which associates every point $f \in C^r(I, I)$ with a closed set $\Lambda(f) \in 2^I$ such that $f(\Lambda(f)) \subset \Lambda(f)$. We say that a map $f \in C^r(I, I)$ is C^r-structurally Λ-stable if there exists a neighborhood $\mathcal{U}(f)$ of the map f in $C^r(I, I)$ such that, for any $g \in \mathcal{U}(f)$, the maps $f|_{\Lambda(f)}$ and $g|_{\Lambda(g)}$ are topologically conjugate.

We consider the cases where a role of the set $\Lambda(f)$ is played either by the entire interval I (this corresponds to C^r-structural stability) or the set of nonwandering points (this corresponds to the so called C^r-structural Ω-stability).

Note that, parallel with structural Λ-stability, it might be interesting to study Λ-stability regarded as the stability of the set $\Lambda(f)$, i.e., to test the map $\Lambda: f \to \Lambda(f)$ for continuity or upper semicontinuity at the point $f \in C^r(I, I)$.

In what follows, we assume that the spaces $C^r(I, I)$, $r \geq 0$, are equipped with metric

$$\rho_r(f, g) = \sum_{0 \leq i \leq r} \max_{x \in I} | D^i f(x) - D^i f(x) |,$$

where

$$D^i f = \frac{d^i f}{dx^i}.$$

In what follows, main attention is paid to the problems of C^2-structural stability and C^1-structural Ω-stability for the following reason: It is clear that the class $C^0(I, I)$ contains no maps that are C^0-structurally stable: Indeed, by small C^0-perturbations of the map in a neighborhood of a fixed point, one can always change at least the qualitative behavior of trajectories in this neighborhood. On the other hand, if a map from the class $C^1(I, I)$ possesses a critical point, then there are maps close to this map in $C^1(I, I)$ which possess an interval of critical points. Therefore, these maps are not topologically equivalent to the original map. A similar situation is also possible for maps from the class C^2 with degenerate critical points. At the same time, it may happen that either $\mathrm{NW}(f)$ contains no critical points of $f \in C^1(I, I)$ or all critical points of the map f are periodic, i.e., isolated in $\mathrm{NW}(f)$. In this case, it seems reasonable to study the problem of C^1-structural Ω-stability of the map f.

Suppose that $f \in C^2(I, I)$ is a unimodal map. The map f cannot be C^2-structurally stable if $f''(c) = 0$ or if it possesses a nonhyperbolic periodic trajectory. If there exists a point $x \in I$ such that $c \in \omega_f(x)$, then the map f may also be C^2-structurally unstable. Structural stability is also impossible in the case where $f^j(c) \in \mathrm{Per}(f)$ for some $j \geq 0$. At the same time, if the indicated possibilities are excluded, then the map f is C^2-structurally stable. Moreover, the results established in the previous chapter imply the following assertion:

Theorem 7.8. *Assume that a unimodal map $f \in C^2(I, I)$ satisfies the conditions*

(a) $f'(x) \neq 0$ for $x \in I \setminus \{c\}$ and $f''(c) \neq 0$;

(b) the set $\mathrm{Per}(f)$ *does not contain nonhyperbolic orbits;*

(c) $c \notin \mathrm{NW}(f)$ and $f^i(c) \notin \mathrm{Per}(f)$ for all $i \geq 1$.

Then f is C^2-structurally stable.

Proof. Since the critical point is unique and $c \notin \mathrm{NW}(f)$, one can indicate $i \geq 0$ and a periodic interval L such that $f^i(c) \in L$. Since $f^i(c) \notin \mathrm{Per}(f)$, the point $f^i(c)$ belongs to the domain of immediate attraction of a certain attracting cycle (by the condition, the map f does not have any nonhyperbolic periodic orbits). By Theorem 6.3, the set $\mathrm{NW}(f)$ is hyperbolic, i.e., there exist $C > 0$ and $\lambda > 1$ such that, for any point $x \in \mathrm{NW}(f)$, either $|D^n f(x)| \geq C\lambda^n$ or $|D^n f(x)| \leq C^{-1}\lambda^{-n}$ for all $n \geq 0$.

By Theorem 5.6, f has finitely many attracting cycles. Let $B_0(f)$ denote the union of the domains of immediate attraction of all attracting cycles of the map f. Then $B_0(f)$ consists of finitely many open intervals. We choose an integer number n such that

$$c \in B_n(f) = \bigcup_{0 \le i \le n} f^{-i}(B_0(f))$$

and the inequality $|Df^{-n}(x)| \le \mu < 1$ holds for any single-valued branch of the map f^{-n} for all x which do not belong to the set $B_n(f)$. Hence, for any map $\tilde{f}$ sufficiently close to the map f in $C^1(I, I)$, the set $B_n(\tilde{f})$ and the constant $\tilde{\mu}$ are close to $B_n(f)$ and μ, respectively. This implies that the maps $f|_{NW(f)}$ and $\tilde{f}|_{NW(\tilde{f})}$ are topologically equivalent and the maps f and $\tilde{f}$ are topologically conjugate.

Corollary 7.4. *Let $f \in C^3(I, I)$ be a unimodal map, let $Sf(x) < 0$ for $x \in I \setminus \{c\}$, and let $|Df(x)| > 1$ for $x \in \partial I$. If $c \notin$ Per(f) and there exists a point $\beta \in$ Per(f) such that $|f'(\beta) f'(f(\beta)) \ldots f'(f^{n-1}(\beta))| < 1$, where n is the period of the point β, then f is C^2-structurally stable.*

Note that a theorem similar to Theorem 7.8 is true for an arbitrary map from the class $C^2(I, I)$ (Jakobson [1]).

Theorem 7.9. *The set of C^2-structurally stable maps is dense in the space $C^1(I, I)$ with metric ρ_1.*

For the complete proof of Theorem 7.9, see (Jakobson [1]). Here, we prove this theorem only for unimodal maps.

First, we show that the collection of maps which have attracting cycles is dense in $C^1(I, I)$. Assume that $c \in \overline{\text{Per}(f)}$, where $f \in C^1(I, I)$, and that c is the point of maximum of the map f. There is a neighborhood U of the point c such that $|Df(x)| < \varepsilon/3$ for $x \in U$. By the assumption, the neighborhood U contains a periodic point β of the map f. Without loss of generality, we can assume that $f(\beta_1) < f(\beta)$ for any $\beta_1 \in$ orb(β), $\beta_1 \ne \beta$. Thus, one can find points $x, x' \in U$ such that $x < c < x'$ and $(x, x') \cap$ orb$(\beta) = \{\beta\}$. Under these conditions, the function f can be replaced in the interval (x, x') by a function $\tilde{f}$ such that $\rho_1(f, \tilde{f}) < \varepsilon$, $\tilde{f}(x) = f(x)$, $\tilde{f}(x') = f(x')$, $\tilde{f}(\beta) = f(\beta)$, and the point β is a unique extremum point of the map $\tilde{f}$.

Note that the maps from the class C^2 are dense in the space $C^1(I, I)$ with metric ρ_1. Hence, we can assume that $f \in C^2(I, I)$.

If $c \notin \overline{\text{Per}(f)}$, then either there is $i \ge 0$ for which $f^i(c)$ belongs to a periodic homterval or $c \in U$, where U is a wandering interval. In the first case, the required assertion is obvious. In the second case, one can assume that U is the maximal wandering interval that contains the point c. It follows from the results of the previous chapter that there exists a critical point c_1 which lies in the interval $\overline{\mathrm{Per}(f)}$, which contradicts the assumption that the extremum point c is unique.

The set of C^2-maps with nondegenerate critical points is dense in the space $C^1(I, I)$. Therefore, by using the reasoning presented above, we have actually proved that the set of C^2-unimodal maps whose single critical point lies in the domain of attraction of an attracting cycle is dense in the space of C^1-unimodal maps.

The fact that maps without nonhyperbolic periodic orbits are typical is established by using the Sard theorem.

Thus, C^2-structurally stable maps form a dense subset of the space $C^1(I, I)$ with metric ρ_1. The answer to the question as to whether C^2-structurally stable maps are dense in the space $C^2(I, I)$ with metric ρ_2 remains unclear even in the case of unimodal maps.

Note that the argument presented above yields the following assertion for structural Ω-stability:

Theorem 7.10. *Assume that a unimodal map $f \in C^1(I, I)$ satisfies the conditions*

(a) $f'(x) \neq 0$ for $x \in I \backslash \{c\}$;

(b) the set Per (f) *does not contain nonhyperbolic orbits;*

(c) for any $i \geq 0$, either $f^i(c) \notin \mathrm{NW}(f)$ or $f^i(c)$ is a periodic point isolated in Per (f).

Then f is C^1-structurally Ω-stable.

Consider the problem of Ω-stability. It is clear that Ω-stable maps cannot have wandering intervals vanishing under perturbations of a map; moreover, the elements of the spectral decomposition of this map should not undergo significant changes under these perturbations.

Theorem 7.11. *Let $f \in C^3(I, I)$ be a unimodal map, let $Sf(x) < 0$ for $x \in I \backslash \{c\}$, and let $f''(c) \neq 0$. Assume that f has no semiattracting periodic orbits and $\partial(C_{m^*}^{(0)}) \cap \mathrm{Per}(f) = \varnothing$. Then the map $\Omega : f \to \mathrm{NW}(f)$ in the space $C^3(I, I)$ with metric ρ_3 is continuous at the point f.*

Proof. Note that the conditions of Theorem 7.11 are, in fact, necessary and sufficient conditions for the C^3-Ω-stability of maps with negative Schwarzian. Both cases where f is not C^3-Ω-stable are displayed in Fig. 37 (where the graph of the map $f^{p_{m^*}}$ is depicted in a neighborhood of the cycle of intervals $A^*_{p_{m^*}}$ which contains the point c). In these cases, the trajectories of almost all points of the interval I (with respect to the

Lebesgue measure) are attracted by an invariant set which is not an attractor: In the first case, this is a semiattracting periodic orbit formed by an attractor "coupled" with a repeller. In the second case, this is the set $C_{m*}^{(0)}$ which is, in this case, a repeller. Note that, in both cases, $m^* < \infty$.

If $m^* < \infty$ and the conditions of Theorem 7.11 are satisfied, then it follows from the results obtained in the previous section that the sets Φ_m^* and $\Phi_m^{(0)}$ undergo small changes under small C^3-perturbations of the map f for all $m \le m^*$. This means that, under the conditions of the theorem, the set $\mathrm{NW}(f)$ cannot become much larger.

On the other hand, $\overline{\mathrm{Per}(f)} = \mathrm{NW}(f)$ for maps with negative Schwarzian. Therefore, under the conditions of the theorem, for any $\varepsilon > 0$, one can choose a finite ε-net formed by hyperbolic periodic orbits in the set $\mathrm{NW}(f)$. Under sufficiently small perturbations of the map f, the periodic orbits of the indicated net do not vanish and the corresponding periodic orbits of the perturbed map $\tilde{f}$ form a 2ε-net in the set $\mathrm{NW}(\tilde{f})$.

If $m^* = \infty$, then the set

$$\Phi_\infty^* = \bigcap_{m \ge 1} \Phi_m^*$$

contains no intervals. Therefore, the required assertion follows from the argument used in the case $m^* < \infty$.

Note that, for maps satisfying the conditions of theorems presented in this section, we have $\overline{\mathrm{Per}(f)} = \mathrm{NW}(f)$. The following theorem demonstrates that this is a typical property of smooth maps:

Theorem 7.12 (Young [1]). *Let* $r \ge 0$ *and* $f \in C^r(I, I)$. *Then, for any* $\varepsilon > 0$, *one can find a map* $g \in C^r(I, I)$ *such that* $\rho_r(g, f) < \varepsilon$ *and* $\mathrm{NW}(g) = \overline{\mathrm{Per}(f)}$.

Proof. We consider only the case of unimodal maps. If $\mathrm{NW}(g) \ne \overline{\mathrm{Per}(f)}$, then $c \notin \mathrm{NW}(f)$ and $f(c) \in \mathrm{NW}(f)$ (see Section 5.2); moreover, there exists a neighborhood U of the point c such that $f^i(U) \cap U = \varnothing$ for any $i \ge 1$. The map g is defined as follows: For $x \in I \setminus U$, we set $g(x) = f(x)$. For $x \in U$, the map g is defined so that c remains its unique point of extremum, $g(U) \subset f(U)$, and $g(c) \notin \partial(f(U))$. (Note that $f(c) \in \partial(f(U))$ because $f(c) \in \mathrm{NW}(f)$). Clearly, the quantity $\rho_r(g, f)$ can be made as small as desired and, by construction, we have $g(c) \notin \mathrm{NW}(f)$ and, therefore,

$$\mathrm{NW}(g) = \overline{\mathrm{Per}(g)} = \overline{\mathrm{Per}(f)}.$$

8. ONE-PARAMETER FAMILIES OF UNIMODAL MAPS

1. Bifurcations of Simple Invariant Sets

If a dynamical system describes a real process or phenomenon, then, as a rule, its properties depend on parameters. Any variation of the parameters inevitably results in a certain perturbation of the trajectories of a dynamical system under consideration. It is worth noting that small changes in the parameters may lead to significant changes in the structure of dynamical systems, i.e., to bifurcations or qualitative changes in the behavior of trajectories. In many cases, it is quite useful to know the values of the parameters for which "small errors" are admissible and the qualitative behavior of trajectories is not affected as well as the values of the parameters for which these "small errors" significantly distort the original dynamical picture.

Here, we consider the simplest case of one-parameter families of maps. As becomes clear from our subsequent presentation, these families are characterized by all types of bifurcations typical of one-dimensional maps.

Let f_λ be a family of maps from the class $C^r(I, I)$, $r \geq 0$, and let λ be a parameter that takes values from an interval Λ. We say that a value $\lambda_0 \in \Lambda$ of the parameter λ is *regular* if there is $\varepsilon > 0$ such that the maps f_λ and f_{λ_0} are topologically conjugate for any $\lambda \in (\lambda_0 - \varepsilon, \lambda_0 + \varepsilon)$. Denote the set of regular values of the parameter by Λ_R. The set $\Lambda_B = \Lambda \setminus \Lambda_R$ is called the set of bifurcation values of the parameter.

Bifurcations of cycles are the simplest type of bifurcations. Their investigation can be reduced to the study of the local behavior of maps in the neighborhood of points that form a cycle.

For one-parameter families of smooth maps, there are several typical bifurcations of periodic trajectories. One of these has already been encountered in Chapter 1, where we studied the family $f_\lambda \colon x \to \lambda x(1-x)$. Indeed, as the value of the parameter λ increases from 0 to $\lambda^* \approx 3.57$, one observes the successive appearance of attracting cycles of periods 1, 2, 2^2, $2^3, \ldots$. These bifurcations of cycles can be described as follows: If λ_n is the bifurcation value of the parameter corresponding to the appearance of a cycle B of period 2^n, then the cycle B is attracting for $\lambda_n < \lambda < \lambda_{n+1}$ and its multiplier varies from

$+1$ (for $\lambda = \lambda_n$) to -1 (for $\lambda = \lambda_{n+1}$). For $\lambda > \lambda_{n+1}$, we have $\mu(B) < -1$. Therefore, the cycle B becomes repelling. The period of the attracting cycle B' that appears for $\lambda > \lambda_{n+1}$ is twice as large as the period of B). This cycle is attracting for $\lambda_{n+1} < \lambda \leq \lambda_{n+2}$ and

$$\lim_{\lambda \to \lambda_{n+1}} \mu(B') = 1.$$

As λ increases, this process is repeated again and again.

For $\lambda > \lambda^* \approx 3.57$, the map $x \to \lambda x(1-x)$ has cycles of periods that are not powers of two. For $\lambda = 4$, this map has cycles of all periods. It is clear that the period doubling bifurcation cannot be responsible for the appearance of all these cycles. Thus, it cannot result in the appearance of cycles with odd periods. In general, the bifurcation that generates cycles of odd periods (including fixed points) can be described as follows: For $\lambda < \lambda_0$, the map f_λ has an interval $\mathcal{J}$ which does not contain fixed points of the map f_λ^n (i.e., $f_\lambda^n(x) \neq x$ for $x \in \mathcal{J}$). For $\lambda = \lambda_0$, the curve $y = f_\lambda^n(x)$ touches the line $y = x$ at a point $x_0 \in \mathcal{J}$, i.e., we observe the appearance of a fixed point x_0 of the map f_λ^n (its multiplier is equal to $+1$). For $\lambda > \lambda_0$, this fixed point decomposes into two fixed points one of which is attracting and the other one is repelling.

It is worth noting that these two types of bifurcations are substantially different. In fact, period doubling bifurcations are local and qualitative changes in the behavior of trajectories are observed only in a small neighborhood of the cycle (mild bifurcation). Bifurcations of the second type (bifurcations of creation of cycles) arrest the motion of points from the domain $\{x < x_0\}$ to the domain $\{x > x_0\}$ near the point $x = x_0$ as soon as the indicated lines touch each other and lead to global (i.e., not only in the neighborhood x_0) qualitative changes in the behavior of a system (rigid bifurcation).

Following Guckenheimer [1], we now formulate the conditions which lead to bifurcations of cycles, in the form of two theorems.

Theorem 8.1. *Let $f_\lambda : I \to I$ be a family of C^2-maps with smooth dependence on the parameter $\lambda \in (\lambda_1, \lambda_2)$, let β_0 be a fixed point of the map f_{λ_0}, $\lambda_0 \in (\lambda_1, \lambda_2)$, and let $f'_{\lambda_0}(\beta_0) = 1$. If*

1) $f''_{\lambda_0}(\beta_0) > 0$ and

2) $\frac{d}{d\lambda} f_\lambda(\beta_0)_{\lambda=\lambda_0} < 0$,

then there exist $\varepsilon > 0$ and $\delta > 0$ such that

(a) for $\lambda \in (\lambda_0 - \delta, \lambda_0)$, the map f_λ has no fixed points in the interval $(\beta_0 - \varepsilon, \beta_0 + \varepsilon)$;

(b) for $\lambda \in (\lambda_0, \lambda_0 + \delta)$, the map f_λ has two fixed points in the interval $(\beta_0 - \varepsilon, \beta_0 + \varepsilon)$; one of these points is attracting and the other one is repelling.

The statement of the theorem remains valid if both 1) and 2) are replaced by the inverse inequalities. If only one of these inequalities is replaced by the inverse inequality, then fixed points appear as λ decreases. In other words, fixed points appear or disappear as λ increases in accordance with the sign of the product $f''_\lambda(x)\frac{d}{d\lambda}f_\lambda(x)$ for $\lambda = \lambda_0$ and $x = \beta_0$.

Proof. Consider the function $h(x, \lambda) = f_\lambda(x) - x$. We have

$$\frac{dh}{d\lambda} \neq 0 \quad \text{and} \quad \frac{dh}{dx} = 0$$

at the point (β_0, λ_0). By the implicit function theorem, there exists a smooth function $\lambda = \varphi(x)$ such that $\lambda_0 = \varphi(\beta_0)$ and $h(x, \varphi(x)) = 0$ in a certain neighborhood of the point β_0. By differentiating the last identity two times, we obtain

$$\frac{\partial^2 h}{\partial x^2} + \frac{\partial h}{\partial \lambda}\frac{d^2\varphi}{dx^2} = 0.$$

Since

$$\frac{d^2\varphi}{dx^2} = -\left(\frac{\partial h}{\partial \lambda}\right)^{-1}\frac{\partial^2 h}{\partial x^2} \neq 0$$

for $x = \beta_0$, the curve $\lambda = \varphi(x)$ lies on the one side of the tangent at the point β_0. The last statement of the theorem follows from the fact that $\frac{\partial}{\partial x}\left(\frac{\partial h}{\partial x}\right) \neq 0$ at the point (β_0, λ_0).

Theorem 8.2. *Let $f_\lambda : I \to I$ be a family of C^3-maps with smooth dependence on the parameter $\lambda \in (\lambda_1, \lambda_2)$, let β_0 be a fixed point of the map f_{λ_0}, $\lambda_0 \in (\lambda_1, \lambda_2)$, and let $f'_{\lambda_0}(\beta_0) = -1$. If*

1) $\dfrac{d^3 f_\lambda^2(x)}{dx^3} < 0$ *and*

2) $\dfrac{d}{d\lambda}\left(f'_\lambda(x)\right) < 0,$

for $\lambda = \lambda_0$ and $x = \beta_0$, then there are $\varepsilon > 0$ and $\delta > 0$ such that

(a) for $\lambda \in (\lambda_0 - \delta, \lambda_0)$, the map f_λ has exactly one fixed point in the interval $(\beta_0 - \varepsilon, \beta_0 + \varepsilon)$ and this fixed point is attracting;

(b) for $\lambda \in (\lambda_0, \lambda_0 + \varepsilon)$, there are three fixed points of the map f_λ in the interval $(\beta_0 - \varepsilon, \beta_0 + \varepsilon)$; moreover, the middle point is a repelling fixed point of the map f_λ and the other two points form an attracting cycle of period two.

If inequality 2) has the opposite sign, then the assertions of the theorem remain true but the cycle of period two appears as λ decreases. If we change the sign in inequality 1), then it is necessary to replace the word "attracting" by "repelling", and vice versa.

Proof. Since $f'_{\lambda_0}(\beta_0) = -1$, we can write

$$\frac{d^2}{dx^2}\left(f^2_{\lambda_0}(x)\right) = 0$$

for $x = \beta_0$. Consider the function $h(x, \lambda) = f^2_\lambda(x) - x$. For $x = \beta_0$ and $\lambda = \lambda_0$, we have

$$h = 0, \qquad \frac{\partial h}{\partial x} = 0, \qquad \frac{\partial^2 h}{\partial x^2} = 0,$$

and

$$\frac{\partial^3 h(x, \lambda)}{\partial x^3} = \frac{1}{3}\left(-\frac{\partial^3}{\partial x^3} f_\lambda(x) - \frac{3}{2}\left[\frac{\partial^2}{\partial x^2} f_\lambda(x)\right]^2\right)\Bigg|_{(\beta_0, \lambda_0)} = \frac{1}{3}\, Sf_\lambda(x)\big|_{(\beta_0, \lambda_0)} \neq 0,$$

where, as above, Sf_λ denotes the Schwarzian of f_λ.

By the implicit function theorem, there exists a function $x = \varphi(\lambda)$ such that $\beta_0 = \varphi(\lambda_0)$ and $f_\lambda(\varphi(\lambda)) = \varphi(\lambda)$ for all λ close to λ_0. Consider the function

$$\tilde{h}(x, \lambda) = \frac{h(x, \lambda)}{x - \varphi(\lambda)}.$$

For $x = \beta_0$ and $\lambda = \lambda_0$, we have

$$\tilde{h} = 0, \quad \frac{\partial \tilde{h}}{\partial x} = 0, \quad \frac{\partial^2 h}{\partial x^2} \neq 0, \quad \text{and} \quad \frac{\partial \tilde{h}}{\partial \lambda} \neq 0.$$

By applying the reasoning used in the proof of Theorem 8.1 to $\tilde{h}(x, \lambda)$, we arrive at the required assertions.

Note that the Schwarzian appears in the proof of Theorem 8.2 as a natural characteristic of the period doubling bifurcation.

The formulations of Theorems 8.1 and 8.2 presented above are adjusted to the case of fixed points. To cover the case of bifurcations of cycles of period n, one must replace there f_λ by f_λ^n.

It is worth noting that condition 1) of Theorem 8.2 is always satisfied for quadratic maps (as well as for general maps with negative Schwarzian). Hence, these maps are characterized by a single type of bifurcations, namely, by period doubling bifurcations in the course of which an attracting cycle of period n becomes repelling and generates an attracting cycle of period $2n$.

2. Properties of the Set of Bifurcation Values. Monotonicity Theorems

We find it reasonable to anticipate the investigation of arbitrary smooth one-parameter families of one-dimensional maps by the analysis of the behavior of some very simple families of maps, e.g., of the family of quadratic maps or more general families of maps with negative Schwarzian.

Let $f_\lambda(x) = \lambda f(x)$, where $f_\lambda\colon [0, 1] \to [0, 1]$ is an S-unimodal map such that $f(0) = f(1) = 0$ and $\lambda \in \Lambda = (0, 1/f(c))$. In addition, we require that the inequality $f''(x) < 0$ must hold for all $x \in (0, 1)$. The importance of this assumption is clarified in what follows. Note that the family of quadratic maps often encountered earlier satisfies the indicated conditions.

For the family f_λ, let Λ_R be the set of regular values of the parameter λ (defined in the previous section). It follows from the definition of the set Λ_R that it is open in Λ.

As shown in Chapter 6, for maps from the family f_λ, the probabilistic limit set $\mathcal{A}(f_\lambda)$ (i.e., the smallest set which contains the ω-limit sets of almost all points with respect to the Lebesgue measure) is either an attracting (or semiattracting) cycle or a cycle of intervals in which the map f_λ possesses the mixing property or coincides with the set $\omega_{f_\lambda}(c)$ (in this case, $\omega_{f_\lambda}(c)$ is a Cantor set and $c \in \omega_{f_\lambda}(c)$; see Theorem 6.9); moreover, in all cases, we have $\omega_{f_\lambda}(c) \subset \mathcal{A}(f_\lambda)$. Thus, the range of the parameter Λ can be split into the following mutually disjoint subsets:

$$\Lambda_0 = \{\lambda \in \Lambda \mid \mathcal{A}(f_\lambda) \text{ is a cycle}\},$$

$$\Lambda_1 = \{\lambda \in \Lambda \mid \mathcal{A}(f_\lambda) \text{ is a cycle of intervals}\},$$

$$\Lambda_2 = \{\lambda \in \Lambda \mid \mathcal{A}(f_\lambda) \text{ is a Cantor set}\}.$$

Let us now formulate several hypotheses concerning the problem of alternation of regular and stochastic behavior in the family f_λ.

Proposition 8.1. *Λ_R is a subset of Λ_0 and $\Lambda_1 \cup \Lambda_2$ is a subset of Λ_B.*

Note that the proof of Proposition 8.1 is closely connected with the investigation of the problem of structural stability of maps with negative Schwarzian.

Let Λ_{dx} be the set of values of the parameter $\lambda \in \Lambda$ for which f_λ possesses an invariant measure absolutely continuous with respect to the Lebesgue measure.

Proposition 8.2. $\text{mes}\, \Lambda_{dx} > 0$.

For a special case, this assertion was formulated and proved by Yakobson.

Theorem 8.3. *Let f be a C^3-map which is sufficiently close in $C^3(I)$ to the map $x \to x(1-x)$. Then, for the family $x \to \lambda f(x)$, the Lebesgue measure of the set Λ_{dx} is positive; moreover, the point $\lambda = 4$ is a density point of this set.*

The proof of this theorem can be found in Jakobson [4].

Note that it follows from Proposition 8.2 and the inclusion $\Lambda_{dx} \subset \Lambda_1 \cup \Lambda_2$ that $\text{mes}\,(\Lambda_1 \cup \Lambda_2) > 0$. It is thus interesting to find the measures of the sets Λ_1 and Λ_2 (it is known that both these sets are uncountable) and to check the validity of the inclusion $\Lambda_{dx} \subset \Lambda_1$.

Let f and g be S-unimodal maps. We say that the map f is not simpler than g if f is semiconjugate to g, i.e., there exists a monotone continuous map $h: I \to I$ such that $g \circ h = h \circ f$ (see Section 2.4). In this case, h maps the trajectory of a point x of the map f into the trajectory of the point $h(x)$ of the map g. Therefore, if f is not simpler than g, then the kneading invariants satisfy the inequality $\nu_f \le \nu_g$ (recall that the points of extremum are assumed to be the points of maximum).

Proposition 8.3. *The dynamics of the map f_λ becomes more complicated as λ increases, i.e., if $\lambda_1 \ge \lambda_2$, then f_{λ_1} is not simpler than f_{λ_2}.*

This proposition is completely proved only for the families of quadratic maps. In this case, it is a consequence of the following theorem (see Milnor [1] and Jonker [2]):

For a quadratic map $f: x \to Ax^2 + Bx + C$, $A \ne 0$, we define its "discriminant" by the formula $\Delta f = B^2 - 4AC - 2B$.

Theorem 8.4 (monotonicity theorem). *Let f and g be quadratic maps. If $\Delta f < \Delta g$, then, for any $n \ge 1$, the number of fixed points and the number of extremum points of the map f^n do not exceed the number of fixed points and the number of extremum points of the map g^n, respectively.*

The proof of this theorem is based on the following statements (see Milnor [1]):

I. f and g are linearly conjugate (i.e., $f = h^{-1} \circ g \circ h$, where h is a linear function) if and only if $\Delta f = \Delta g$.

II. f possesses an invariant interval if and only if $\Delta f \in [-1, 8]$.

III. If the extremum points of f and g are periodic and form equivalent cycles, then f and g are linearly conjugate.

The main problem encountered in proving Theorem 8.4 is connected with the proof of the third statement. Although this statement seems to be obvious, the proof suggested by Milnor [1] requires the transition to the complex plane.

In conclusion, we present the formulation of another monotonicity theorem (Matsumoto [1]) for general families of smooth unimodal maps.

Theorem 8.5. *Let* $f_\lambda(x) = \lambda f(x)$, $f \in C^2(I, I)$, $f(0) = f(1) = 0$, *and* $f'(x) < 0$ *for all* $x \in \operatorname{int} I$. *If* f_λ *has a cycle of odd period* k, *then, for any* $\mu > \lambda$, *the map* f_μ *also possesses a cycle of period* k. *Moreover, if* f *is an S-unimodal map, then this assertion holds for any* $k \neq 2^i$, $i = 1, 2, \ldots$.

3. Sequence of Period Doubling Bifurcations

Consider a family of continuous unimodal maps $f_\lambda = \lambda f(x)$, where $f\colon [0, 1] \to [0, 1]$, $f(0) = f(1) = 0$, $f(c) = 1$ (c is the point of extremum of the function f), and $\lambda \in [0, 1]$. Denote

$$\lambda[n] = \inf\{\lambda \in [0, 1] \mid f_\lambda \text{ has a cycle of period } n\}.$$

Then, for any $\lambda < \lambda[n]$, the map f_λ has no cycles of period n. Therefore, $\lambda[n]$ may be called the value of the parameter for which a cycle of period n appears in the family f_λ. By the theorem on coexistence of cycles, the following statement is true:

Theorem 8.6. *Let* $f_\lambda \in C^0(I, I)$. *Then the inequality* $\lambda[n_1] \leq \lambda[n_2]$ *holds for any* n_1 *and* n_2 *such that* $n_1 \triangleleft n_2$.

In this section, we consider families of maps of the indicated type without any further comments and explicitly mention only additional restrictions imposed on the maps.

If a map f belongs to the class C^1, then cycles of different periods appear in the family f_λ for different values of the parameter, i.e., $\lambda[n] \neq \lambda[m]$ if $n \neq m$ (Milnor and Thurston [1]; see also Chapter 2). In particular, this is true for the families of maps with negative Schwarzian.

Theorem 8.7. *Let f_λ be a family of maps with negative Schwarzian. Then*

$$\lambda[1] < \lambda[2] < \lambda[4] < \ldots < \lambda[5\cdot 2] < \lambda[3\cdot 2] < \ldots < \lambda[5] < \lambda[3].$$

Theorem 8.7 implies that infinite sequences of period doubling bifurcations may appear in families of maps with negative Schwarzian.

At the same time, Theorem 8.8 demonstrates that infinite sequences of period doubling bifurcations are impossible for one-parameter families of unimodal maps whose Schwarzian is equal to zero, i.e., for maps composed of two pieces of linear-fractional functions.

Theorem 8.8. *Let f_λ be a one-parameter family of unimodal maps whose Schwarzian is equal to zero such that, for some $\lambda_0 \in \Lambda$, the map f_{λ_0} has cycles whose periods are not equal to powers of two. Then there exists an integer number $n \geq 0$ such that $\lambda[2^{n+j}] = \lambda[2^n]$ for all $j > 0$.*

Proof. Suppose that the assertion of the theorem is not true. Then one can indicate a sequence n_i, $i = 0, 1, \ldots$, such that $\lambda[2^{n_0}] < \lambda[2^{n_1}] < \ldots < \lambda[2^{n_i}] < \ldots$ and there exists $\lambda_\infty = \lim\limits_{i\to\infty} \lambda[2^{n_i}] \leq \lambda_0$. Hence, the central branch of the phase diagram of the map $f_{\lambda_\infty}: I \to I$ consists of infinitely many vertices formed by the cycles of intervals $A^*_{p_m}$, $m \geq 1$, $p_m = 2^m$. In view of the fact that unimodal maps whose Schwarzian is equal to zero have no wandering intervals, any neighborhood $U_\varepsilon(c) = (c-\varepsilon, c+\varepsilon)$, $\varepsilon > 0$, of the point c contains infinitely many intervals from $A^*_{p_m}$.

Let $f_{\lambda_\infty}(x)$ be a symmetric function, i.e., if $f_{\lambda_\infty}(x_0) = f_{\lambda_\infty}(x_1)$, then $|Df_{\lambda_\infty}(x_0)| = |Df_{\lambda_\infty}(x_1)|$. Hence, either $|Df_{\lambda_\infty}(x)| > 1$ for any $x \in I\backslash\{c\}$ (the map f_{λ_∞} is expanding) or $\inf\limits_{x\in I}\left|Df^{p_2}_{\lambda_\infty}(x)\right| > 1$ because

$$\inf_{x\in[c,c_{p_2}]}\left|Df^{p_2}_{\lambda_\infty}(x)\right| = \left|Df^{p_2}_{\lambda_\infty}(c_{p_2})\right|,$$

where c_{p_2} is the nearest right $f^{p_2}_{\lambda_\infty}$-preimage of the point c.

If $f_{\lambda_\infty}(x)$ is an asymmetric function, then there exists a neighborhood $U_\varepsilon(c)$ such that either

$$|Df_{\lambda_\infty}(x)| > |Df_{\lambda_\infty}(x')|, \text{ or } |Df_{\lambda_\infty}(x)| < |Df_{\lambda_\infty}(x')|, \text{ or } |Df_{\lambda_\infty}(x)| = |Df_{\lambda_\infty}(x')|$$

for any $x \in (c-\varepsilon, c)$ and $x' \in (c, c+\varepsilon)$ such that $f(x_0) = x'$. For the neighborhood $U_\varepsilon(c)$, there exists $m < \infty$ such that $A^*_{p_m} = [x_m, y_m] \subset U_\varepsilon(c)$ and, hence, either

$$\left|Df^{p_m}_{\lambda_\infty}(x)\right| > \left|Df^{p_m}_{\lambda_\infty}(x')\right|, \text{ or } \left|Df^{p_m}_{\lambda_\infty}(x)\right| < \left|Df^{p_m}_{\lambda_\infty}(x')\right|, \text{ or } \left|Df^{p_m}_{\lambda_\infty}(x)\right| = \left|Df^{p_m}_{\lambda_\infty}(x')\right|$$

for all $x \in [x_m, c)$ and $x' \in (c, y_m]$. The case where $\left|Df^{p_m}_{\lambda_\infty}(x)\right| = \left|Df^{p_m}_{\lambda_\infty}(x')\right|$ for $f^{p_m}_{\lambda_\infty}\big|_{A^*_{p_m}}$ has already been considered above ($f^{p_m}_{\lambda_\infty}\big|_{A^*_{p_m}}$ is symmetric).

In the other cases, the map $f^{p_m}_{\lambda_\infty}\big|_{A^*_{p_m}}$ may take the shape displayed in Fig. 45.

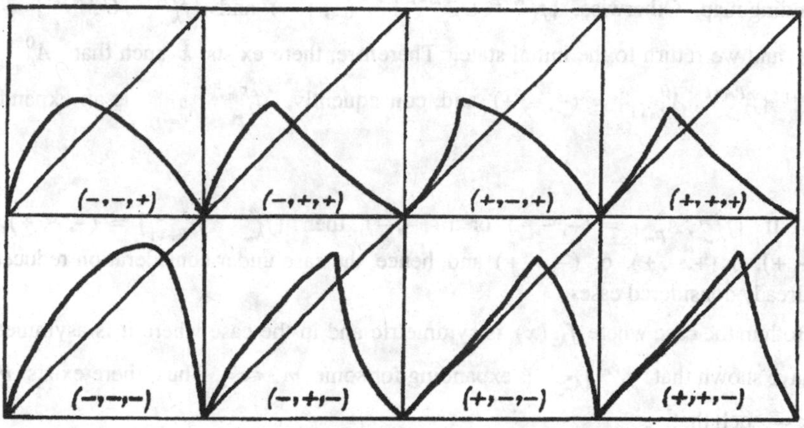

Fig.45

Without loss of generality, we can assume that x_m is a fixed point of the map $f^{p_m}_{\lambda_\infty}: I \to I$. Then all possibilities depicted in Fig. 45 can be characterized by the expression

$$\{f^{p_m}_{\lambda_\infty}, A^*_{p_m}\} = \left(\operatorname{sign} D^2 f^{p_m}_{\lambda_\infty}\big|_{[x_m, c)}, \operatorname{sign} D^2 f^{p_m}_{\lambda_\infty}\big|_{[c, y_m)}, \operatorname{sign}\left(Df^{p_m}_{\lambda_\infty}(x_m) + Df^{p_m}_{\lambda_\infty}(y_m)\right)\right),$$

which may take the values $(-,-,+)$, $(-,+,+)$, $(+,-,+)$, $(+,+,+)$, $(-,-,-)$, $(-,+,-)$, $(+,-,-)$, and $(+,+,-)$.

One can easily show that the map $f^{p_m}_{\lambda_\infty}\big|_{A^*_{p_m}}$ is expanding if $\{f^{p_m}_{\lambda_\infty}, A^*_{p_m}\}$ is equal to $(+,-,-)$ or to $(+,+,-)$.

In the remaining cases, we have the following picture:

1. If $\{f^{p_m}_{\lambda_\infty}, A^*_{p_m}\} = (-,-,+)$ or $(+,-,+)$, then $\{f^{p_{m+1}}_{\lambda_\infty}, A^*_{p_{m+1}}\} = (+,-,-)$ or $(+, +,-)$ and, hence, the map $f^{p_{m+1}}_{\lambda_\infty}\big|_{A^*_{p_{m+1}}}$ is expanding.

2. If $\{f_{\lambda_\infty}^{p_m}, A_{p_m}^*\} = (-,+,+)$ or $(+,+,+)$, then $\{f_{\lambda_\infty}^{p_{m+1}}, A_{p_{m+1}}^*\} = (-,-,-)$. In this case, we have $\{f_{\lambda_\infty}^{p_{m+1}}, U_{c_p}^{(p_{m+1})}\} = (-,-,+)$, where $U_{c_p}^{(p_{m+1})} \subset f_{\lambda_\infty} A_{p_{m+1}}^*$ is a neighborhood of the right preimage of the point c (this point is denoted by c_p). The map $f_{\lambda_\infty}^{p_{m+1}}\big|_{f_{\lambda_\infty} A_{p_{m+1}}^*}$ can be taken as the original map. By the assumption, the middle part of the phase diagram of this map consists of infinitely many vertices formed by cycles of intervals of periods p_{m+i}, $i = 2, 3, \ldots$. We denote these intervals by $A_{p_{m+i}}^0$. Then, for any $i \geq 2$ and some $j \in \overline{1, p_{m+i+1} - 1}$, we have $f_{\lambda_\infty}^j A_{p_{m+k}}^* = A_{p_{m+k}}^0$. Thus, if $A_{p_{m+2}}^0 \supset U_{c_p}^{(p_{m+1})}$, then $\{f_{\lambda_\infty}^{p_{m+2}}, A_{p_{m+2}}^0\} = (+,-,+)$ or $(+,+,-)$ and, hence, $f_{\lambda_\infty}^{p_{m+2}}\big|_{A_{p_{m+2}}^0}$ is an expanding map. Otherwise, $\{f_{\lambda_\infty}^{p_{m+2}}, U_{c_p}^{(p_{m+1})}\} = (+,+,-)$ and $\{f_{\lambda_\infty}^{p_{m+3}}, U_{c_p}^{(p_{m+1})}\} = (-,-,+)$ and we return to the initial state. Therefore, there exists k such that $A_{p_{m+k}}^0 \supset U_{c_p}^{(p_{m+1})}$ $\{f_{\lambda_\infty}^{p_{m+k}}, A_{p_{m+k}}^0\} = (+,-,+)$ and, consequently, $f_{\lambda_\infty}^{p_{m+k}}\big|_{A_{p_{m+1}}^0}$ is an expanding map.

3. If $\{f_{\lambda_\infty}^{p_m}, A_{p_m}^*\} = (-,+,-)$ or $(-,-,-)$, then $\{f_{\lambda_\infty}^{p_{m+1}}, A_{p_{m+1}}^*\} = (-,-,+)$, or $(+,+,+)$, or $(+,-,+)$, or $(-,+,+)$ and, hence, the case under consideration reduces to the already considered cases.

Both in the case where $f_{\lambda_\infty}(x)$ is symmetric and in the case where it is asymmetric, we have shown that $f_{\lambda_\infty}^{p_{m'}}\big|_{A_{p_{m'}}^*}$ is expanding for some $m' < \infty$. Thus, there exists $m' < m'' < \infty$ such that

$$\inf_{x \in A_{p_{m''}}^*} \left| Df^{p_{m''}}(x) \right| \geq \sqrt{2}$$

and, therefore, $A_{p_{m''}}^*$ contains no periodic intervals whose periods are greater than $p_{m''}$. Hence, the number of vertices in the phase diagram of the map $f_{\lambda_\infty} : I \to I$ is finite.

Thus, in a one-parameter family of piecewise smooth unimodal maps that are not C^1-smooth, the sequence of period doubling bifurcations can be finite at most at one point.

Numerical results demonstrate that the dynamics of maps in the family $f_\lambda = \lambda f$ becomes more complicated as the parameter λ increases. Thus, it follows from Theorem 8.4 that, for the family $x \to \lambda x(1-x)$, the topological entropy and kneading invariant are monotone functions of the parameter. By the same theorem, if a family $f_\lambda(x) = \lambda f(x)$ of convex maps with negative Schwarzian is characterized by the property that, for some $\lambda_0 \in \Lambda$, the map f_{λ_0} has a cycle of period $m \neq 2^k$, $k \in \mathbb{N}$, then, for any $\lambda \geq \lambda_0$, the map f_λ also possesses a cycle of period m, i.e., bifurcations of cycles exhibit the property of monotonicity.

It was conjectured that the families of unimodal maps with negative Schwarzian must be characterized by the property of monotonicity of bifurcations of cycles and by the monotone dependence of the topological entropy and the kneading invariant on the parameter. However, it was shown that if f is not a convex function, then, for the family $f_\lambda = \lambda f$, it may happen that $h(f_\lambda)$ and $\nu(f_\lambda)$ are nonmonotone functions of λ and no monotonicity of bifurcations of cycles is observed.

Theorem 8.9. *There exists a unimodal map f with negative Schwarzian such that, for the family $f_\lambda = \lambda f$, no monotonicity of bifurcations of cycles is observed and $h(f_\lambda)$ and $\nu(f_\lambda)$ are nonmonotone functions of λ.*

Proof. Let us construct a unimodal map f with negative Schwarzian such that the family $f_\lambda = \lambda f$ has the following properties: As the parameter λ increases from 0 to some $\lambda_0 > 0$, one observes the appearance of cycles of all periods. As λ increases further, all cycles (except fixed points) first disappear (for some $\lambda_1 > 0$) and then appear again. The required map f is given by the equality

$$f(x) = \begin{cases} g_\mu(x) = \mu^2 x(1-x)(1-\mu x(1-x)), & x \in [0, x_0], \\ g(x) = \int\limits_x^{x_0} (ax^2 + bx + c)^{-2} dx + g_\mu(x_0), & x > x_0, \end{cases}$$

where

$$\mu \in (3, 4), \quad a = \frac{1}{4}\left[\gamma + \frac{3}{2}\beta^2\alpha^{-1}\right]\alpha^{-3/2},$$

$$b = \frac{\beta}{2}\alpha^{-3/2} - \frac{x_0}{2}\left(\gamma + \frac{3}{2}\beta^2\alpha^{-1}\right)\alpha^{-3/2},$$

$$c = \alpha^{-1/2} + \frac{x_0^2}{4}\left(\gamma + \frac{3}{2}\beta^2\alpha^{-1}\right)\alpha^{-3/2} - \frac{x_0}{2}\beta\alpha^{-3/2},$$

$$\alpha = |\, g_\mu'(\alpha)\,|, \quad \beta = g_\mu(x_0), \quad \gamma = g_\mu''(x_0), \quad x_0 = \frac{1}{2} - \delta, \quad 0 < \delta < \frac{1}{2}.$$

We choose $x_0 < 1/2$ close to $1/2$ and select constants a, b, and c such that $f\colon \mathbb{R}^+ \to \mathbb{R}^+$ is a unimodal C^3-map with negative Schwarzian (see Fig. 46). Note that the last assertion can be readily verified by using the following criterion of negativity for Schwarzians: $Sg(x) < 0$ in a given interval if and only if the function $|g'(x)|^{-1/2}$ is concave in this interval.

For μ close to 4, the map $x \to \mu^2 x(1-x)(1-\mu x(1-x))$ has cycles of all periods. Moreover, for any μ of this sort, one can find sufficiently small $\delta_0 = \delta_0(\mu)$ such that

$$x_0 g'_\mu(x_0) \geq -g_\mu(x_0) \tag{8.1}$$

for all $\delta \leq \delta_0$.

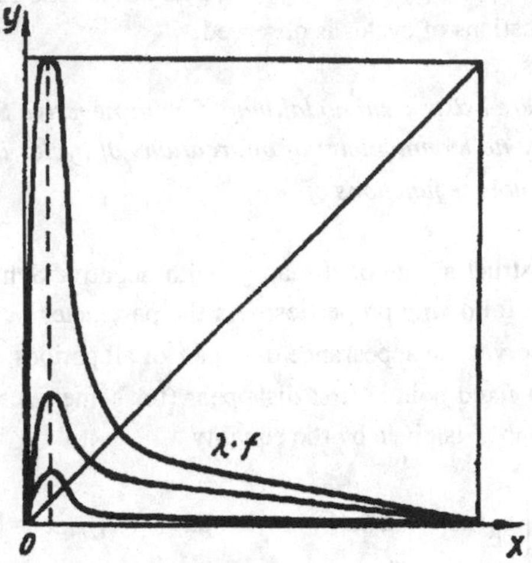

Fig. 46

We fix μ and δ for which the map $f: \mathbb{R}^+ \to \mathbb{R}^+$ has cycles of all periods and condition (8.1) is satisfied. In this case, cycles of all periods appear in the one-parameter family λf as λ changes from 0 to 1. Further, by virtue of (8.1), for

$$\lambda = \lambda_1 = \frac{x_0}{f(x_0)} = \frac{x_0}{g_\mu(x_0)} > 1,$$

the map λf possesses an attracting fixed point other than the fixed point 0. Note that $Sf_{\lambda_1} < 0$ and the map f_{λ_1} is unimodal. Therefore, by virtue of Theorem 5.3, this map may have only fixed points.

It is clear that, for the family of maps constructed above, we observe not only the violation of monotonicity of bifurcations of cycles but also the nonmonotone dependence of the entropy and kneading invariant on the parameter. Similarly, one can construct a family of unimodal maps with negative Schwarzian for which the topological entropy and kneading invariant regarded as functions of the parameter may have arbitrarily many intervals of monotonicity.

Note that the family of maps constructed above is defined for $x \in \mathbb{R}^+$ and $\lambda \in (0, \infty)$. Clearly, it is possible to construct a family of maps with the properties indicated in Theorem 8.9 but defined for $x \in [0, 1]$ and $\lambda \in (0, 1)$.

According to Theorem 8.5, the property of monotonicity of bifurcations for families of maps with negative Schwarzian is guaranteed by the convexity of maps from this families. The following assertion makes the result of Theorem 8.5 more precise:

Theorem 8.10. *Let* $f_\lambda = \lambda f$ *be a family of unimodal convex maps with negative Schwarzian. Then*

$$\lim_{i\to\infty} \frac{\lambda\left[(2i-1)\,2^n\right] - \lambda\left[(2i+1)\,2^n\right]}{\lambda\left[(2i+1)\,2^n\right] - \lambda\left[(2i+3)\,2^n\right]} = \gamma_n(f) > 1.$$

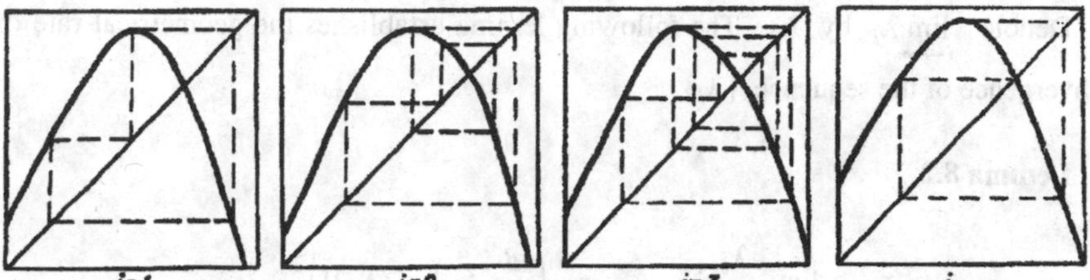

Fig. 47

Our proof of Theorem 8.10 is based on a hypothesis formulated somewhat later.

First, we consider the case $n = 0$. In the following lemma, we use the concept of cycles of minimal type (or, simply, minimal cycles) introduced in Chapter 3:

Lemma 8.1. *There exists a monotonically decreasing sequence* $\{\lambda_i, i \geq 1\}$ *of values of the parameter* λ *such that the point* c *belongs to the minimal cycle of the map* f_{λ_i} *of period* $2i+1$.

Proof. The arrangement of points of minimal cycles of periods $2i+1$, $i \geq 1$, on the real line is known. In Fig. 47, we display the arrangement of points of these cycles for $i = 1, 2, 3$ and the graph of the limit function $f_{\lambda_0}(x)$, where $\lambda_0 = \lim_{i\to\infty} \lambda_i$.

In proving the lemma, we assume that $\lambda > \lambda[2]$. Consider a point $c_+(\lambda) \in (c, 1)$ such that $f_\lambda(c_+(\lambda)) = c$. The restriction of the map f_λ^2 to the interval $[c, c_+(\lambda)]$ is a homeomorphism which covers $[c, c_+(\lambda)]$. Therefore, the map $\varphi_\lambda : [c, c_+(\lambda)] \to [c, c_+(\lambda)]$ such that

$$\varphi_\lambda(x) = \left(f_\lambda^2\right)^{-1}(x)$$

is well defined.

Since f_λ is a map with negative Schwarzian, $\varphi_\lambda(x)$ is a monotone strictly increasing function and the fixed point $x^* = x^*(\lambda)$ of the map f_λ lying in the interval $(c, c_+(\lambda))$ is a globally stable fixed point of φ_λ.

Consider functions $c_k(\lambda)$, $k \geq 0$, defined on the interval $[\lambda[2], 1]$ by the equality $c_k(\lambda) = \varphi_\lambda^k(c)$. Due to the monotonicity of φ_λ, we have $c_0(\lambda) < c_1(\lambda) < c_2(\lambda) < \ldots < c_k(\lambda) < x^*(\lambda)$ for $\lambda \in [\lambda[2], 1]$. Moreover, $c_k(\lambda) \to x^*(\lambda)$ as $k \to \infty$. Now let $z(\lambda) = f_\lambda^3(c)$. Then $z(\lambda[2]) > x^*(\lambda[2])$ and $z(1) = 0 < c$. Since the functions $z(\lambda)$, $x^*(\lambda)$, and $c_k(\lambda)$, $k \geq 0$, are continuous, there exists a decreasing sequence $\lambda_1 > \lambda_2 > \lambda_3 > \ldots$ such that $z(\lambda_i) = c_{i-1}(\lambda_i)$, $i = 1, 2, 3, \ldots$, which is equivalent to the assertion of the lemma.

Denote $\lim\limits_{i \to \infty} \lambda_i$ by λ_0. The following lemma establishes the geometrical rate of convergence of the sequence $\{\lambda_i\}$.

Lemma 8.2.

$$\lim_{i \to \infty} \frac{\lambda_{i-1} - \lambda_i}{\lambda_i - \lambda_{i+1}} = \left(\frac{d}{dx} f_{\lambda_0}\left(x^*(\lambda_0)\right)\right)^2.$$

Proof. By using the mean value theorem, we obtain

$$c_{i-1}(\lambda_i) - x^*(\lambda_0) = f_{\lambda_i}^3(c) - f_{\lambda_0}^3(c) = \frac{d}{d\lambda} f_\mu^3(c)\,(\lambda_i - \lambda_0), \tag{8.2}$$

where $\mu = \lambda_0 + \theta(\lambda_i - \lambda_0)$, $\theta \in (0, 1)$.

On the other hand,

$$c_{i-1}(\lambda_i) - x^*(\lambda_i) = \frac{d}{dx} f_{\lambda_i}^2\left(x^* + \theta_1\left(c_i(\lambda_i) - x^*(\lambda_i)\right)\right)\left(c_i(\lambda_i) - x^*(\lambda_i)\right), \tag{8.3}$$

where $\theta_1 \in (0, 1)$.

Denote $c_i(\lambda) - x^*(\lambda)$ by $\Delta_i(\lambda)$. Then

$$\lambda_i - \lambda_0 \sim \Delta_{i+1}(\lambda_i)\left(\frac{d}{dx} f_{\lambda_i}^2\left(x^* + \theta\Delta_{i+1}\right)\right)\left(\frac{d}{d\lambda} f_{\lambda_0}^3(c)\right)^{-1}$$

for large i. By using relations (8.2) and (8.3) and assuming that

$$\frac{d}{d\lambda} f_\lambda^3(c) \neq 0 \quad \text{for} \quad \lambda = \lambda_0,$$

we arrive at the required result. Indeed,

$$\lim_{i \to \infty} \frac{\lambda_{i-1} - \lambda_i}{\lambda_i - \lambda_{i+1}} = \frac{d}{dx} f^2_{\lambda_0}\left(x^*(\lambda_0)\right) = \left(\frac{d}{dx} f_{\lambda_0}\left(x^*(\lambda_0)\right)\right)^2.$$

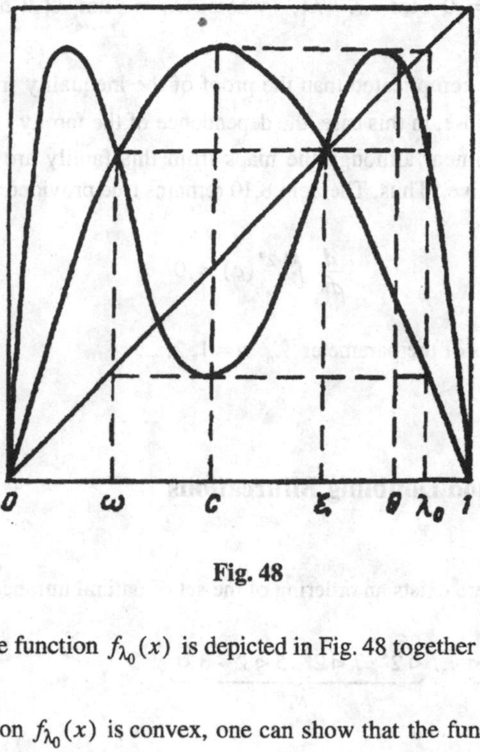

Fig. 48

The graph of the function $f_{\lambda_0}(x)$ is depicted in Fig. 48 together with the graph of the function $f^2_{\lambda_0}(x)$.

Since the function $f_{\lambda_0}(x)$ is convex, one can show that the function $f^2_{\lambda}(x)$ is convex in the interval $[\xi, \lambda_0]$. Indeed,

$$(f^2_{\lambda_0})''(x) = f''_{\lambda_0}(f_{\lambda_0}(x))[f'_{\lambda_0}(x)]^2 + f'_{\lambda_0}(f_{\lambda_0}(x))f''_{\lambda_0}(x).$$

Since any map with negative Schwarzian has at most one inflection point in each interval of monotonicity, it suffices to prove that $f''_{\lambda_0}(x)$ is negative at the points $x = \xi$ and $x = \lambda_0$. For $x = \xi$, we can write

$$(f^2_{\lambda_0})''(\xi) = f''_{\lambda_0}(\xi)f'_{\lambda_0}(\xi)(1 + f'_{\lambda_0}(\xi)) < 0$$

because the convexity of f_{λ_0} implies the inequality $f'_{\lambda_0}(\xi) < -1$. Similarly, for $x = \lambda_0$,

$$(f^2_{\lambda_0})''(\lambda_0) = f''_{\lambda_0}(\omega)[f'_{\lambda_0}(\lambda_0)]^2 + f'_{\lambda_0}(\omega)f''_{\lambda_0}(\lambda_0) < 0.$$

To prove the theorem for $n = 1$, one must consider the map f_λ^2 in the interval $[\xi, \lambda]$, where ξ is the fixed point of the map f_λ other than 0. In this case, it is necessary to prove the inequality

$$\frac{d}{d\lambda} f_\lambda^6(c) = 0 \quad \text{for} \quad \lambda = \lambda_1, \quad \text{where} \quad \lambda_1 = \lim_{i\to\infty} \lambda[2(2i+1)].$$

This problem is more complicated than the proof of the inequality from the hypothesis considered above because, in this case, the dependence of the family $F_\lambda = f_\lambda^2|_{[\xi,\lambda]}$ on the parameter is not linear although the maps from this family are convex and their Schwarzians are negative. Thus, Theorem 8.10 remains true provided that the inequality

$$\frac{d}{d\lambda} f_\lambda^{3\cdot 2^n}(c) \neq 0$$

holds for proper values of the parameter λ_n, $n = 1, 2, \ldots$.

4. Rate of Period Doubling Bifurcations

As already known, there exists an ordering of the set of natural numbers

$$\underbrace{1 \triangleleft 2 \triangleleft 4 \triangleleft 8 \triangleleft}_{N^*} \ldots \underbrace{\triangleleft\, 2^k\cdot 7 \triangleleft 2^k\cdot 5 \triangleleft 2^k\cdot 3 \triangleleft}_{N_k} \ldots$$

$$\underbrace{\triangleleft\, 2\cdot 7 \triangleleft 2\cdot 5 \triangleleft 2\cdot 3 \triangleleft}_{N_1} \ldots \underbrace{\triangleleft\, 7 \triangleleft 5 \triangleleft 3}_{N_0}$$

such that if a continuous map $f: I \to I$ (or $f: \mathbb{R} \to \mathbb{R}$) has a cycle of period m, then it also has a cycle of period n for any $n \triangleleft m$. Hence, for any family $f_\lambda(x)$ of continuous maps of the real line into itself, the order of appearance of cycles is specified by the indicated ordering of natural numbers.

There are many families of maps for which one can observe not only single bifurcations but also infinite sequences of bifurcations of cycles (as the parameter changes within a certain finite interval). Among maps of this sort, one can mention convex unimodal maps with negative Schwarzian and, in particular, quadratic maps. By analyzing the behavior of these maps, one can establish some "universal" properties of sequences of bifurcations of cycles.

In studying the family $x \to \lambda x(1-x)$, we observe an infinite sequence of period doubling bifurcations as the parameter λ increases from $\lambda = 3$ to $\lambda = 3.57$ (as a result of this sequence of bifurcations, the map has cycles of periods 2^n, $n = 0, 1, 2, \ldots,$ for $\lambda > 3.57$). Note that it follows from the theorem on coexistence of periods of cycles that, in any family of smooth maps, the appearance of infinitely many cycles is a result of period doubling. Moreover, families of maps are characterized by the universal order of the appearance of cycles of new periods and, in addition, for a broad class of families, the sequence of bifurcation values of the parameter converges with certain universal rate (for all families from a given class, this rate is the same).

To clarify these observations, we consider the following family of quadratic maps: $g_\mu(x) = 1 - \mu x^2$, $x \in [-1, 1]$, $\mu \in [0, 2]$. The first period doubling bifurcation occurs at $\mu = \mu_0 = 0.75$: The fixed point $\beta_1(0.75) = 2/3$ generates a cycle of period two. The subsequent bifurcation values corresponding to the appearance of cycles of periods 2^n, $n = 2, 3, 4, \ldots,$ are equal to $\mu_1 = 1.25$, $\mu_2 = 1.3681\ldots$, $\mu_3 = 1.3940\ldots, \ldots,$ respectively. As $n \to \infty$, the sequence μ_n approaches the value $\mu_\infty = 1.40155\ldots$ for which the map $f_{\mu_\infty}\colon [-1, 1] \to [-1, 1]$ has cycles of all periods equal to powers of two and has no cycles of other periods. The ratio

$$\delta_n = \frac{\mu_n - \mu_{n-1}}{\mu_{n+1} - \mu_n}$$

takes values $\delta_1 = 4.23$, $\delta_2 = 4.55$, $\delta_3 = 4.65$, $\delta_4 = 4.664$, $\delta_5 = 4.668$, $\delta_6 = 4.669, \ldots$. As in the case of the family $f_\lambda(x) = \lambda x(1 - x)$ (see Chapter 1), the limit of the sequence δ_n as $n \to \infty$ is equal to $\delta = 4.6692\ldots$.

The value of the quantity α, which characterizes the sizes of appearing cycles, also coincides with the corresponding value for the family f_λ, i.e., if β'_{2^n} is the first point of the cycle of period 2^n (which appears for $\mu > \mu_n$) to the right of $x = 0$ and

$$\beta''_{2^n} = g_\mu^{2^{n-1}}(\beta'_{2^n}),$$

then

$$\alpha_n = \frac{\beta'_{2^n} - \beta''_{2^n}}{\beta'_{2^{n+1}} - \beta''_{2^{n+1}}} \to \alpha = 2.502\ldots \quad \text{as } n \to \infty.$$

The phenomenon of universality means that the sequences δ_n and α_n determined for different one-parameter families of maps (not only for quadratic maps but also for the families $\lambda \sin x$, $\lambda x(1-x)^2$, etc.) converge, for all these families, to the same values δ and α, respectively. This phenomenon was discovered and investigated by Feigenbaum in 1978 (see Feigenbaum [1, 2]); almost simultaneously, similar results were obtained by Grosmann and Thomae [1].

In order to explain the phenomenon of universality, we consider the set G formed by unimodal maps $\psi \in C^1(I, I)$, where $I = [-1, 1]$, such that $\psi(0) = 1$ and $\psi(1) < 0$.

For any $\psi \in G$, we define

$$T\psi(x) = \frac{1}{a}\psi^2(ax),$$

where $a = \psi(1)$. The nonlinear operator $T: G \to G$ is called the transformation of doubling.

Let $G^\omega \subset G$ be the set of analytic functions from G. We want to determine fixed points of the map $T: G^\omega \to G^\omega$, i.e., the solutions of the functional equation $T\psi = \psi$ in the set of analytic functions G^ω. The degree of degeneracy of the critical point of the function $\psi \in C^\omega$ is invariant under the action of T. Therefore, the form of the solutions of the indicated functional equation depends on the degree of degeneracy. We require that the critical point of any function which is a fixed point of the operator T must be nondegenerate. By Theorem 8.14, this function must be even.

Proposition 8.4. *There exists an even analytic function*

$$\psi_0(x) = 1 - 1.52763\ldots \cdot x^2 + 0.104815\ldots \cdot x^4 - 0.0267057\ldots \cdot x^6 + \ldots,$$

which is a fixed point of the operator of doubling $T\psi(x) = \psi^2(ax)/a$, *where* $a = a(\psi_0) = \psi_0(1) = -1/\alpha = -0.3995\ldots$.

Let $\mathcal{H}$ denote the Banach space of functions $\psi(z)$ analytic and bounded in a certain complex neighborhood of the interval I and real-valued on the real axis. Let $\mathcal{H}_0$ be the subspace of $\mathcal{H}$ formed by the functions satisfying the conditions $\psi(0) = 1$, $\psi'(0) = 0$, and $\psi''(0) \neq 0$.

Proposition 8.5. *There exists a neighborhood* $U(\psi_0)$ *of the point* ψ_0 *in* $\mathcal{H}_0$ *such that* $T \in C^\infty(U(\psi_0), \mathcal{H}_0)$. *The operator* $DT(\psi_0)$ *is hyperbolic and possesses a one-dimensional unstable subspace and a stable subspace of codimensionality one. The eigenvalue of* $DT(\psi_0)$ *in the unstable subspace is equal to* $\delta = 4.6692\ldots$.

Let $\Sigma_0 \subset G$ be the "surface" formed by the maps ψ whose derivative at a fixed point $x_0 = x_0(\psi) \in [0, 1]$ is equal to -1 and $S\psi(x_0) < 0$.

Proposition 8.6. *An unstable local "manifold" defined in a neighborhood of* ψ_0 *can be extended to a global unstable "manifold"* $W^u(\psi_0)$ *which transversally crosses the surface* Σ_0; *the "manifold"* $W^u(\psi_0)$ *consists of maps with negative Schwarzian.*

At present, all known proofs of these propositions are computer-assisted (see Wul, Sinai, and Khanin [1]).

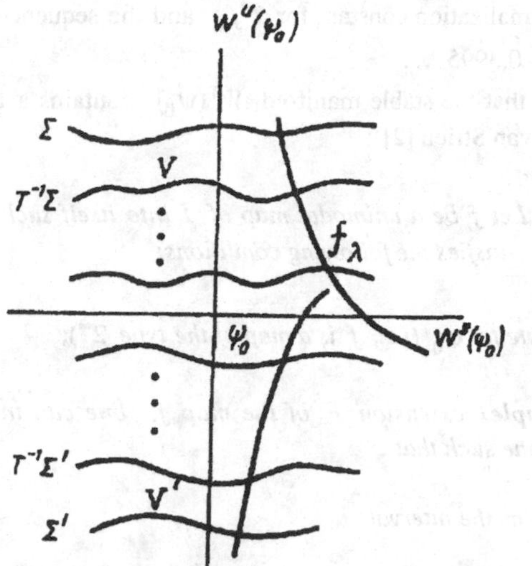

Fig. 49

By using Propositions 8.4–8.6, one can explain the phenomenon of universality, which is known as Feigenbaum universality. In the neighborhood $U(\psi_0)$, we have the following picture (Fig. 49): If a one-parameter family f_λ transversally crosses the stable manifold $W^s(\psi_0)$, then it transversally crosses the surfaces $T^{-n}\Sigma_0$ for sufficiently large n. The points f_{λ_n} of intersection of f_λ with $T^{-n}\Sigma_0$ correspond to period doubling bifurcations of cycles of periods 2^n and the point f_{λ_∞} corresponds to the accumulation point of the set of bifurcation values, i.e., $\lambda_\infty = \lim_{n\to\infty} \lambda_n$. For large n, the distance between $T^{-(n+1)}\Sigma_0$ and $W^s(\psi_0)$ is about δ times less than the distance between $T^{-n}\Sigma_0$ and $W^s(\psi_0)$. Hence, the bifurcation values of the parameter λ of the family f_λ satisfy the relation

$$\lambda_\infty - \lambda_n \sim c_0\delta^{-n},$$

where c_0 depends on the family of maps.

Let β_n be the point of the 2^n-periodic cycle of the map f_{λ_n} whose distance from the point of extremum is minimal and let β_0 be the fixed point of the map $T^n f_{\lambda_n}$ from [0, 1]. Then we have

$$\beta_n = \beta_0 \prod_{i=1}^{n} \alpha_i,$$

where α_i is a renormalization constant for $T^i f_{\lambda_n}$ and the sequence $\{\alpha_i\}$ converges to the value $-1/\alpha = -0.3995\ldots$.

Sullivan proved that the stable manifold $W^s(\psi_0)$ contains a fairly broad class of functions (see, e.g., van Strien [2]):

Theorem 8.11. *Let f be a unimodal map of I into itself such that $f(-1) = f(1) = -1$. Assume that f satisfies the following conditions:*

(i) f is conjugate to ψ_0 (i.e., f is a map of the type 2^∞);

(ii) for the complex extension F of the map f, one can indicate a disk in the complex plane such that

(a) it contains the interval I,

(b) it contains the unique critical point of the map F, and

(c) under the map F, its boundary is mapped into the outside of the disk.

Then f belongs to the stable manifold $W^s(\psi_0)$.

By using the properties of the operator T, one can construct the unstable manifold $W^u(\psi_0)$ numerically. Thus, a construction of this sort was suggested by Wul, Sinai, and Khanin [1].

As indicated above, Feigenbaum universality is observed for a broad class of one-parameter families of smooth unimodal maps. It is thus interesting to clarify the conditions under which an individual family f_λ exhibits the phenomenon of universality.

First, for a given family f_λ, it is desirable to establish simple conditions guaranteeing the monotonicity of the sequence of bifurcations similar to that observed for the family of quadratic maps $\lambda x(1-x)$. As follows from Theorems 8.5 and 8.7, for the family λf, the required property is apparently guaranteed by the analyticity, convexity, and negativity of the Schwarzian of the map f. Second, one can apply Theorem 8.11 to require that the map $\lambda_\infty f$ be similar to a quadratic map in a sense of Douady and Hubbard (i.e., that $\lambda_\infty f$ satisfy the conditions of Theorem 8.11).

Let $\mathcal{F}_\lambda$ be the space of families with smooth dependence on the parameter λ. We now study the phenomenon of Feigenbaum universality for families of maps from the space $\mathcal{F}_\lambda$. The following description of the doubling operator in the space of analytic functions in the vicinity of its fixed point ψ_0 seems to be quite reasonable:

The stable manifold $W^s(\psi_0)$ splits the neighborhood $U(\psi_0)$ into two parts (Fig. 49). We define the fundamental domain V of the operator T^{-1} as the domain bounded by the surfaces $\Sigma = \{f \mid f^2(0) = 0\}$ and $T^{-1}\Sigma$ (Fig. 50). The fundamental domain V' of the operator T^{-1} is bounded by $\Sigma' = \{f \mid f^2(0) = -1\}$ and $T^{-1}\Sigma'$ (Fig. 51). The stable manifold $W^s(\psi_0)$ separates the maps with simple structure (i.e., with finite sets of nonwandering points and topological entropy equal to zero) from the maps with complicated structure (with infinite sets of nonwandering points and positive topological entropy).

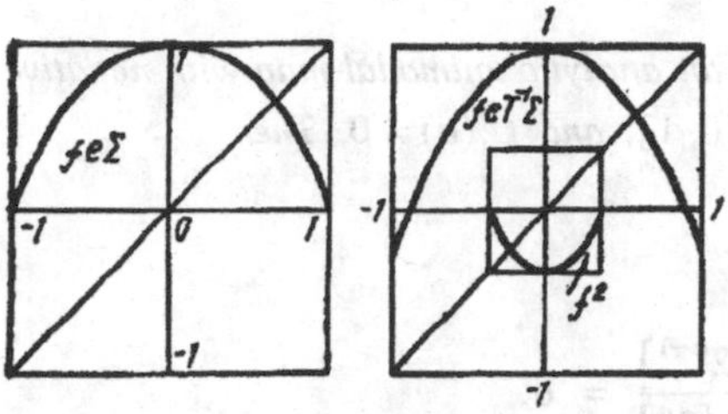

a b

Fig. 50

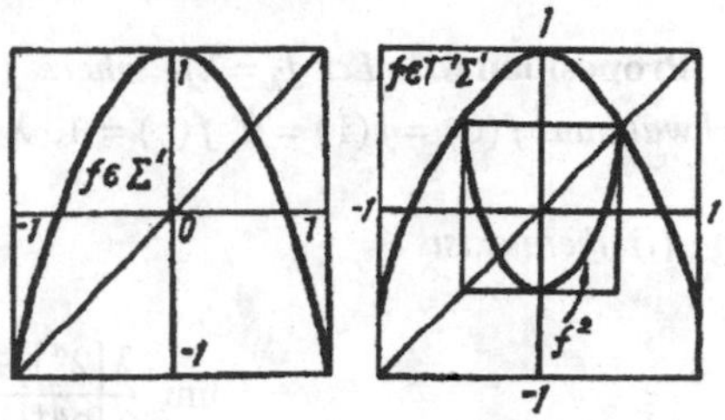

a b

Fig. 51

If a family $g_\lambda \in \mathcal{F}_\lambda$ transversally crosses the manifold $W^s(\psi_0)$ then, for all sufficiently large n, it transversally crosses the surfaces $T^{-n}\Sigma$ and, hence, we observe the phenomenon of Feigenbaum universality. Moreover, for large n, the family g_λ transversally crosses the surfaces $T^{-n}\Sigma'$. The values of the parameter λ_n' corresponding to the intersections of g_λ with $T^{-n}\Sigma'$ are points of bifurcations of creation of cycles of intervals of period 2^{n+1}, and the accumulation point of the set of bifurcation values $\lambda_\infty = \lim\limits_{n\to\infty} \lambda_n'$ corresponds to the intersection of the family g_λ with $W^s(\psi_0)$. Since the distance between $T^{-n}\Sigma'$ and $W^s(\psi_0)$ is proportional to δ^{-n}, we have $\lambda_n' - \lambda_\infty \sim c\delta^{-n}$.

The value λ_n' corresponds to the appearance of a trajectory homoclinic to the minimal cycle of period 2^n in the family g_λ (i.e., to a cycle from the block N_*). This means that homoclinic trajectories appear with the same rate δ as the corresponding cycles (but in the inverse order).

Consider another universal property of families of maps, which is a direct consequence of Feigenbaum universality. It characterizes bifurcations (creation) of cycles whose periods are not powers of two. By Theorem 8.10, for families λf of unimodal convex maps with negative Schwarzian, each block N_k is characterized by a certain asymptotic rate $\gamma_k = \gamma_k(f)$ of creation of cycles of periods $(2n+1)2^k$ as $n \to \infty$. However, this theorem does not imply that the sequence $\{\gamma_k\}$ converges as $k \to \infty$. If we assume that λf transversally crosses $W^s(\psi_0)$ for $\lambda = \lambda_\infty$ and use the fact that $\lambda f \in T^{-k}U'$ for $\lambda = \lambda[(2n+1)2^k]$, $n \geq 1$, then we arrive at the following conclusion:

For the family $f_\lambda = \lambda f$, we have $\gamma_k \to \gamma$ as $k \to \infty$, i.e., the rate of the process of creation of cycles of periods $(2n+1)\,2^k$ (in the blocks N_k) is asymptotically constant for large k. Numerical experiments corroborate these conclusions and give approximate values of the asymptotic rate. Thus, according to Kolyada and Sivak [1] and Geisel and Nierwetberg [1], $\gamma = 2.9480\ldots$. It should also be noted that the rate of creation of cycles of periods $(2n+1)\,2^k$, where n is fixed and $k \to \infty$ is equal to δ.

Proposition 8.7. *Let $f_\lambda = \lambda f$, where f is an analytic unimodal map with negative Schwarzian, $f(0) = f(1) = 0$, $f(c) = 1$, $\lambda \in [0, 1]$, and $f''(c) \neq 0$. Then*

(i) there exists

$$\lim_{n\to\infty} \frac{\lambda[2^n] - \lambda[2^{n-1}]}{\lambda[2^{n+1}] - \lambda[2^n]} = \delta,$$

where $\delta = 4.669201\ldots$;

(ii) there exists

$$\lim_{m\to\infty} \lim_{n\to\infty} \frac{\lambda[(2m+1)2^n] - \lambda[(2m-1)2^{n-1}]}{\lambda[(2m+1)2^{n+1}] - \lambda[(2m+1)2^n]} = \gamma,$$

where $\gamma = 2.94805\ldots$;

(iii) there exists

$$\lim_{n\to\infty} \frac{\lambda'[2^n] - \lambda'[2^{n-1}]}{\lambda'[2^{n+1}] - \lambda'[2^n]} = \delta,$$

where $\delta = 4.669201\ldots$ *and*

$$\lambda'[2^n] = \lim_{m\to\infty} \lambda[(2m+1)2^{n-1}], \quad n \geq 1;$$

(iv) the family f_λ has no other bifurcation values of the parameter in the interval $[0, \lambda_\infty)$, where

$$\lambda_\infty = \lim_{n\to\infty} \lambda[2^n] = \lim_{n\to\infty} \lambda'[2^n].$$

If a map f belongs to $W^s(\psi_0)$, then it has periodic points of all periods 2^n, $n = 0, 1, 2, \ldots$, and has no periodic points of other periods. The set of nonwandering points

NW(f) is equal to Per$(f) \cup K(f)$, where $K(f)$ is a closed uncountable minimal set of the map f (i.e., a 2^∞-type quasiattractor) (see Misiurewicz [2]).

Properties of the maps from $W^s(\psi_0)$ are characterized by the following theorem proved by Paluba:

Theorem 8.12 (Paluba [1]). *Assume that f and g belong to the stable "manifold" $W^s(\psi_0)$. Then the sets $K(f)$ and $K(g)$ are topologically conjugate and the conjugating homeomorphism h belongs to C^{Lip} in a sense that, for any $x \in K(f)$ and $y \in [-1, 1]$, there exists a Lipschitz constant γ_0 such that*

$$|h(x) - h(y)| \leq \gamma_0 |x - y|.$$

The fact that $\operatorname{mes} K(f) = 0$ for any $f \in W^s(\psi_0)$ is an important consequence of Theorem 8.12 (because it is clear that $\operatorname{mes} K(\psi_0) = 0$).

5. Universal Properties of One-Parameter Families

Let f_λ be a one-parameter family of smooth unimodal maps. For $\lambda = \lambda^*$, we assume that the central branch of the phase diagram of the map f_{λ^*} consists of infinitely many vertices (see Chapter 5). Then the map f_{λ^*} possesses an infinite sequence of periodic intervals $I_1 \supset I_2 \supset \ldots \supset I_m \supset \ldots$, which contain the point of extremum. In this case, for any $m > 1$, p_{m-1} is a divisor of p_m and the set

$$K = \bigcap_{m=1}^{\infty} \bigcap_{i=1}^{p_m} f_{\lambda^*}^i (I_m)$$

is a quasiattractor. The case where $p_{m+1}/p_m = 2$ for any $m \geq 1$ was studied in the previous section. In particular, it was mentioned that $\lim\limits_{m \to \infty} |I_m|/|I_{m+1}| = 2.502\ldots$ for $f_{\lambda^*} \in W^s(\psi_0)$, where $|I_m|$ is the length of the interval I_m.

It is natural to expect that universal properties are exhibited by the family f_λ not only for $p_{m+1}/p_m = 2$ but also for $p_{m+1}/p_m = k > 2$.

As in the previous section, we consider the set G^ω of analytic unimodal maps f: $[-1, 1] \to [-1, 1]$ such that $f(0) = 1$. The set G^ω can be decomposed into infinitely many mutually disjoint subsets as follows: $G^\omega = \bigcup_{i \geq 1} G^{(2i)}$, where each $G^{(2i)}$ is formed by the maps satisfying the conditions

$$\frac{d^r f}{dx^r}(0) = 0 \quad \text{for} \quad r = 1, 2, \ldots, 2i-1 \quad \text{and} \quad \frac{d^{2i} f}{dx^{2i}}(0) \neq 0.$$

An operator T_k, $k \geq 2$, is introduced by the formula

$$(T_k f)(x) = \frac{1}{\alpha} f^k(\alpha x), \quad \alpha = f^k(0), \quad x \in [-1, 1].$$

If, for a map $f \in G^{\omega}$, we have $m^*(f) < \infty$, then, for some $j \geq 1$, the operator T_k is not well defined for the function $\left(T_k^j f\right)(x)$ (T_k^j denotes the jth iteration of T_k). At the same time, one can easily give examples of maps from G^{ω} for which all iterations of the operator T_k are well defined.

Let

$$\mathcal{E}(T_k) = \{ f \in G^{\omega} \mid m^*(f) = \infty \text{ and } p_{m_{i+1}} / p_{m_i} = k$$
$$\text{for some subsequence } m_1 < m_2 < m_3 < \ldots \}.$$

As above, it is not difficult to show that $T_k^j f \in G^{\omega}$ for all $j \geq 0$ if and only if $f \in \mathcal{E}(T_k)$.

We say that maps f and g from the class G^{ω} are of the same type if $m^*(f) = m^*(g)$ and the types of the cycles of intervals $A^*_{p_m}(f)$ and $A^*_{p_m}(g)$ in the phase diagrams of the maps f and g coincide for all m (all relevant definitions can be found in Chapter 5). By using the concept of maps of the same type, we can decompose the set $\mathcal{E}(T_k)$ into classes of maps of the same type. The class of maps from $\mathcal{E}(T_k)$ that consists of maps of the same type as $f \in \mathcal{E}(T_k)$ is denoted by $\mathcal{E}_f(T_k)$ (or simply $\mathcal{E}_f$ if this does not lead to ambiguity).

Lemma 8.3. *If* $k = 2$ *or* $k = 3$, *then* $\mathcal{E}_f(T_k) = \mathcal{E}(T_k)$ *for any* $f \in \mathcal{E}(T_k)$. *If* $k = 4$, *then* $\mathcal{E}(T_k)$ *splits into uncountably many classes of maps of the same type.*

Proof. If a permutation $\pi = (t_0, t_1, \ldots, t_{n-1})$ determines the type of a cycle of intervals of a unimodal map, then this permutation π is cyclic and the map $\pi : \{0, 1, \ldots, n-1\} \to \{0, 1, \ldots, n-1\}$ with $\pi(i) = t_i$ is unimodal. Permutations of this sort are called U-permutations. If π is a U-permutation, then one can easily construct an example of a unimodal map with a cycle of intervals of the type π.

Suppose that a unimodal map f has a cycle of intervals A. If c is the point of maximum (minimum) of the map f, then the map $\pi : \{0, 1, \ldots, n-1\} \to \{0, 1, \ldots, n-1\}$, defined by the permutation $\pi(A)$ also has the maximum (minimum). The statement of the lemma for $k=2$ and $k=3$ follows from the fact that the only U-permutations of lengths 2 and 3 with maximum are $\pi_2 = (1, 0)$ and $\pi_3 = (1, 2, 0)$, respectively.

For $k \geq 4$, the situation is absolutely different. Thus, one can always find two different U-permutations of length k, e.g., $\pi'_k = (1, 2, \ldots, k-1, 0)$ and

$$\pi''_k = (k/2, k-1, k-2, \ldots, k/2+1, k/2-1, k/2-2, \ldots, 1, 0)$$

for even k or

$$\pi''_k = ((k+1)/2, k-1, k-2, \ldots, (k+1)/2+1, (k+1)/2-1, (k+1)/2-2, \ldots, 1, 0)$$

for odd k. Note that π'_k and π''_k have no nontrivial divisors (for definitions, see Chapter 7).

It is clear that maps $f, g \in \mathcal{E}(T_k)$ have the same type if and only if

$$\pi\left(A^*_{p_2}\left(T^i_k f\right)\right) = \pi\left(A^*_{p_2}\left(T^i_k g\right)\right)$$

for all $i \geq 0$. For any $i \geq 0$, the permutation $\pi\left(A^*_{p_2}\left(T^i_k f\right)\right)$ can be equal to any U-permutation of length k. In particular, it can be equal to π'_k or π''_k. Thus, for any sequence $\{\pi^{(0)}, \pi^{(1)}, \pi^{(2)}, \ldots\}$, where $\pi^{(i)}$ is equal to π'_k or π''_k, $i = 0, 1, 2, \ldots$, one can find a map $f \in \mathcal{E}(T_k)$ such that $\pi\left(A^*_{p_2}\left(T^i_k f\right)\right) = \pi^{(i)}$.

Hence, the cardinality of the set of classes of maps of the same type lying in $\mathcal{E}(T_k)$ is not less than the cardinality of continuum because the set of infinite sequences over a two-letter alphabet has the cardinality of continuum.

For any map $f \in \mathcal{E}(T_k)$, there are three possibilities, namely,

(a) $T_k(f) \in \mathcal{E}_f$;

(b) $T^i_k f \notin \mathcal{E}_f$ for $i = 1, 2, 3, \ldots, n-1$ and $T^n_k f \in \mathcal{E}_f$;

(c) $T^i_k f \notin \mathcal{E}_f$ for all $i \geq 1$.

In each of these cases, for the class $\mathcal{E}_f$, we can, respectively, write

(a) $T_k(\mathcal{E}_f) \subset \mathcal{E}_f$,

(b) $T^i_k(\mathcal{E}_f) \cap \mathcal{E}_f = \varnothing$ for $i = 1, 2, \ldots, n-1$ and $T^n_k(\mathcal{E}_f) \subset \mathcal{E}_f$,

(c) $T^i_k(\mathcal{E}_f) \cap \mathcal{E}_f = \varnothing$ for all $i \geq 1$.

In case (a), it is natural to say that the class $\mathcal{E}_f$ is a fixed class of the operator T_k. In case (b), we say that this class is periodic with period n and, in case (c), we say that it is aperiodic.

Lemma 8.4. *For $k \geq 4$, the operator T_k has periodic classes of all periods.*

Proof. Assume that a map $f \in \mathcal{E}(T_k)$ is such that $\pi\left(A^*_{p_2}(T^i_k f)\right) = \pi'_k$ for $i = nj$ and $\pi\left(A^*_{p_2}(T^i_k f)\right) = \pi''_k$ for $i \neq nj$, $n > 0$, $j = 0, 1, 2, \ldots$, where π'_k and π''_k are the permutations defined in the proof of Lemma 8.3. By virtue of Theorem 2.6, this map exists. Hence, $\mathcal{E}_f$ is a periodic class of the operator T_k with period n.

Lemma 8.5. *For any $k \geq 2$, the number of fixed classes of the operator T_k is finite.*

The proof follows from the fact that, for any $k \geq 2$, there are finitely many different permutations of length k (including finitely many unimodal permutations of length k).

The following assertion is an immediate consequence of the definition of periodic classes of the operator T_k:

Lemma 8.6. *If $\mathcal{E}_f$ is an n-periodic class of the operator T_k, then $\mathcal{E}_f$, $T_k(\mathcal{E}_f)$, $\ldots$, $T_k^{n-1}(\mathcal{E}_f)$ are fixed classes of the operator T_{kn}.*

Corollary 8.1. *For any $n \geq 1$, the operator T_k has finitely many periodic classes of period n.*

The theory of Feigenbaum universality is based, in particular, on the assumption that the operator $T_2 : \mathcal{E}(T_2) \to \mathcal{E}(T_2)$ possesses a unique fixed point $f^* \in \mathcal{E}(T_2)$ globally stable in the space $\mathcal{E}(T_2) \cap G^{(2)}$ and such that $(f^*)''(0) \neq 0$. There are several known methods for proving the existence of the fixed point of the operator T_2 with the indicated properties. We also note that there are papers devoted to the investigation of the spectrum of the operator $DT_2(f^*)$.

In what follows, unless otherwise stated, we always assume that $f \in G^{(2)}$ and T_k: $G^{(2)} \to G^{(2)}$ (in other words, we assume that $f''(0) \neq 0$ for all maps $f \in G^{\omega}$ under consideration). It is not difficult to show that, for any $f \in G^{(2)}$, we have either $T_k f \in G^{(2)}$ or $T_k f \notin G^{\omega}$. In order not to introduce new notation for the intersections of the indicated classes of maps of the same type with the space $G^{(2)}$, we use the same notation both for these objects and for the original classes. It is worth noting that the reasoning presented below is also applicable to the investigation of the operator T_k in the spaces $G^{(2i)}$, $i = 2, 3, \ldots$.

Let $k \geq 2$ and let $\mathcal{E}_f$ be a fixed class of the operator T_k. Suppose that the operator T_k has the following properties:

Property 1. *The operator* T_k *possesses a fixed point* $f^* \in \mathcal{E}_f$ *and this point is a globally attracting fixed point of the operator* T_k *in* $\mathcal{E}_f = \mathcal{E}_{f^*}$.

Property 2. *The operator* $D\,T_k(f^*)$ *has only one simple real eigenvalue* $\delta = \delta(T_k, f^*)$ *which is greater than one; the other eigenvalues belong to the interior of the unit disk.*

Consider the following "surfaces":

$$\Sigma_0 = \{f \in G^{(2)} \mid f(1) = 0\}$$

and

$$\Sigma_1 = \{f \in G^{(2)} \mid f(-1) = f(1) = -1\}.$$

Property 3. *The unstable manifold* $W^u(T_k, f^*)$ *of the operator* T_k *which crosses* f^* *and corresponds to the eigenvalue* $\delta(T_k, f^*)$ *has dimensionality one (i.e., it is a one-parameter family of maps from* $G^{(2)}$*). This family transversally crosses the surfaces* $\mathcal{E}_{f^*}(T_k)$ *(i.e., the stable "manifold" of* T_k*),* Σ_0*, and* Σ_1.

Let F_λ, $\lambda \in [0, 1]$, be a family of maps from $G^{(2)}$ which is sufficiently close to the family $W^u(T_k, f^*)$ and transversally crosses the surface $\mathcal{E}_{f^*}(T_k)$ as $\lambda = \lambda_\infty$. Then, at least for sufficiently large n, there exists a unique value λ_n close to λ_∞ and such that $F_{\lambda_n} \in T_k^{-n}(\Sigma_0)$ and a unique value $\bar{\lambda}_n$ such that $F_{\bar{\lambda}_n} \in T_k^{-n}(\Sigma_1)$. Without loss of generality, we can assume that $\lambda_n < \bar{\lambda}_n$. In this case, $\lambda_\infty \in (\lambda_n, \bar{\lambda}_n)$. If $\lambda \in (\lambda_n, \bar{\lambda}_n)$, then for the phase diagram of the map F_λ, we have $m^*(F_\lambda) \geq n$. Hence, the length of the central branch of the phase diagram of maps from the family F_λ increases with n and the values λ_n and $\bar{\lambda}_n$ correspond to bifurcations in the phase diagram.

Properties 1–3 of the operator T_k immediately imply that

$$\lim_{n\to\infty} \frac{\lambda_n - \bar{\lambda}_n}{\lambda_{n+1} - \bar{\lambda}_{n+1}} = \delta(T_k, f^*).$$

This relation enables us to estimate the measure of the set of values of the parameter for which the map F_λ possesses a cycle of intervals of period k^n. Furthermore, the constant $\alpha(T_k, f^*) = (f^*)^k(0)$ determines the rate of decrease in the sizes of cycles of intervals of periods k^n as n increases. This rate is asymptotically equal to α^{-1}.

It is convenient to represent the family $W^u(T_k, f^*)$ corresponding to the eigenvalue $\delta(T_k, f^*)$ of the operator T_k in the form $W_\lambda(x) = f^*(x) + \lambda\psi(x)$, where $\psi(0) = 0$ and $\psi(1) = 1$.

The following assertion indicates that the constants $\alpha(T_k, f^*)$ and $\delta(T_k, f^*)$ substantially depend on the behavior of the trajectory of the point 0 under the action of the map f^* (i.e., under the action of the relevant fixed point of the operator T_k; see Kolyada and Sivak [2]):

Lemma 8.7. *The following equalities hold:*

$$\alpha(T_k, f^*) = \left[\prod_{i=1}^{k-1} (f^*)'((f^*)^i(0))\right]^{-1},$$

$$\delta(T_k, f^*) = \alpha^{-2}\left[L_0 + \sum_{i=1}^{k-1}\left\{L_i \prod_{j=1}^{i} ((f^*)'((f^*)^j(0)))^{-1}\right\}\right],$$

where

$$L_s = \psi((f^*)^s(\alpha)) \prod_{i=1}^{k-s-1} \left\{\left[(f^*)'((f^*)^{k-i}(\alpha))\right]\left[(f^*)'((f^*)^{k-i}(0))\right]^{-1}\right\}$$

for $s = 0, 1, \ldots, k-2$ *and* $L_{k-1} = \psi((f^*)^{k-i}(\alpha))$.

Proof. We have $T_k f^* = f^*$. By differentiating this identity two times, we obtain

$$(T_k f^*)''(0) = (f^*)''(0).$$

By using the chain rule for differentiation of composite functions and the assumption that $(f^*)''(0) \neq 0$, we immediately arrive at the first equality of the lemma.

To required relation for $\delta(T_k, f^*)$ can be established by using the fact that

$$\delta(T_k, f^*) = \frac{\frac{d}{d\lambda} T_k(W_\lambda(x))}{\frac{d}{d\lambda} W_\lambda(x)}$$

for $\lambda = \lambda_\infty$. Therefore,

$$\delta(T_k, f^*) = \Psi(x, f^*, \psi)$$

for some function Ψ, and this equality can be regarded as a functional equation for finding the map $\psi(x)$ provided that the function f^* and the constant $\delta(T_k, f^*)$ are known. Since $\delta(T_k, f^*)$ is a number, i.e., does not depend on x, the right-hand side of the indicated equality also does not depend on x. By setting $x = 1$ and using the representation for $\alpha(T_k, f^*)$, we arrive at the second equality of Lemma 8.7.

Note that if $\alpha(T_k, f^*)$ is sufficiently small (e.g., for large k), then the constants L_s in the representation of $\delta(T_k, f^*)$ are close to the values $\psi((f^*)^s(0))$.

Now assume that the fixed points f_i^* of a sequence of operators T_{k_i}, $i \geq 1$, converge to a map f_∞^* from the space $G^{(2)}$. Then, for large i, the constants $\alpha(T_{k_i}, f_i^*)$ and $\delta(T_{k_i}, f_i^*)$ substantially depend on the values of the map f_∞^* at the points of the trajectory of the point 0.

As an illustration of this assertion, we consider the case where $k_i = i$, $i = 2, 3, \ldots$, and f_i^* is a fixed point of the operator T_i for which the permutation $\pi(A_{p_2}^*(f_i^*))$ is equal to π_k'' from Lemma 8.3. Fixed points of this sort are called minimal. (For odd $i > 1$, cycles with permutations of the indicated type and the limiting function f_∞^* are depicted in Fig. 47.)

Theorem 8.13. *Let* f_i^* *be minimal fixed points of the operators* T_i, $i = 2, 3, \ldots$, *respectively. Assume that*

$$f_i^* \to f_\infty^* \quad \text{as} \quad i \to \infty \tag{8.4}$$

in the metric of C^0. *Then*

(a)

$$\lim_{i\to\infty} \frac{\alpha(T_i, f_i^*)}{\alpha(T_{i+1}, f_{i+1}^*)} = (f_\infty^*)'(x^*) \overset{\text{def}}{=} \gamma,$$

where x^* *is the fixed point of the map* f_∞^*, *and*

(b)

$$\lim_{i\to\infty} \frac{\delta(T_{i+1}, f_{i+1}^*)}{\delta(T_i, f_i^*)} = [(f_\infty^*)'(x^*)]^2 \overset{\text{def}}{=} \gamma^2.$$

Proof. The required equalities follow from the structure of trajectories of the point 0 of the minimal fixed points f_i^* and from the representations of α and δ in Lemma 8.7. Indeed, for large i, the right-hand sides of equalities for $\alpha(T_i, f_i^*)$ differ by the number of multipliers close to $\gamma_i = (f_\infty^*)'(x_i^*)$, where x_i^* is the fixed point of the map

f_i^*. Hence, by using Lemma 8.7, we arrive at the equalities of Theorem 8.13.

A similar result for another sequence of fixed points of the operators T_i was established by Eckmann, Epstein, and Wittwer [1]. They considered the sequence of maximal fixed points $\tilde{f}_i^*$ of the operators T_i (a fixed point $\tilde{f}_i^*$ of the operator T_i is called maximal if $\pi(A_{p_2}^*, (\tilde{f}_i^*)) = \pi_i$, where π_i is defined in Lemma 8.3 and equal to $(1, 2, 3, \ldots, i-1, 0)$). Moreover, Eckmann, Epstein, and Wittwer [1] proved that, in this case, the map $1 - 2x^2$, $x \in [-1, 1]$, is the limiting map $\tilde{f}_i^*$. By using their arguments, one can show that condition (8.4) in Theorem 8.13 is satisfied and $f_\infty^* = 1 - \lambda_\infty x^2$, where λ_∞ is the value of the parameter λ for which $F_\lambda(x) = 1 - \lambda x^2$ satisfies the conditions $F_\lambda(1) < 0$ and $F_\lambda^2(1) = x^*$, where x^* is a fixed point of the map F_λ lying to the right of the origin.

The following assertion seems to be true: If a sequence $\{f_i^*\}_{i=2}^\infty$ of fixed points of the operators T_i has a limiting point f_∞^* in the metric of C^0, then $f_\infty^* = 1 - \lambda x^2$ for some $\lambda \in (0, 2)$.

This enables us to conclude that

$$W^u(T_i, f_i^*) \approx 1 - \lambda x^2 \quad \text{and} \quad \frac{d}{d\lambda} W^u(T_i, f_i^*) \approx x^2$$

for large i if we use parametrization introduced above.

The equalities of Theorem 8.13 are corroborated by the results of numerical calculation of the relevant constants for families of quadratic maps. Below, we present the corresponding results for the constants $\alpha(T_i, f_i^*)$ and $\delta(T_i, f_i^*)$. The numerical value of the constant γ is approximately equal to 1.71.

i	α^{-1}	δ	i	α^{-1}	δ
3	$9.27 \cdot 10^0$	$5.52 \cdot 10^1$	4	$6.26 \cdot 10^0$	$2.18 \cdot 10^1$
5	$2.01 \cdot 10^1$	$2.55 \cdot 10^2$	6	$2.09 \cdot 10^1$	$2.18 \cdot 10^2$
7	$4.91 \cdot 10^1$	$1.44 \cdot 10^3$	8	$6.63 \cdot 10^1$	$2.30 \cdot 10^3$
9	$1.29 \cdot 10^2$	$9.60 \cdot 10^3$	10	$1.97 \cdot 10^2$	$2.10 \cdot 10^4$
11	$3.52 \cdot 10^2$	$7.00 \cdot 10^4$	12	$5.68 \cdot 10^2$	$1.77 \cdot 10^5$
13	$9.78 \cdot 10^2$	$5.35 \cdot 10^5$	14	$1.61 \cdot 10^3$	$1.44 \cdot 10^6$
15	$2.74 \cdot 10^3$	$4.18 \cdot 10^6$	16	$4.57 \cdot 10^3$	$1.15 \cdot 10^7$
17	$7.70 \cdot 10^3$	$3.30 \cdot 10^7$	18	$1.28 \cdot 10^4$	$9.24 \cdot 10^7$
19	$2.21 \cdot 10^4$	$2.56 \cdot 10^8$			

For large i, computation becomes much more complicated because the sizes of cycles of intervals $A^*_{p_2}(f^*_i)$ rapidly decrease (the rate of this process is equal to γ) and the constants $\delta(T_i, f^*_i)$ rapidly increase as i increases. Computations were carried out for the family of quadratic maps $\lambda x(1-x)$, $x \in [0, 1]$, $\lambda \in [0, 4]$, which is equivalent to the family $1-\lambda x^2$, $x \in [0, 1]$, $\lambda \in [0, 2]$.

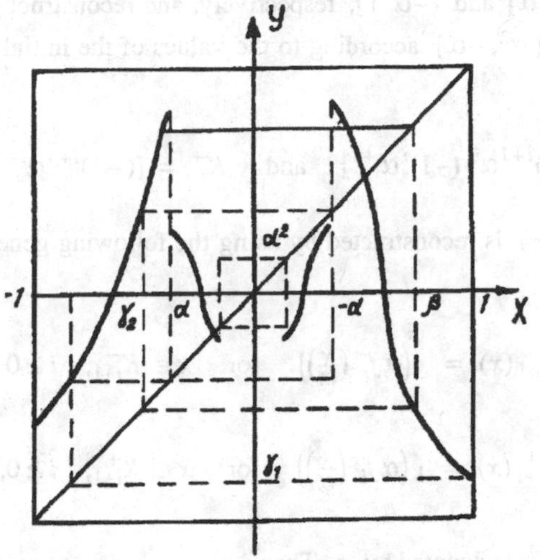

Fig. 52

Let us now study the problem of existence of solutions of the equation $T_k f = f$ in the space of unimodal maps and investigate some properties of these solutions.

First, we describe a method for the construction of solutions of the functional equation $\alpha^{-1} f^k(\alpha x) = f(x)$, where α is a nonzero constant whose absolute value is less than one. To avoid cumbersome explanations, we consider the case $k = 3$ as an example. (Note that a method for the construction of even solutions in the case $k = 2$ was described by Cosnard and Eberhard [2]).

We choose $\alpha \in (-1, 0)$. Generally speaking, the choice of the sign of the constant α depends on the type of a solution to be constructed. Thus, for minimal fixed points f^*_i of the operators T_i, $i = 2, 3, \ldots$, the constants $\alpha(T_i, f^*_i)$ are positive for all even $i > 2$ and negative for $i = 2$ and all odd $i > 2$. For maximal fixed points $\tilde{f}^*_i$, the constants $\alpha(T_i, \tilde{f}^*_i)$ are negative for all $i \geq 2$.

Let $k = 3$ and let $\alpha \in (-1, 0)$ be an arbitrary fixed number. We define $J_0 = [\alpha, -\alpha]$, $J_1 = [\beta, 1]$, and $J_2 = [\gamma_1, \gamma_2]$, where $-1 < \gamma_1 < \gamma_2 < \alpha < 0 < -\alpha < \beta < 1$. We construct a map f^* such that the intervals J_0, J_1, and J_2 form a cycle of intervals of period 3, $f^*(J_1) = J_2$, and $f^*(J_2) = J_0$. If we want to construct a unimodal map f^*, then it is necessary to require that $f^*(1) = \gamma_1$, $f^*(\beta) = \gamma_2$, $f^*(\gamma_1) = \alpha$, and $f^*(\gamma_2) = -\alpha$ (Fig. 52).

On the interval $[-1, \alpha]$, we define a continuous monotonically increasing function f_0^- such that $f_0^-(\gamma_1) = \alpha$, $f_0^-(\gamma_2) = -\alpha$, and $f_0^-(\alpha) > \beta$. Similarly, on $[-\alpha, 1]$, we define a continuous monotonically decreasing map f_0^+ such that $f_0^+(-\alpha) > \beta$, $f_0^+(\beta) = \gamma_2$, and $f_0^+(1) = \gamma_1$. Let us now construct a solution which coincides with f_0^- and f_0^+ in the intervals $[-1, \alpha]$ and $[-\alpha, 1]$, respectively, and reconstruct $f^*(x)$ in the intervals $[\alpha, -\alpha^2]$ and $[\alpha^2, -\alpha]$ according to the values of the initial functions f_0^- and f_0^+. Denote

$$K_i^- = [(-1)^{i+1}\alpha^i, (-1)^i\alpha^{i+1}] \quad \text{and} \quad K_i^+ = [(-1)^{i+1}\alpha^{i+1}, (-1)^i\alpha^i].$$

The solution $f^*(x)$ is reconstructed by using the following general recurrence relations:

$$\begin{aligned} f_{i+1}^-(x) &= g\left(\alpha f_i^+\left(\tfrac{x}{\alpha}\right)\right) \quad \text{for} \quad x \in K_{i+1}^-, \quad i \geq 0, \\ f_{i+1}^+(x) &= g\left(\alpha f_i^-\left(\tfrac{x}{\alpha}\right)\right) \quad \text{for} \quad x \in K_{i+1}^+, \quad i \geq 0, \end{aligned} \tag{8.5}$$

where $(f_0^+)^{-1} \circ (f_0^-)^{-1}$ is denoted by g. Then

$$f^*(x) = \begin{cases} 1, & x = 0, \\ f_i^+(x), & x \in K_i^+, \quad i \geq 0, \\ f_i^-(x), & x \in K_i^-, \quad i \geq 0, \end{cases} \tag{8.6}$$

One can easily check that (8.6) is a solution of the equation $T_3 f = f$. To guarantee the continuity of the solution under consideration, it is necessary that $f_1^-(\alpha) = f_0^-(\alpha)$ and $f_1^+(-\alpha) = f_0^+(-\alpha)$. As follows from the recurrence relations (8.5), for the initial functions f_0^- and f_0^+, these equalities are equivalent to

$$\begin{aligned} f_0^-(\alpha) &= g(\alpha f_0^+(1)), \\ f_0^+(\alpha) &= g(\alpha f_0^-(-1)). \end{aligned} \tag{8.7}$$

Since conditions (8.7) establish the correspondence between f_0^- and f_0^+ only at finitely many points, one can always find initial functions satisfying these conditions. The following lemma demonstrates that conditions (8.7) do not guarantee the continuity of the map $f^*(x)$:

Lemma 8.8. *If conditions (8.7) are satisfied for continuous initial functions, then the map f^* is continuous on $[-1, 1]\backslash\{0\}$. At the point 0, the map f^* may be discontinuous.*

Proof. To prove that the map f^* is continuous on $[-1, 1]\backslash\{0\}$, it suffices to show that $f_i^+(x) = f_{i+1}^+(x)$ for $x \in K_i^+ \cap K_{i+1}^+$ and $f_i^-(x) = f_{i+1}^-(x)$ for $x \in K_i^- \cap K_{i+1}^-$. It follows from (8.5) that the equality $f_i^+(x) = f_{i+1}^+(x)$ is equivalent to the equality

$$g\left(\alpha f_{i-1}^-\left(\frac{x}{\alpha}\right)\right) = g\left(\alpha f_i^-\left(\frac{x}{\alpha}\right)\right).$$

Since the map g is monotone, the required equality for f_i^+ reduces to the analogous equality for f_{i-1}^- and the equality for f_i^- reduces to the equality for f_{i-1}^+. Hence, the required assertion can be obtained from conditions (8.7) by induction.

In order to prove that the map f^* may be discontinuous at the point 0, we consider the map $\alpha g: [+\alpha, -\alpha] \to [+\alpha, -\alpha]$. Note that $\alpha g(\alpha) = \alpha$ and αg monotonically increases in the interval $[+\alpha, -\alpha]$.

It follows from (8.5) that

$$f_i^+(x) = \alpha^{-1}(\alpha g)^i\left(\alpha f_0^{\pm}\left(\frac{x}{\alpha_i}\right)\right), \quad x \in K_i^+,$$

$$f_i^-(x) = \alpha^{-1}(\alpha g)^i\left(\alpha f_0^{\mp}\left(\frac{x}{\alpha_i}\right)\right), \quad x \in K_i^-,$$

where $f_0^{\pm}$ and $f_0^{\mp}$ denote either f_0^+ or f_0^- (depending on i). These equalities imply that $f_i^+(x)$ and $f_i^-(x)$ approach 1 if and only if the map αg has no fixed points in the interval $[\alpha, -\alpha]$ except α. It is clear that one can easily find initial functions f_0^- and f_0^+ for which this condition is not satisfied.

As follows from the proof of Lemma 8.8, for the solution $f^*(x)$ to be continuous, it suffices to require that conditions (8.7) be satisfied and the initial functions f_0^- and f_0^+ be convex upward. In this case, αg has no fixed points other than α because it is convex downward and $\alpha g(-\alpha) < -\alpha$. Note that the requirement of convexity of the initial functions restricts the choice of γ_1, γ_2, and β but the construction of the required solution is still possible.

Lemma 8.9. *Assume that initial functions f_0^- and f_0^+ and the functions $(f_0^-)^{-1}$ and $(f_0^+)^{-1}$ belong to C^2. Let f_0^- and f_0^+ be convex upward and satisfy conditions (8.7) and the conditions*

$$\frac{d^i f_0^-}{dx^i}(\alpha) = \frac{d^i f_1^-}{dx^i}(\alpha)$$

and

$$\frac{d^i f_0^-}{dx^i}(-\alpha) = \frac{d^i f_1^+}{dx^i}(-\alpha)$$

for $i = 1, 2$. *If, in addition,* $(f_0^-)'(\gamma_1)(f_0^+)'(1) < \alpha^{-1}$, *then* $f^* \in C^2([-1, 1])$. *Moreover,*

$$\frac{d^2 f^*}{dx^2}(0) = 0.$$

Proof. As in the proof of Lemma 8.8, by using the recurrence relations (8.5) and the conditions of Lemma 8.9, one can easily establish the existence and continuity of the corresponding derivatives everywhere in the interval $[-1, 1]$ except 0 because, in this case, the proof of the existence and continuity of these derivatives can be reduced to the proof of their existence and continuity at the points α and $-\alpha$ and in the intervals $[-1, \alpha]$ and $[-\alpha, 1]$, i.e., to the conditions of the lemma.

Since initial functions are convex upward, we have $f^*(0) = 1$. Let us show that $(f^*)'(0) = 0$.

It follows from relations (8.5) that

$$(f_i^+)'(x) = g'\left(\alpha f_{i-1}^-\left(\tfrac{x}{\alpha}\right)\right)(f_{i-1}^-)'\left(\tfrac{x}{\alpha}\right), \quad x \in K_i^+,$$

$$(f_i^-)'(x) = g'\left(\alpha f_{i-1}^+\left(\tfrac{x}{\alpha}\right)\right)(f_{i-1}^+)'\left(\tfrac{x}{\alpha}\right), \quad x \in K_i^-.$$

Denoting $(f_i^+)'(x)$ by z_i, $(f_i^-)'(x)$ by z_{i-1}, and $g'\left(\alpha f_{i-1}^-\left(\frac{x}{\alpha}\right)\right)$ by ε_i, we obtain a recurrence relation of the form $z_i = \varepsilon_i z_{i-1}$.

For large i, the argument of the function g' is close to α (because $f^*(0) = 1$ and the map f^* is continuous). Hence, $\varepsilon_i \to g'(\alpha)$ as $i \to \infty$. Since the function g is convex upward and continuously differentiable, we have $-1 < g'(\alpha) < 0$. One can easily show that, in this case, all sequences satisfying the recurrence relation established above converge to zero independently of their initial values and, consequently, the derivative of the map f^* at the point 0 exists and is continuous.

For the second derivative, relations (8.5) imply that

$$(f_i^+)''(x) = g''\left(\alpha f_{i-1}^-\left(\tfrac{x}{\alpha}\right)\right)\left[(f_{i-1}^-)'\left(\tfrac{x}{\alpha}\right)\right]^2 + g'\left(\alpha f_{i-1}^-\left(\tfrac{x}{\alpha}\right)\right)\tfrac{1}{\alpha}(f_{i-1}^-)''\left(\tfrac{x}{\alpha}\right),$$

$$(f_i^-)''(x) = g''\left(\alpha f_{i-1}^+\left(\tfrac{x}{\alpha}\right)\right)\left[(f_{i-1}^+)'\left(\tfrac{x}{\alpha}\right)\right]^2 + g'\left(\alpha f_{i-1}^+\left(\tfrac{x}{\alpha}\right)\right)\tfrac{1}{\alpha}(f_{i-1}^+)''\left(\tfrac{x}{\alpha}\right).$$

In this case, we set

$$z_i = (f_i^+)''(x), \quad z_{i-1} = (f_{i-1}^-)''\left(\frac{x}{\alpha}\right), \quad \text{and} \quad k_i = \frac{1}{\alpha}\, g'\left(\alpha f_{i-1}^-\left(\frac{x}{\alpha}\right)\right)$$

and denote the first term on the right-hand side of the first equality by ε_i. This gives the recurrence relation $z_i = k_i z_{i-1} + \varepsilon_i$. As shown above, f^* belongs to C^1 and $(f^*)'(0) = 0$. Therefore, $\varepsilon_i \to 0$ as $i \to \infty$. The results established above also imply that $k_i \to \alpha^{-1} g'(\alpha)$ as $i \to \infty$. By the condition of the lemma, we have

$$g'(\alpha) = \left[(f_0^-)'(\gamma_1)\,(f_0^+)'(1)\right]^{-1} > \alpha$$

and, consequently, $0 < \alpha^{-1} g'(\alpha) < 1$. It is easy to see that all sequences satisfying this recurrence relation converge to zero, whence it follows that $(f^*)''(0) = 0$ and, therefore, $f^* \in C^2([-1, 1])$.

By using the proof of Lemma 8.9, one can easily show that if initial functions are convex upward, belong to C^r, $r \geq 1$, together with their inverse functions, and satisfy conditions (8.7), the equalities

$$\frac{d^i f_0^-}{dx^i}(\alpha) = \frac{d^i f_1^-}{dx^i}(\alpha)$$

and

$$\frac{d^i f_0^+}{dx^i}(-\alpha) = \frac{d^i f_1^+}{dx^i}(-\alpha)$$

for all $i \leq r$, and the inequality

$$\left|(f_0^-)'(\gamma_1)\,(f_0^+)'(1)\right| > |\alpha^{-r}|,$$

then $f^* \in C^r([-1, 1])$. In this case,

$$\frac{d^i f^*}{dx^i}(0) = 0$$

for all $i \leq r$.

Thus, the equation $T_k f = f$ may have infinitely many different unimodal solutions with degenerate critical point from the class C^r. If $(f_0^+)'(1) = -\infty$, then (under the corresponding conditions of continuity and compatibility imposed on the initial functions) we can construct solutions of the equation $T_k f = f$ from the class $C^\infty([-1, 1])$

with infinitely degenerate critical point. Since the existence of solutions with nondegenerate critical point is usually postulated in the theory of universal behavior for families of one-dimensional maps, we do not present a detailed description of the procedure used to construct C^∞-solutions. We also note that, in view of relations (8.5) and compatibility conditions imposed on the initial functions, the application of this procedure to the construction of solutions with nondegenerate critical point leads to a system of functional equations which is in no case simpler than the original equation.

Consider some properties of solutions with nondegenerate critical point.

Lemma 8.10. *Let f be a C^r-unimodal map which solves the equation $T_k f = f$ and let $(f^*)''(0) \neq 0$. Then $\frac{d^i f^*}{dx^i}(0) = 0$ for all odd $i \leq r$.*

Proof. For simplicity, we consider only the case $k = 3$. In all other cases, the proof is similar.

First, we differentiate the identity $(T_3 f^*)(x) = f^*(x)$ i times. This gives

$$\frac{d^i}{dx^i} f^*(x) = \sum_{j=0}^{i-1} c_j \frac{d^j}{dx^j}\left[\tilde{g}'(f^*(\alpha x))\right] \frac{d^{i-j}}{dx^{i-j}}\left[f^*(\alpha x)\right], \tag{8.8}$$

where $\tilde{g} = (f^*)^2$. For $i = 2$ and $x = 0$, we obtain $(f^*)''(0) = \alpha \tilde{g}'(1)(f^*)''(0)$ (here, we have used the facts that $f^*(0) = 1$ and $(f^*)'(0) = 0$). Therefore, the inequality $(f^*)''(0) \neq 0$ implies that $\tilde{g}'(1) = \alpha^{-1}$.

Assume that i is odd and greater than 2. Then, for all $j > 0$, we can write

$$c_j \frac{d^j}{dx^j}\left[\tilde{g}'(f^*(\alpha x))\right] \frac{d^{i-j}}{dx^{i-j}}\left[f^*(\alpha x)\right] = 0 \quad \text{for} \quad x = 0.$$

This equality can be proved by induction on i. Actually, assume that $\frac{d^s f^*}{dx^s}(0) = 0$ for all odd $s < i$ and that i is odd. If $i - j$ is odd, then $\frac{d^{i-j}}{dx^{i-j}} f^*(\alpha x) = 0$ for $x = 0$. Otherwise, j is odd and then $\frac{d^j}{dx^j}\left[\tilde{g}'(f^*(\alpha x))\right] = 0$ for $x = 0$ because

$$\frac{d^j}{dx^j}\left[\tilde{g}'(f^*(\alpha x))\right] = \sum_{l=0}^{j-1} c_l \frac{d^l}{dx^l}\left[\tilde{g}''(f^*(\alpha x))\right] \frac{d^{j-l}}{dx^{j-l}}\left[f^*(\alpha x)\right],$$

i.e., we arrive at an equality similar to (8.8) for $j < i$. Acting as above, we apply the assumption of induction to some terms and transform the other terms into an expression of the form $c_0 \tilde{g}^{(\xi)}(f^*(\alpha x))(f^*)'(\alpha x)$, where $\tilde{g}^{(\xi)}$ denotes the ξth derivative of the map

$\tilde{g}$. This expression is also equal to zero for $x = 0$ because $(f^*)'(0) = 0$ and, hence, equality (8.8) can be rewritten in the form

$$\frac{d^i f^*}{dx^i}(0) = \tilde{g}'(f^*(\alpha x)) \frac{d^{i-1}}{dx^{i-1}}((f^*)'(\alpha x))\Big|_{x=0} = \alpha^{i-2} \frac{d^i f^*}{dx^i}(0).$$

This equality implies the required assertion because $\alpha^{i-2} \neq 1$ for $i > 2$.

For analytic solutions of equations $T_k f = f$, Lemma 8.10 yields the following theorem:

Theorem 8.14. *Assume that a map* $f^* : [-1, 1] \to [-1, 1]$ *belongs to the class* $G^{(2)}$ *and is a fixed point of the operator* T_k, $k \geq 2$. *Then* $f^*(x)$ *is an even function, i.e.,* $f^*(-x) = f^*(x)$ *for any* $x \in [-1, 1]$.

Note that if Properties 1–3 of the operator T_k in the vicinity of its fixed point f^* were proved, then the assertion of Theorem 8.14 would follow from the fact that any fixed class $\mathcal{E}(T_k)$ of the operator T_k contains even functions (parallel with functions of other types) and the evenness of maps is preserved under the action of the operator T_k.

Theorem 8.14 enables us to restrict the investigation of the operators T_k to the subspace of even functions $\tilde{G}^{(2)}$ of the space $G^{(2)}$. By using the Schauder theorem, Lanford [1], [2] proved that the equation $T_2 f = f$ is solvable in the space $\tilde{G}^{(2)}$. It is not difficult to show that the existence of fixed points of various types for the operators T_k with $k > 2$ can be proved by using the same method as in the indicated works. (Recall that, according to Lemma 8.3, all fixed points of the operator T_2 are of the same type.) Thus, we can regard the assertion that, for any $k \geq 2$, any fixed class $\mathcal{E}_f(T_k)$ of the operator T_k contains a fixed point of this operator as an established fact.

It is worth noting that the result of Paluba mentioned in the previous section (Theorem 8.12; Paluba [1]) is an analog of the Hermann theorem on smoothness of the conjugation of a diffeomorphism of a circle with the corresponding rotation. It seems likely that Paluba's result can be generalized to the case of arbitrary stable manifolds $\mathcal{E}_{f^*}(T_k)$.

The results of numerical simulation demonstrate that the constant $\delta = 4.6692\ldots$ characterizes the properties of period doubling bifurcations not only in one-dimensional dynamical systems. Sequences of period doubling bifurcations exhibiting the property of Feigenbaum universality were detected as a result of numerical analysis of the Lorenz model, the Hénon map, and some other systems. For these systems, we observe local expansion in one direction and contraction in all other directions. The dynamics of systems of this sort is fairly similar, in a certain sense, to the dynamics of one-dimensional systems. In particular, in this case, we also observe an infinite sequence of period doubling bifurcations appearing at the same rate as in the one-dimensional case.

For families of two-dimensional area-preserving maps, period doubling bifurcations are characterized by the same asymptotic law $\lambda_\infty - \lambda_n \sim \text{const}\, \delta^{-n}$ but with $\delta = 8.72\ldots$.

In many-dimensional case, parallel with period doubling bifurcations, we observe period tripling bifurcations, period quadrupling bifurcations, etc. It is natural to expect that these bifurcations may also exhibit universal properties for some families of maps. Thus, for period tripling bifurcations in a family $f(z, \mu)$ of maps of $\mathbb{C}^1$ into itself depending on the complex parameter μ, we have $\mu_\infty - \mu_n \sim \text{const}\, \delta_3^{-n}$, where $\delta_3 = 4.600 + 8.981\,i$ (see Wul, Sinai, and Khanin [1]).

REFERENCES

Adler, R., Konheim, A., and McAndrew, M.

[1] *Topological entropy,* Trans. Amer. Math. Soc. **114** (1965), 309–319.

Agronsky, S. J., Bruckner, A. M., Ceder, J. G., and Pearson, T. L.

[1] *The structure of ω-limit sets for continuous functions,* Real Analysis Exchange **15** (1989/90), 483–510.

Agronsky, S. J., Bruckner, A. M., and Laczkovich, M.

[1] *Dynamics of typical continuous functions,* J. London Math. Soc. **40** (1989), 227–243.

Aleksandrov, P. S.

[1] Introduction to the Theory of Sets and General Topology, Nauka, Moscow, 1977 *(in Russian).*

Alekseev, V. M.

[1] Symbolic Dynamics, 11th Mathematical School, Institute of Mathematics, Ukrainian Academy of Sciences, Kiev, 1976, pp. 5–210 *(in Russian).*

Aliev, S. Ya., Ivanov, A. F., Maistrenko, Yu. L., and Sharkovsky, A. N.

[1] *Singular perturbations of difference equations with continuous time,* Preprint No. 84.33, Inst. Mat. Akad. Nauk Ukrain. SSR, Kiev, 1984 *(in Russian).*

Allwright, D.

[1] *Hypergraphic functions and bifurcations in recurrence relations,* SIAM J. Appl. Math. **34** (1978), 687–691.

Alseda, L. and Fedorenko, V.

[1] *Simple and complex dynamics for circle maps,* Publicacions Matematiqus, UAB, 37, 1993, 305–316.

Alseda, L., Llibre, J., and Misiurewicz, M.

[1] Combinatorial Dynamics and Entropy in Dimension One, Advanced Series in Nonlinear Dynamics, 5, World Scientific, Singapore, 1993.

Alseda, L., Llibre, J., and Serra, R.

[1] *Minimal periodic orbits for continuous maps of the interval,* Trans. Amer. Math. Soc. **286** (1984), 595–627.

Anosov, D. V., Aranson, S. Kh., Bronshtein, I. U., and Grines, V. Z.

[1] Smooth Dynamical Systems, VINITI Series on Contemporary Problems in Mathematics. Fundamental Directions, vol. 1, VINITI, Moscow, 1985, pp. 151–242 *(in Russian).*

Arneodo, A., Ferrero, P., and Tresser, S.

[1] *Sharkovskii's order for the appearance of superstable cycles in one parameter families of simple real maps: an elementary proof,* Commun. Pure Appl. Math. **XXXVII** (1984), 13–17.

Arnold, V. I., Afraimovich, V. S., Il'yashenko, Yu. S., and Shilnikov, L. P.

[1] Theory of Bifurcations, VINITI Series on Contemporary Problems in Mathematics. Fundamental Directions, vol. 5, VINITI, Moscow, 1986, pp. 5–218 *(in Russian); **English transl.:*** Dynamical Systems V: Bifurcation Theory and Catastrophe Theory, Encyclopedia of Mathematical Sciences, Springer, 5, 1994.

Babenko, K. I. and Petrovich, V. Yu.

[1] *On demonstrative numerical calculations,* Preprint No. 133, Keldysh Institute of Applied Mathematics, Academy of Sciences of the USSR, Moscow, 1983 *(in Russian).*

Barkovsky, Yu. S. and Levin, G. M.

[1] *On the limiting Cantor set,* Uspekhi Mat. Nauk **35** (1980), No. 2, 201–202; ***English transl.:*** Russian Math. Surveys **35** (1980).

Barna, B.

[1] *Über die Iteration relleer Funktionen,* Publ. Math. Debrecen. I: **7** (1960), 16–40; II: **13** (1966), 169–172; III: **22** (1975), 269–278.

Birkhoff, G. D.

[1] Dynamical Systems, Amer. Math. Soc., Colloquium Publications, IX, Providence, R. I., 1966.

Block, L.

[1] *Homoclinic points of mapping of the interval,* Proc. Amer. Math. Soc. **72** (1978), 576–580.

[2] *Simple periodic orbits of continuous mappings of the interval,* Trans. Amer. Math. Soc. **254** (1979), 391–398.

[3] *Stability of periodic orbits in the theorem of Sharkovskii,* Proc. Amer. Math. Soc. **81** (1981), 333–336.

Block, L. and Coppel, W. A.

[1] *Stratification of continuous maps of an interval,* Australian National University, Canberra (Res. rep. No. 32), 1984.

[2] *Dynamics in one dimension,* Springer Lect. Notes Math. **1513** (1992).

Block, L. and Franke, J. E.

[1] *The chain recurrent set for maps of the interval,* Proc. Amer. Math. Soc. **87** (1983), 723–727.

Block, L., Guckenheimer, J., Misiurewicz, M., and Young, L. S.

[1] *Periodic points and topological entropy of one-dimensional maps,* Springer Lect. Notes Math. **819** (1980), 18–34.

Block, L. and Hard, D.

[1] *Stratification of the space of unimodal interval maps,* Ergod. Theory Dynam. Syst. **3** (1983), 553–539.

Blokh, A. M.

[1] *Limiting behavior of one-dimensional dynamical systems,* Uspekhi Mat. Nauk **37** (1982), No. 1, 137–138; ***English transl.:*** Russian Math. Surveys **37** (1982).

[2] *Sensitive maps of an interval,* Uspekhi Mat. Nauk **37** (1982), No. 1, 189–190; ***English transl.:*** Russian Math. Surveys **37** (1982).

[3] *On the spectral decomposition for piecewise-monotone maps of an interval,* Uspekhi Mat. Nauk **37** (1982), No. 3, 175–176; ***English transl.:*** Russian Math. Surveys **37** (1982).

[4] *Decomposition of dynamical systems on an interval,* Uspekhi Mat. Nauk **38** (1983), No. 5, 179–180; ***English transl.:*** Russian Math. Surveys **38** (1983).

Blokh, A. M. and Lyubich, M. Yu.

[1] *Attractors of transformations of an interval,* Funk. Anal. Prilozh. **21** (1987), 70–71; ***English transl.:*** Funct. Anal. Appl. **21** (1987).

Bowen, R.

[1] *Entropy for group endomorphisms and homogeneous spaces,* Trans. Amer. Math. Soc. **153** (1971), 401–414.

[2] *Periodic points and measures for axiom A on diffeomorphisms,* Trans. Amer. Math. Soc. **154** (1971), 377–397.

[3] Methods of Symbolic Dynamics, Mir, Moscow, 1979 *(Russian translation).*

Bowen, R. and Franks, J.

[1] *The periodic points of maps of the disk and interval,* Topology **15** (1976), 337–342.

Bronshtein, I. U. and Burdaev, V. P.

[1] *Chain recurrence and extensions of dynamical systems,* Algebraic Invariants of Dynamical Systems. Mat. Issled., **55** (1980), 3–11 *(in Russian).*

Bunimovich, L. A., Pesin, Ya. B., Sinai, Ya. G., and Yakobson, M. V.

[1] *Ergodic Theory of Smooth Dynamical Systems,* VINITI Series on Contemporary Problems in Mathematics. Fundamental Directions, vol. 2, VINITI, Moscow, 1985, pp. 113–1231 *(in Russian).*

Burkart, U.

[1] *Interval mapping graphs and periodic points of continuous functions,* J. Combin. Theory, Ser. B. **32** (1982), 57–68.

Campanino, M. and Epstein, H.

[1] *On the existence of Feigenbaum's fixed point,* Commun. Math. Phys. **79** (1981), 261–302.

Collet, P. and Eckman, J.-P.

[1] Iterated Maps on the Interval as Dynamical Systems, Birkhäuser, Boston, 1980.

Collet, P., Eckman, J.-P., and Lanford, O. E.

[1] *Universal properties of maps on an interval,* Commun. Math. Phys. **76** (1980), 211–254.

Coppel, W. A.

[1] *Sharkovskii-minimal orbits,* Math. Proc. Cambr. Phil. Soc. **93** (1983), 397–408.

[2] *Maps of an interval,* Preprint No. 26 Institute of Mathematics and Its Applications, University of Minnesota, Minneapolis, Minnesota (1983).

Cosnard, M. Y. and Eberhard, A.

[1] *Sur les cycles d'une application continue de la variable reale,* Sem. Anal. Num. 274, USMG Lab. Math. Appl. Grenoble, (1977).

[2] *On the use of renormalizations technics in the study of discrete population in one dimension,* Lect. Notes Biomath. **49** (1983), 217–226.

Coven, E. M. and Hedlund, G. A.

[1] *Continuous maps of the interval whose periodic points form a closed set,* Proc. Amer. Math. Soc. **79** (1980), 127–133.

[2] *$\overline{P} = \overline{R}$ for maps of the interval,* Proc. Amer. Math. Soc. **79** (1980), 316–318.

Coven, E. M. and Nitecki, Z.

[1] *Non-wandering sets of the powers of maps of the interval,* Ergod. Theory Dynam. Syst. **1** (1981), 9–31.

Derrida, B., Gervois, A., and Pomeau, Y.

[1] *Iterations of endomorphisms on the real axis and representations of numbers,* Ann. Inst. Henri Poincaré **29** (1978), 305–356.

Dinaburg, E. I.

[1] *Connection with various entropy characterizations of dynamical systems,* Izv. Akad. Nauk SSSR **35** (1971), 324–366 *(in Russian).*

Dobrynsky, V. A.

[1] *Typicalness of dynamical systems with a stable prolongation,* Dynamical Systems and Problems of Stability of Solutions of Differential Equations, Inst. Mat. Akad. Nauk Ukrain. SSR, Kiev, 1973, pp. 43–53 *(in Russian).*

Dobrynsky, V. A. and Sharkovsky, A. N.

[1] *Generality of dynamical systems almost all trajectories of which are stable under constantly acting perturbations,* Dokl. Akad. Nauk SSSR **211** (1973), 273–276; ***English transl.:*** Soviet Math. Dokl. **211** (1973).

Du, B. S.

[1] *The minimal number of periodic orbits guaranteed in Sharkovskii's theorem,* Bull. Austral. Math. Soc. **31** (1985), 89–103.

Eckman, J.-P., Epstein, H., and Wittwer, P.

[1] *Fixed points of Feigenbaum's type for the equation* $f^p(\lambda x) = \lambda f(x)$, Commun. Math. Phys. **93** (1984), 495–516.

Fatou, P.

[1] *Sur les equations fonctione,* Bull. Soc. Math. de France **47** (1919), 161–270; **48** (1920), 33–95, 208–314.

Fedorenko, V. V.

[1] *A partially ordered set of types of periodic trajectories of one-dimensional dynamical systems,* Dynamical Systems and Differential–Difference Equations, Inst. Mat. Akad. Nauk Ukrain. SSR, Kiev, 1986, pp. 90–97 *(in Russian).*

[2] *Continuous maps of the interval exhibiting a closed set of almost periodic points,* Studies on Theoretical and Application Problems in Mathematics, Inst. Mat. Akad. Nauk Ukrain. SSR, Kiev, 1986, p. 79 *(in Russian)*

[3] *Simple one-dimensional dynamical systems,* Differential Equations and Their Applications to Nonlinear Boundary-Value Problems, Inst. Mat. Akad. Nauk Ukrain. SSR, Kiev, 1987 *(in Russian).*

[4] *One-dimensional dynamical systems stable in the sense of Lyapunov on every minimal set,* Dynamical Systems and Turbulence, Inst. Mat. Akad. Nauk Ukrain. SSR, Kiev, 1989 *(in Russian).*

[5] *Dynamics of interval maps with zero topological entropy on the sets of chain-recurrent points,* Dynamical Systems and Nonlinear Phenomena, Inst. Mat. Akad. Nauk Ukrain. SSR, Kiev, 1990 *(in Russian).*

[6] *Classification of simple one-dimensional dynamical systems,* Preprint No. 91.5, Inst. Mat. Akad. Nauk Ukrain. SSR, Kiev, 1991 *(in Russian).*

Fedorenko, V. V. and Sharkovsky, A. N.

[1] *On the existence of periodic and homoclinic trajectories,* Abstracts of V All-Union Conference on Qualitative Theory of Differential Equations, Stiinça, Kishinev, 1979, pp. 174–175 *(in Russian).*

[2] *Continuous maps of the interval whose periodic points form a closed set,* Studies of Differential and Differential–Difference Equations, Inst. Mat. Akad. Nauk Ukrain. SSR, Kiev, 1980, pp. 137–145 *(in Russian).*

Fedorenko, V. V., Sharkovsky, A. N., and Smital, J.

[1] *Characterizations of weakly chaotic maps of the interval,* Proc. Amer. Math. Soc. **110** (1990), No. 1, 141–148.

Fedorenko, V. V. and Smital, J.

[1] *Maps of the interval Ljapunov stable on the set of non-wandering points,* Acta Math. Univ. Comenianae **LX** (1991), No. 1, 11–14.

Feigenbaum, M.

[1] *Quantitative universality for a class of nonlinear transformations,* J. Stat. Phys. **19** (1978), 25–52.

[2] *The universal metric properties of nonlinear transformations,* J. Stat. Phys. **21** (1979), 669–706.

[3] *Universality in the behavior of nonlinear systems,* Uspekhi Fiz. Nauk **141** (1983), No. 2, 343–374 *(in Russian).*

Foguel, S. R.

[1] *The ergodic theory of Markov processes,* Van Nostrand Math. Studies **21** (1969).

Geisel, T. and Nierwetberg, J.

[1] *Universal fine structure of the chaotic region in period doubling systems,* Phys. Rev. Lett. **47** (1981), 975–978.

Grosmann, S. and Thomae, S.

[1] *Invariant distributions and stationary correlation functions of one-dimensional discrete processes,* Zeitschr. f. Naturforschg. **32a** (1977), 1353–1363.

Guckenheimer, J.

[1] *On the bifurcation of maps on the interval,* Invent. Math., **39** (1977), 165–178.

[2] *Sensitive dependence on initial conditions for one-dimensional maps,* Commun. Math. Phys. **70** (1979), 133–160.

[3] *Limit set of S-unimodal maps with zero entropy,* Commun. Math. Phys. **110** (1987), 655–659.

Gumowsky, I. and Mira, C.

[1] *Dynamique Chaotique,* Cepadues, Toulouse, 1980.

[2] *Recurrences and discrete dynamical systems,* Springer Lect. Notes in Math. **809** (1980).

Henry, B.

[1] *Escape from the unit interval under the transformation $x \to \lambda x(1-x)$,* Proc. Amer. Math. Soc. **41** (1973), 146–150.

Ho, C. and Morris, C.

[1] *A graph-theoretic proof of Sharkovskii's theorem on the periodic points of continuous functions,* Pacif. J. Math. **92** (1981), 361–370.

Jakobson, M. V.

[1] *On smooth maps of a circle into itself,* Mat. Sbornik **85** (1971), 163–188; ***English transl.:*** Math. USSR Sb. **85** (1971).

[2] *On the properties of dynamical systems generated by mappings of the form* $x \to Axe^{-x}$, Simulation of Biological Communities, Dalnevostoch. Nauch. Tsentr Akad. Nauk SSSR, Vladivostok, 1975, pp. 141–162 *(in Russian).*

[3] *Topological and metrical properties of one-dimensional endomorphisms,* Dokl. Akad. Nauk SSSR **293** (1978), 866–869 *(in Russian).*

[4] *Absolutely continuous invariant measures for one-parameter families of one-dimensional maps,* Commun. Math. Phys. **81** (1981), 39–48.

Jankova, K. and Smital, J.

[1] *A characterization of chaos,* Bull. Austral. Math. Soc. **34** (1986), 283–292.

Johnson, S. D.

[1] *Singular measures without restrictive intervals,* Commun. Math. Phys. **110** (1987), 185–190.

Jonker, L.

[1] *Periodic orbits and kneading invariants,* Proc. London Math. Soc. **39** (1979), 428–450.

[2] *A monotonicity theorem for the family* $f_\lambda(x) = a - x^2$, Proc. Amer. Math. Soc. **85** (1982), 434–436.

Jonker, L. and Rand, S.

[1] *Bifurcations in one dimension. I, II,* Invent. Math. **62**, 347–365; **63**, 1–16 (1981).

Julia, G.

[1] *Memoire sur L'Iteration des Fonctions Rationelles,* J. Math. Pures Appl. **4** (1918), 47–245.

Kaplan, H.

[1] *A cartoon-assisted proof of Sharkowskii's theorem,* Ann. J. Phys **55** (1987), 1023–1032.

Keller, G. and Nowicki, T.

[1] *Fibonacci maps revisited*, Preprint, Erlangen, (1992).

Kloeden, P. E.

[1] *On Sharkovsky's cycle coexistence ordering,* Bull. Austral. Math. Soc. **20** (1979), 171–178.

Kolyada, S. F.

[1] *Maps of an interval with zero Schwarzian derivative,* Differential–Difference Equations and Their Applications, Inst. Mat. Akad. Nauk Ukrain. SSR, Kiev, 1985, pp. 47–57 *(in Russian).*

[2] *On the measure of quasiattractors of one-dimensional smooth maps,* Preprint No. 86.35, Inst. Mat. Akad. Nauk Ukrain. SSR, Kiev, 1986 *(in Russian).*

[3] *One-parameter families of maps of the interval with negative Schwarzian derivative, in which the monotonicity of bifurcations breaks down,* Ukrain. Mat. Zh. **41** (1989), 258–261; ***English transl.:*** Ukrain. Math. J. **41**, 1989, 230–232.

Kolyada, S. F. and Sharkovsky, A. N.

[1] *On the measure of repellers of one-dimensional smooth maps,* Differential–Functional and Difference Equations, Inst. Mat. Akad. Nauk Ukrain. SSR, Kiev, 1981, pp. 16–22 *(in Russian).*

Kolyada, S. F. and Sivak, A. G.

[1] *Universal constants for one-parameter maps,* Oscillation and Stability of Solutions of Differential–Difference Equations, Inst. Mat. Akad. Nauk Ukrain. SSR, Kiev, 1982, pp. 53–60 *(in Russian).*

[2] *A certain class of functional equations and universal behavior of the families of one-dimensional maps,* Differential–Difference Equations and Problems of Mathematical Physics, Inst. Mat. Akad. Nauk Ukrain. SSR, Kiev, 1984, pp. 33–37 *(in Russian).*

Kolyada, S. F. and Snoha, L.

[1] *On ω-limit sets of triangular maps,* Real Analysis Exchange **18** (1992–93), 115–130.

Krylov, N. M. and Bogolyubov, N. N.

[1] *General Theory of Measure in Nonlinear Mechanics,* Zapiski Kaf. Mat. Fiz. Inst. Stroit. Mekh. Akad. Nauk Ukrain. SSR **3** (1937), 55–112 (see also **Bogolyubov, N. N.**, Selected Works, vol. 1, Naukova Dumka, Kiev, 1969) *(in Russian).*

Kuchta, M. and Smital, J.

[1] *Two points scrambled set implies chaos,* Proceedings of the European Conference on Iteration Theory (Caldas de Malavella, Spain, 1987), World Scientific (1989), pp. 427–430.

Kuratowski, K.

[1] Topology, Academic Press, New York, 1968.

Lanford, O. E.

[1] *Remarks on the accumulation of period-doubling bifurcations,* Springer Lect. Notes in Physics **116** (1980), 340–342.

[2] *A shorter proof of the existence of the Feigenbaum fixed point,* Commun. Math. Phys. **96** (1984), 521–538.

Lasota, A. and Yorke, J.

[1] *On the existence of invariant measures for piecewise monotonic transformations,* Trans. Amer. Math. Soc. **186** (1973), 481–488.

Leonov, N. N.

[1] *Map of the line onto itself,* Radiofiz., **2** (1959), 942–956 *(in Russian).*

[2] *Piecewise linear map,* Radiofiz., **3** (1960), 496–510 *(in Russian).*

[3] *On discontinuous pointwise transformation of the line into itself,* Dokl. Akad. Nauk SSSR, **143** (1962), 1038–1041 *(in Russian).*

Li, T. Y., Misiurewicz, M., Pianigiani. G., and Yorke, J.

[1] *No division implies chaos,* Trans. Amer. Math. Soc. **273** (1982), 191–199.

Li, T. Y. and Yorke, J.

[1] *Periodic three implies chaos,* Amer. Math. Mon. **82** (1975), 285–992.

Lyubich, M. and Milnor, J.

[1] *The dynamics of the Fibonacci polynomial,* J. Amer. Math. Soc. **6** (1993), 425–457.

Maistrenko, Yu. L. and Sharkovsky, A. N.

[1] *Turbulence and simple hyperbolic systems,* Preprint No. 84.2, Inst. Mat. Akad. Nauk Ukrain. SSR, Kiev, (1984) *(in Russian).*

Mané, R.

[1] *Hyperbolicity, sinks, and measure in one-dimensional dynamics,* Commun. Math. Phys. **100** (1985), 495–524.

Matsumoto, S.

[1] *Bifurcation of periodic points of maps of the interval,* Bull. Sci. Math. **107** (1983), 49–75.

May, R. M.

[1] *Simple mathematical models with very complicated dynamics,* Nature **261** (1976), 459–467.

Melo, W. de and Strien, S. J. van

[1] *A structure theorem in one-dimensional dynamics,* Preprint No. 86.29, Delft University of Technology, 1986.

[2] One-Dimensional Dynamics, Springer, Berlin, 1993.

Metropolis, M., Stein, M. L., and Stein, P. R.

[1] *On finite limit sets for transformations of unit interval,* J. Combin. Theory A **15** (1973), 25–44.

Milnor, J.

[1] *The monotonicity theorem for real quadratic maps,* Morphologic Complexer Grenzen, Universität Bremen, Bremen, 1985, pp. 10–12.

[2] *On the concept of attractor,* Commun. Math. Phys. **99**, 177–195; **102**, 517–519 (1985).

Milnor, J. and Thurston, W.

[1] *On iterated maps of the interval. I. The kneading matrix.. II. Periodic points,* Preprint, Princeton, 1977.

[2] *On iterated maps of the interval,* Lecture Notes in Math. **1342** (1988), 465–563.

Misiurewicz, M.

[1] *Horseshoes for mappings of an interval,* Bull. Acad. Pol. Sci, Ser. Sci. Math., Astron. Phys. **27** (1979), 167–169.

[2] *Structure of mapping of an interval with zero entropy,* Publ. Math. Inst. Hautes Etud. Sci. **53** (1981), 1–16.

[3] *Absolutely continuous measures for certain maps of an interval,* Publ. Math. Inst. Hautes. Etud. Sci. **53** (1981), 17–51.

[4] *Attracting Cantor set of positive measure for a* C^∞ *map of an interval,* Ergod. Theor. Dynam. Syst. **2** (1982), 405–415.

[5] *Jumps of entropy in one dimension,* Fund. Math. **132** (1989), 215–226.

Misiurewicz, M. and Nitecki, M.

[1] *Combinatorial patterns for maps of the interval,* Mathematica Gottingensis. **35** (1989).

Misiurewicz, M. and Szlenk, W.

[1] *Entropy of piecewise monotone mappings,* Studia Math. **67** (1980), 45–63.

Mori, M.

[1] *The extension of Sharkovskii's results and topological entropy in unimodal transformations,* Tohoky Math. J. **4** (1981), 133–152.

Myrberg, P. J.

[1] *Iteration der rellen polynome zweiten grades,* Ann. Acad. Sci. Fenn. Ser. A. **I: 256** (1958), 1–16; **II: 268** (1959), 1–10; **III: 336** (1963), 1–18.

Myshkis, A. D. and Lepin, A. Ya.

[1] *Existence of the invariant set consisting of two points for some continuous self-maps of an interval,* Uchen. Zapiski Belorussk. Univ. Ser. Fiz.-Mat. **32** (1957), 29–32 *(in Russian).*

Nemytsky, V. V. and Stepanov, V. V.

[1] Qualitative Theory of Differential Equations, 2nd edition, Gostekhizdat, Moscow–Leningrad, 1949 *(in Russian)*; ***English transl.:*** Princeton University Press, Princeton, N. J., 1960.

Nitecki, Z.

[1] *Maps of the interval with closed periodic set,* Proc. Amer. Math. Soc. **8 5** (1982), 451–456.

Paluba, W.

[1] *The Lipschitz condition for the conjugacies of Feigenbaum-like mappings,* Preprint No. 4/87, Uniwersytet Warszawski, Instytut Matematyki, Warszawa, 1987.

Peitgen, H.-O. and Richter, P. H.

[1] The Beauty of Fractals, Springer-Verlag, Berlin–Heidelberg, 1986.

Poincaré, H.

[1] *Sur les equations lineaires and differentielles ordinaires et aux differences finies,* Amer. J. Math. **7** (1885), 213–217; 237–258.

Preston, C.

[1] *Iterates of maps on an interval,* Springer Lect. Notes Math. **999** (1983).

Pulkin, S. P.

[1] *On iterations of functions of one independent variable,* Izv. Akad. Nauk SSSR, Ser. Math., **6** (1942) *(in Russian).*

[2] *Oscillation sequences of iterations,* Dokl. Akad. Nauk SSSR, **22**, No. 6 (1950) *(in Russian).*

Romanenko, E. Yu. and Sharkovsky, A. N.

[1] *Asymptotic properties of the semigroup of maps of an interval,* Dynamical Systems and Differential–Difference Equations, Inst. Mat. Akad. Nauk Ukrain. SSR, Kiev, 1986, pp. 85–90 *(in Russian).*

Schwartz, A.

[1] *A generalization of a Poincaré Bendixon theorem to closed two-dimensional manifolds,* Amer. J. Math. **85** (1963).

Schwarz, H. A.

[1] *Über einige Abbildungsaufgaben,* J. für reine und angewandte Mathematik **70** (1868), 105–120.

Shapiro, A. P. and Luppov, S. P.

[1] Recurrence Equations in Population Biology, Nauka, Moscow, 1983 *(in Russian).*

Sharkovsky, A. N.

[1] *Coexistence of cycles of a continuous transformation of a line into itself,* Ukrain. Mat. Zh. **16** (1964), 61–71 *(in Russian)*; ***English transl.:*** Int. J. Bifurcation and Chaos. **5** (1995), 1263–1273.

[2] *Non-wandering points and the center of a continuous map of a line into itself,* Dokl. Akad. Nauk Ukrain. SSR, Ser. A (1964), No. 7, 865–868 *(in Ukrainian).*

[3] *On the cycles and structure of a continuous map,* Ukrain. Mat. Zh. **17** (1965), No. 3, 104–111 *(in Russian).*

[4] *On a certain classification of fixed points,* Ukrain. Mat. Zh. **17** (1965), No. 5, 80–95 *(in Russian)*; ***English transl.:*** Amer. Math. Soc. Translations. **97**, No. 2 (1970), 159–179.

[5] *On attracting and attractable sets,* Dokl. Akad. Nauk SSSR **160** (1965), 1036–1038 *(in Russian)*; ***English transl.:*** Soviet Math. Dokl. **6** (1965), 268–270.

[6] *On a continuous mapping on a set of the ω-limit points,* Dokl. Akad. Nauk Ukrain. SSR, Ser. A (1965), No. 11, 1407–1410 *(in Ukrainian).*

[7] *On ω-limit sets of dynamical systems with discrete time,* Doctoral Degree Thesis, Inst. Mat. Akad. Nauk Ukrain. SSR, Kiev, 1966 *(in Russian).*

[8] *Behavior of a map in the neighborhood of an attracting set,* Ukrain. Mat. Zh. **18** (1966), No. 2, 60–83 *(in Russian)*; ***English transl.:*** Amer. Math. Soc. Translations. **97**, No. 2 (1970), 227–258.

[9] *Continuous mapping on the set of limit points of an iteration sequence,* Ukrain. Mat. Zh. **18** (1966), 127–130 *(in Russian).*

[10] *Partially ordered system of attracting sets,* Dokl. Akad. Nauk SSSR, **170** (1966), 1276–1278; ***English transl.:*** Soviet Math. Dokl. **7**, No. 5 (1966), 1384–1386.

[11] *About one Birkhoff's theorem,* Dokl. Akad. Nauk Ukrain. SSR, Ser. A (1967), No. 5, 429–432 *(in Russian).*

[12] *Attracting sets without cycles,* Ukrain. Mat. Zh. **20** (1968), No. 1, 136–142 *(in Russian).*

[13] *On the problem of isomorphism of dynamical systems,* Proceedings of V ICNO (Kiev), vol. 2, Naukova Dumka, Kiev, 1970, pp. 541–545 *(in Russian).*

[14] *Stability of trajectories and dynamical systems as a whole,* Proceedings of IX Summer Mathematical Workshop, 2nd edition, Naukova Dumka, Kiev, 1976, pp. 349–360 *(in Russian).*

[15] *The structural theory of differentiable dynamical systems and weakly non-wandering points,* Abhandlungen der Wissenschaften der DDR, Abteilung Mathematik, Naturwissenschaften, Technik VII Intern. Konf. über Nichlinear Schwingungen, Berlin, Band 1, 2, No. 4, 1977, pp. 193–200.

[16] *Difference equations and dynamics of population numbers,* Preprint No. 82, Inst. Mat. Akad. Nauk Ukrain. SSR, Kiev, 1982 *(in Russian).*

[17] *On some properties of discrete dynamical systems,* Colloque Intern. du C.N.R.S. No. 322 sur la Theorie de L'Iteration et ses Applications, Univ. Paul Sebatier, Toulouse, 1982, pp. 153–158.

[18] *"Dry" turbulence,* Nonlinear and Turbulent Processes in Physics, in 3 volumes (ed. by R. Z. Sagdeev), vol. 3, Gordon and Breach, Chur–London–Paris–New-York, 1984, pp. 1621–1626.

[19] *Classification of one-dimensional dynamical systems,* Proceedings of the European Conference on Iteration Theory (Caldas de Malavella, Spain, 1987), World Scientific, 1989, pp. 42–55.

Sharkovsky, A. N. and Dobrynsky, V. A.

[1] *Non-wandering points of dynamical systems,* Dynamical Systems and Problems of Stability of Solutions of Differential Equations, Inst. Mat. Akad. Nauk Ukrain. SSR, Kiev, 1973, pp. 165–174 *(in Russian).*

Sharkovsky, A. N. and Fedorenko, V. V.

[1] *Types of return for simple dynamical systems,* Preprint No. 91.2, Inst. Mat. Akad. Nauk Ukrain. SSR, Kiev, 1991 *(in Russian).*

Sharkovsky, A. N. and Ivanov, A. F.

[1] *C^∞-maps of an interval exhibiting attracting cycles with arbitrary large period,* Ukrain. Mat. Zh. **35** (1983), 537–539; ***English transl.:*** Ukrain. Math. J. **35** (1983).

Sharkovsky, A. N., Maistrenko, Yu. L., and Romanenko, E. Yu.

[1] *Dry turbulence,* The Lavrentyev Hearings on Mathematics, Physics and Mechanics, II All-Union Conference (Kiev, 1985), Inst. Mat. Akad. Nauk Ukrain. SSR, Kiev, 1985, pp. 210–212 *(in Russian).*

[2] Difference Equations and Their Applications, Naukova Dumka, Kiev, 1986; ***English transl.:*** Kluwer, Dordrecht, 1993.

[3] *Asymptotic periodicity of solutions of difference equations with continuous time,* Ukrain. Mat. Zh. **39** (1987), No. 1; ***English transl.:*** Ukrain. Math. J. **39** (1987).

Sharkovsky, A. N. and Romanenko, E. Yu.

[1] *Asymptotic behavior of solutions of difference–differential equations,* Qualitative Approaches to the Study of Nonlinear Differential Equations and Nonlinear Oscillations, Inst. Mat. Akad. Nauk Ukrain. SSR, Kiev, 1981, pp. 171–199 *(in Russian).*

Sibirsky, K. S.

[1] Introduction to Topological Dynamics, Academy of Sciences of Moldavian SSR, Kishinev, 1970 *(in Russian), **English transl.:*** Noordhoff, Leiden, 1975.

Sinai, Ya. G.

[1] *Stochasticity of dynamical systems,* Nonlinear Waves, Nauka, Moscow, 1979, pp. 192–221 *(in Russian).*

Singer, D.

[1] *Stable orbits and bifurcations of maps on the interval,* SIAM J. Appl. Math, **35** (1978), 260–267.

Sivak, A. G.

[1] *On a measure of attraction domain of semiattracting cycles of unimodal maps,* Differential–Difference Equations and Their Applications, Inst. Mat. Akad. Nauk Ukrain. SSR, Kiev, 1985, pp. 57–59 *(in Russian).*

[2] *Stability and bifurcations of periodic intervals of one-dimensional maps,* Differential–Difference Equations and Their Applications, Inst. Mat. Acad. Nauk Ukrain. SSR, Kiev, 1985, pp. 60–73 *(in Russian).*

[3] *Limit behavior, stability, and bifurcations of one-dimensional maps,* Preprint No. 86.51, Inst. Mat. Akad. Nauk Ukrain. SSR, Kiev, 1986 *(in Russian).*

Sivak, A. G. and Sharkovsky, A. N.

[1] *Universal order and universal rate of bifurcations of solutions of differential–difference equations,* Rough and Qualitative Studies of Differential and Differential–Functional Equations, Inst. Mat. Akad. Nauk Ukrain. SSR, Kiev, 1983, pp. 98–106 *(in Russian).*

Smale, S.

[1] *Differentiable dynamical systems,* Bull. Amer. Math. Soc. **73** (1967), 747–817.

Smital, J.

[1] *Chaotic functions with zero topological entropy,* Trans. Amer. Math. Soc. **297** (1986), 269–282.

Snoha, L.

[1] *Characterization of potentially minimal periodic orbits of continuous mappings of an interval,* Acta Math. Comenian. **52–53** (1987), 112–124.

Stefan, P.

[1] *A theorem of Sharkovskii on the existence of periodic orbits of continuous endomorphisms of the real line,* Commun. Math. Phys. **54** (1977), 237–248.

Strange Attractors (Collection of papers), Mir, Moscow, 1981 *(Russian translation).*

Strien, S. J. van

[1] *On the bifurcations creating horseshoes,* Springer Lect. Notes in Math. **898** (1981), 316–351.

[2] *Smooth dynamics on the interval (with an emphasis on unimodal maps),* Preprint No. 87.09, Delft University of Technology, Delft, 1987.

[3] *Hyperbolicity and invariant measures for general* C^2 *interval maps satisfying the Misiurewicz condition,* Preprint, Delft University of Technology, Delft, 1987.

Targonski, G.

[1] Topics in Iteration Theory, Vandenhoeck and Ruprecht, Göttingen, 1981.

Ulam, S.

[1] Unsolved Mathematical Problems, Nauka, Moscow, 1964 *(Russian translation).*

Vereikina, M. B. and Sharkovsky, A. N.

[1] *Returnability in one-dimensional dynamical systems,* Rough and Qualitative Studies of Differential and Differential–Functional Equations, Inst. Mat. Akad. Nauk Ukrain. SSR, Kiev, 1985, pp. 35–46 *(in Russian).*

[2] *A set of almost-returning points of a dynamical system,* Dokl. Akad. Nauk Ukrain. SSR, Ser. A. (1984), No. 1, 6–9 *(in Russian).*

Wul, E. B., Sinai, Ya. G., and Khanin, K. M.

[1] *Feigenbaum universality and thermodynamic formalism,* Uspekhi Mat. Nauk **39** (1984), No. 3, 3–37; ***English transl.:*** Russian Math. Surveys **39** (1984).

Xiong, J.

[1] *A set of almost periodic points of continuous self-maps of an interval,* Acta Math. Sinica **2** (1986), No. 1, 73–77.

Ye, X.

[1] *D-function of a minimal set and an extension of Sharkovskii's theorem to minimal sets,* Ergod. Theory Dynam. Syst., **12** (1992), 365–376.

Young, L.-S.

[1] *A closing lemma on the interval,* Invent. Math. **54** (1979), 179–187.

SUBJECT INDEX

NOTATION

$\mathbb{N}$	–	the set of natural numbers
$\mathbb{Z}$ ($\mathbb{Z}^+$)	–	the set of integer (nonnegative integer) numbers
$\mathbb{R}$ ($\mathbb{R}^+$)	–	the set of real (nonnegative real) numbers
I	–	a closed interval of $\mathbb{R}$
$C^0(I, I)$	–	the space of continuous maps of I
2^X	–	the space of closed subsets of the space X
$\overline{A}$	–	the closure of the set A
∂A	–	the boundary of the set A
$\operatorname{int} A$	–	internal points of the set A
$U_\delta(A)$	–	the δ-neighborhood of the set A
$\operatorname{mes} A$	–	the Lebesgue measure of the set A
$f\|_A$	–	the restriction of the map f to the set A
f^n	–	the nth iteration of the map f
$\operatorname{orb}(x)$	–	the trajectory of the point x
$\omega(x)$	–	the set of ω-limit points of the trajectory of the point x
$\Omega(f)$	–	the set of ω-limit points
$\operatorname{NW}(f)$	–	the set of nonwandering points
$\operatorname{NW}^+(f), \operatorname{NW}^-(f)$	–	the sets of unilateral nonwandering points
$\operatorname{CR}(f)$	–	the set of chain recurrent points
$\operatorname{Fix}(f)$	–	the set of fixed points
$\operatorname{Per}(f)$	–	the set of periodic points
$\operatorname{AP}(f)$	–	the set of almost periodic points
$\operatorname{APB}(f)$	–	the set of almost periodic points in the sense of Bohr
$\operatorname{R}(f)$	–	the set of recurrent points
$\operatorname{RR}(f)$	–	the set of regularly recurrent points
$\operatorname{C}(f)$	–	the center
$h(f)$	–	topological entropy
ν_f	–	kneading invariant

Other *Mathematics and Its Applications* titles of interest:

J.-F. Pommaret: *Partial Differential Equations and Group Theory. New Perspectives for Applications.* 1994, 473 pp. ISBN 0-7923-2966-X

Kichoon Yang: *Complete Minimal Surfaces of Finite Total Curvature.* 1994, 157 pp. ISBN 0-7923-3012-9

N.N. Tarkhanov: *Complexes of Differential Operators.* 1995, 414 pp. ISBN 0-7923-3706-9

L. Tamássy and J. Szenthe (eds.): *New Developments in Differential Geometry.* 1996, 444 pp. ISBN 0-7923-3822-7

W.C. Holland (ed.): *Ordered Groups and Infinite Permutation Groups.* 1996, 255 pp. ISBN 0-7923-3853-7

K.L. Duggal and A. Bejancu: *Lightlike Submanifolds of Semi-Riemannian Manifolds and Applications.* 1996, 308 pp. ISBN 0-7923-3957-6

D.N. Kupeli: *Singular Semi-Riemannian Geometry.* 1996, 187 pp. ISBN 0-7923-3996-7

L.N. Shevrin and A.J. Ovsyannikov: *Semigroups and Their Subsemigroup Lattices.* 1996, 390 pp. ISBN 0-7923-4221-6

C.T.J. Dodson and P.E. Parker: *A User's Guide to Algebraic Topology.* 1997, 418 pp. ISBN 0-7923-4292-5

B. Rosenfeld: *Geometry of Lie Groups.* 1997, 412 pp. ISBN 0-7923-4390-5

A. Banyaga: *The Structure of Classical Diffeomorphism Groups.* 1997, 208 pp. ISBN 0-7923-4475-8

A.N. Sharkovsky, S.F. Kolyada, A.G. Sivak and V.V. Fedorenko: *Dynamics of One-Dimensional Maps.* 1997, 272 pp. ISBN 0-7923-4532-0